Engineering Project Management for the Global High-Technology Industry

About the Author

Sammy G. Shina, Ph.D., P.E., is a professor of mechanical engineering at the University of Massachusetts Lowell (UML), and has lectured in the University of Pennsylvania's ExMSE Program and at the University of California Irvine. He is the coordinator of the Design and Manufacturing Certificate, the Quality Engineering Certificate, mechanical engineering senior capstone projects, and co-op education for the College of Engineering at UML. He is the founder of the New England Lead-Free Electronics Consortium, which researches, tests, and evaluates materials and processes for lead-free and RoHS compliance and conversion to nanotechnology.

Dr. Shina is an international consultant, trainer, and seminar provider on project management, quality methods in design and manufacturing, Six Sigma, and design of experiments (DoE), as well as technology supply chains, product design and development, and electronics manufacturing, testing, and automation. He worked for 22 years in high-technology companies developing new products and state-of-the-art manufacturing technologies. Dr. Shina received B.S. degrees in electrical engineering and industrial management from Massachusetts Institute of Technology, an M.S. degree in computer science from Worcester Polytechnic Institute, and a Ph.D. degree in mechanical engineering from Tufts University. He is the author of several best-selling books on concurrent engineering, Six Sigma, green design, and engineering project management, and more than 100 papers.

Engineering Project Management for the Global High-Technology Industry

Sammy G. Shina, Ph.D., P.E.

Mc Graw Hill Education

New York Chicago San Francisco
Athens London Madrid
Mexico City Milan New Delhi
Singapore Sydney Toronto

1 2 3 4 5 6 7 8 9 0 DOC/DOC 1 9 8 7 6 5 4 3

ISBN 978-0-07-181536-9
MHID 0-07-181536-8

Sponsoring Editor
Michael McCabe

Editing Supervisor
Stephen M. Smith

Production Supervisor
Lynn M. Messina

Acquisitions Coordinator
Bridget L. Thoreson

Project Manager
Sheena Uprety,
Cenveo® Publisher Services

Copy Editor
James Dryden

Proofreader
Carol Shields

Indexer
Bev Weiler

Art Director, Cover
Jeff Weeks

Composition
Cenveo Publisher Services

Printed and bound by RR Donnelley.

This book is printed on acid-free paper.

To my wife Jackie,
and our children and grandchildren.

Contents

Preface

Engineering project management is becoming more important as technology companies compete in a worldwide market for customers desiring high-quality and low-cost products. The project manager (PM) has to be a *jack of all trades*, a product champion, a great organizer, a leader, mentor, and motivator of the team; the PM has to be an effective communicator, a salesperson, a financial analyst, and much more. The PM today must be an expert in technology, quality, cost, teamwork, supply chains, and market dynamics. The PM must always balance priorities and make good decisions regarding resource allocation, schedule variability, cost, technology adoption, and risk management.

This book attempts to augment the basic project-management principles of scheduling, tracking, and control of projects with answering many of the questions posed by the role of technology in new product creation. Why do some companies thrive in the technology arena, while others start well but cannot maintain the momentum? Why is it so difficult for companies to enter some markets? What are the options available to companies for setting new product price and performance? What types of organizational structures and methods are needed to successfully manage technical projects? How can company resources and the supply chain be leveraged?

This book attempts to answer these questions by examining product lifecycles, project management types, and where they should be used as well as tools and techniques of quality cost and marketplace. Economic analysis of the project potential and how to best leverage internal resources versus supply chains, as well as risk and rewards of project decisions, are also examined. The book illustrates these principles with examples of current technology-company policies, some drawn from the headlines and some from my own experience. I have an extensive history of managing many development projects, consulting to technology companies, and researching the tools and techniques of new product creation. In addition, long conversations and meetings with many of the creators of project management tools, CEOs, and members of the boards of directors of companies, and several expert-witness litigation cases, have given me a unique perspective of the challenges and concerns of global technology companies.

The book also aims to help the PM to become more successful, using the technical, organizational, financial, leadership, and communications skills covered in this book. Topics presented deal with the historical development of the tools and techniques of project management through the last 40 years and how to successfully use these tools for effectively managing technical projects. The PM can understand the best use of the management structures explained in this book, depending on the lifecycle of the product. The use of financial analysis and tools can effectively augment the PM's plans and decisions. Understanding the use of the global supply chain, its opportunities and risks, can also help the PM in project and product cost formulation and schedule realization as well as advocating decisions to management. The effective use of scheduling tools to plan, track, and control projects is important for the PM in maintaining the product creation schedule and evaluating and managing its risks. The effective communications skills, teamwork, and leadership covered in this book will help the PM navigate successfully through these important but nontechnical issues.

About the Book Organization

This book is intended to introduce newly minted as well as experienced project managers in technology companies to many of the issues regarding the use of project-management tools and techniques and how to effectively apply them for new product creation. It is based on my experience in researching, practicing, consulting on, and teaching project management for the last 40 years.

The approach I use in this book is to start with the historical development of project management tools and then go on to what are the proper conditions for using these tools, why they were created, and how they became widely adopted. The following chapters deal with the step-by-step elements of technology product realization, starting with the technology product lifecycle and the management organization best suited for each phase of the cycle. Technology management from research to advanced development to adoption in new products is explained with examples of organizational structure and timelines needed. Other chapters discuss the marketing aspects of customer expectations and finding the best opportunity for new product success, with tools and examples of using them effectively.

Once the market opportunity for the new product is realized, the hard work of the PM begins with the business plan and economic analysis for the project. Issues of how to leverage internal resources and the supply chain and how to select suppliers are presented. This is followed by the methodology to organize and plan the development project, how to control projects using phases and milestones, tracking a project's progress, and reporting to management. In addition, the value and use of risk management to estimate and mitigate risk are illustrated with the definitions of methods used and case studies from industry.

The final chapters of the book deal with important skill development for the PM, including communications, leadership, and teamwork. I asked experienced professionals who deal with these issues to help me by co-authoring these chapters in order to share their own experiences and insight.

I hope this book will be of value to the neophyte as well as the experienced project managers in technology companies, in particular, in the small- to medium-sized companies that do not have the support staff and the resources necessary to have a well-organized project-management process. It is beneficial to try out some of the principles and tools of project management outlined in this book and meld them into the company culture. The experiences documented here should be helpful to encourage many companies to venture out and develop new world-class products that can make them grow and prosper for the future.

Sammy G. Shina, Ph.D., P.E.

Acknowledgments

The principles of engineering project management discussed in this book have been learned, collected, and practiced through my almost 50 years in industry and academia. After graduation from MIT, I worked in the high-technology industry for 22 years, followed by now 26 years on the faculty of the University of Massachusetts Lowell. At the university, I have worked as a teacher, then as a researcher and a consultant to different companies, increasing my personal knowledge and experience in the fields of engineering project management, design, manufacturing, and quality.

I am indebted to several organizations for supporting and encouraging me during the lengthy time it took to collect my materials, write the chapters, and edit the book—notably the University of Massachusetts Lowell, for its continuing support for my courses, programs, and certificates, especially the chairman of the Department of Mechanical Engineering, John McKelliget, and the ME faculty. They supported me in my research and work on developing the book materials and approved my plans for academic programs and certificates and encouraged me to organize, write, and edit this book.

In addition, I want to give my thanks to Steve Chapman, publisher, and Michael McCabe, senior editor, at McGraw-Hill. Steve was my editor for my previous four books on green design, Six Sigma, and concurrent engineering. Michael was my editor for this book. Mike's humor, encouragement, and good spirit guided me through this book, and for that I am very grateful. I also wish to extend my gratitude to Sheena Uprety of Cenveo Publisher Services for her prompt and efficient editing and production of this book. In addition, I want to thank Marc Wakim of UML for proofing and editing of the book; and Sharon Sambursky of SpectraLink Corporation, who lectured to my classes on the topics of leadership and teamwork; and Srini Swaroop of Raytheon Corporation for his lectures to my classes on risk management. I also want to thank the men who contributed chapters to this book—Robert Campbell, Ralph Jordan, and David Nolte—and who worked together with me on planning and organizing the topics. Each contributor brought with him his own deep experience and skill in his specialty. I also want to thank the many family members who hosted me through the long period of writing the book, including Brenda Shina of St. John's Wood, London, and Nancy Shina Aguirre of Ogden, Utah.

Many colleagues provided review and thoughtful criticism. In particular, I wish to thank Travis Done, Jack Burnham, and Dick Ugolini of United Technologies Aerospace Systems. I also thank Tom Bergeron, president of ISR Systems, United

Technologies Aerospace Systems, for his support and encouragement and for sharing legacy Goodrich/United Technologies approaches to program management.

Finally, many thanks to my family for emotional support during the writing, editing, and production of the book, including my wife Jackie, who edited the book with her superb English, our children—Mike, Gail, Nancy, and Jon—and our grandchildren, who brought me great joy between the many days of writing and editing. I also wish to thank the many students who have attended my classes in engineering project management over many years and peppered me with questions and challenges to explain the many topics, which cleared and refocused my mind. I wish them best success in implementing engineering project management principles and methods in their companies.

Contributor Biographies

David A. Nolte (Chaps. 4 and 9) is a manager with ISR Systems, United Technologies Aerospace Systems, Westford, Massachusetts, where he supports the development and evolution of program management, ACE, and continuous improvement and lean product development culture. Mr. Nolte has over 25 years of progressive management and engineering experience in defense industries, government agencies, and nonprofit organizations. His experience ranges from field and test engineering to program management.

Robert J. Campbell, Jr. (Chap. 10), is a mechanical engineer with a love for the technical discipline of precision machine design and the interpersonal discipline of collaborative development. His father, Robert Sr., taught him that a good engineer does not get lost in his circuit or mechanism, but instead maintains sight of the whole. In the years since, Mr. Campbell has been fortunate to work with and learn from other wise engineers and agile organizations that put this belief into daily practice. As a consulting engineer, he had to adapt those practices to development teams that spanned organizations and continents. Now, through his website engineerunbound.com, he helps engineers and organizations to not only overcome the challenges posed by distributed development and remote collaboration, but to achieve competitive advantage. Mr. Campbell is a licensed professional engineer, with a masters of management science from the University of Massachusetts Lowell and a BS in mechanical engineering from Virginia Tech. With several peers, he holds patents for precision optical and mechanical systems and devices.

Ralph E. Jordan (Chap. 12) is the former director of Massachusetts' Executive Office of Labor and Workforce Development's Office of Professional Development. Presently, he is a visiting lecturer within the University of Massachusetts Lowell's Manning School of Business, where he lectures on professional communications, managing teams and projects, and leadership processes. While relatively new to teaching at the undergraduate level, Mr. Jordan has a long history in training and leading teams in total quality management, business process reengineering, and lean Six Sigma–type initiatives within the high-tech and communications industries. He has held high-level management positions in several high-tech companies. Mr. Jordan spent several years serving Massachusetts as the undersecretary of economic affairs. He has led high-tech business initiatives in the Republic of Korea, Kuwait, and the Republic of Slovenia.

Engineering Project Management for the Global High-Technology Industry

CHAPTER 1

The Engineering Project Lifecycle and Historical Development of Engineering Project Management Tools and Techniques

In this chapter, the historical perspective for the genesis of modern engineering project management for high-technology companies will be reviewed. Emphasis will be placed on the trends of each successive generation, starting with the 1980s and on to the new century. The trends and tools of each decade will be outlined as well as the resulting shifts in total engineering project management experiences. The resulting impact on the organizational structure of modern high-technology companies, on managers, engineers, research and development, and the introduction of highly specialized tools, will be examined. The challenges to the project manager in terms of completing the project on time and within budget, while having the new product meet all design specifications with the lowest manufacturing cost and quality, will be illustrated.

Engineering project management for the global high-technology industry began to be organized in the last 50 years with the advent of the 1970s, as the Japanese technology industry competition began to make a major impact on global U.S. companies' competitive position. Due in part to the oil crisis of the 1970s, U.S. consumers were looking for more energy-efficient smaller cars, and in the process, were pleased to discover the higher-quality and customer-friendly Japanese cars, as compared with their American counterparts. This created a thirst in American companies for all things Japanese and began the focus on improving the new product development cycle and engineering project management. New concepts were adopted widely and began to take effect, including the following:

- Just-in-time (JIT) to reduce inventories and shorten manufacturing cycle time
- Total quality management (TQM) to bring together a set of tools focused on process improvements for the total enterprise

- Quality circles to involve production associates in improving the manufacturing process and their duties and responsibilities
- Partnering with Japanese companies, such as the GM/Toyota partnership, to better understand their manufacturing techniques for auto manufacturing plants in California
- The Taguchi method, which streamlined the difficult topic of design of experiments (DoE) and took it from the preview of advanced-degreed statisticians into the hands of project and design engineers
- Quality function deployment (QFD), which focused on better defining new product specifications using customers' input and competitive analysis (QFD will be further discussed in Chap. 3 in this book)

Several American homegrown design, quality, and cost improvement tools also emerged to meet the Japanese industry challenge. They include the following tools that were especially aimed at the project development cycle:

- Design for manufacturing/assembly (DFM/DFA) to reduce product manufacturing cost
- Concurrent/collaborative engineering (CE) to focus on design project collaboration among the different parts of the organization and shortening the new product development cycle
- Six Sigma (6σ) to merge the quality issues of design and manufacturing

These tools and techniques were developed to augment the Japanese-developed tools for improving new product development quality and reducing cost. They focused on distinctly American cultural and managerial nuances, being quite different from their Japanese counterparts.

A historical listing of these changing trends is summarized in Table 1.1. The chronology of the effect of these trends will be examined for each decade as follows.

1.1 The 1980s

During this decade, companies were focused on increasing profits by matching their global competitors in reducing the cost of new products while at the same time speeding up their development. In addition, companies were intent on incorporating new technology into their products as fast as possible and winning the race for their customers' thirst for state-of-the-art product performance. This resulted in the need for quickly introducing successive new products, with increased technology adoption.

Innovative engineering project management was needed to adopt these new tools for reducing cost, increasing quality, and shortening development time. The specific tools and techniques of choice were as follows (to be explained later):

- DFM tools: Boothroyd-Dewhurst Incorporated (BDI) and GE/Hitachi (GE/Hit) methodologies
- Variability reduction and design optimizing tools: classical DoE and the Taguchi method
- Design for quality tools: 6σ and process capability based on *Cp* and *Cpk* methodologies

Decade	Focus	Customer Expectation	Product Development	Organizational Emphasis
1980s	Speed Margin	Satisfied	Functional silo DFM/remove walls	Serial design
1990s	Design process Market share	Delighted	Project teams Bring in marketing	Concurrent engineering
2000s	Global market size	Customized products	Distributed teams Supply chain	Collaborative engineering
2010s	Grow new markets, globally		Leverage intellectual capital/ resources Design anywhere, build anywhere	

Table 1.1 Changing Historical Trends for Engineering Project Management

1.1.1 Design for Manufacturing

DFM concepts were used for inputting feedback from the manufacturing part of the organization into the design cycle. This would lead to reducing the number of parts in new products, encouraging the reuse of older parts and reducing manufacturing cost. The DFM analysis should be performed early in the design cycle so that recommendations could be fully implemented in new products. Two techniques predominated:

1. The Boothroyd-Dewhurst Incorporated (BDI) is a system for rating parts in an assembly by assigning a number to each part on the basis of part geometry. The numbers resulted in assembly-time estimates and were then added up to determine total assembly time. There are two numbering systems, one for the handling and another for insertion of parts. The resultant estimated assembly time is compared with an ideal design time and a figure of merit (design efficiency) is determined. The ideal design time is based on parts geometries that are symmetrical and easy to insert. The assembly can then be redesigned, using three simple guidelines to help reduce the number of parts and assembly time. The process is continued until the maximum design efficiency is achieved.

   The BDI methodology encourages the ease of assembly by focusing on parts' geometry to make the parts more symmetrical (or alternatively exaggerating asymmetry) and for easier part orientation for subsequent handling by manual or automatic means. In addition, it encourages the proper aligning of parts for ease of insertion into the assembly through geometry or part features. The same principles could be used for other parts or materials. For example, lower cost could be obtained by reducing the number of hole diameters in sheet-metal assembly or moving the parting lines in the case of plastic parts mold design.

   The BDI methodology grew out of universal part-numbering systems used to identify part geometry by a specialized numbering code obtained from tables. A successful BDI analysis results in the reduction of the number of parts and time to assemble these parts. This is expressed through a higher design efficiency of the assembly as compared with an ideal assembly of symmetrical and easy to join parts.

2. The GE/Hit method was developed jointly by the two companies. The GE/Hit method focuses on reducing the number of assembly motions and encouraging design engineers to have parts that assemble together through downward motions, called T-downs. The GE/Hit method is a natural growth of industrial engineering systems that measured assembly times through time and motion studies on the manufacturing floor. It discourages side or upward-assembly motions, tight fits, and micro-positioning as well as complex joining or assembly processes. These non-downward motions are given penalty points and then added up to produce a total estimated assembly time. This total time is compared to an ideal number of downward motions or T-downs. The ratio of the two numbers is the design efficiency.

There are no specific guidelines for increasing the efficiency number for the GE/Hit method. It is adaptable to comparing competing designs or encouraging the design team to explore ideas to increase the design efficiency. The GE/Hit method was not adopted as widely as the BDI methodology, given that it does not encourage the reduction of parts, or provide specific instructions to increase the design efficiency. Table 1.2 is a comparison of the two design efficiency measurement systems.

Both techniques above use symbolic or numerical methods to label individual parts based on previous techniques for part-number classifications (BDI) or time and motion assembly analysis (GE/Hit). They allocate penalty points to each part that was not ideal (either by geometry or assembly motion). By adding all of the parts in an assembly, a design efficiency (or assembly efficiency) score is generated that compares the assembly design to one with ideal parts or down motions. The goal is to have the project manager (PM) set a desired efficiency number, which is significantly higher than the company's current product portfolio as well as benchmarking the competition's design efficiency. In summary, these DFM/DFA methodologies provided the following process for reducing cost:

- Evaluate new or alternate designs or assemblies, including competitors' designs.

BDI	GE/Hitachi
Emphasis on part geometry	Emphasis on the assembly process
Includes guidelines for auto/robotic assembly	Focus on manual assembly
Compares to ideal design	Compares alternatives
Focus on reducing part numbers	Does not penalize excess parts
Measures assembly time from DFM tables	Normalizes assembly time to down motions

TABLE 1.2 Comparison of DFM Design Efficiency Techniques

- Use assembly motions or part geometries to generate a design efficiency number as compared with an ideal or alternate design.
- Improve the design efficiency by targeting an efficiency goal. A numerical goal or a percentage improvement can be set based on the design efficiency of current products or the competition's designs.
- A redesign is encouraged if the new design does not meet the design efficiency goals.

Many comparisons have been made of the two methods (BDI, GE/Hit), as well as methods-time measurement (MTM) studies, to calculate the design efficiency and the total assembly time for global products. Different teams from University of Massachusetts Lowell graduate students compared the same design using the same methodology (BDI or GE/Hit). Their design-efficiency and assembly-time estimates were well within 20 percent among the different teams using the same method. When the teams compared the same assemblies by using alternative methods (BDI or GE/Hit), the differences in estimating the design efficiency were quite large. The difference narrowed to within 20 percent when the ideal number of parts (as prescribed by the BDI method) was reached. This indicates that a crucial step is needed for the GE/Hit method to reduce the number of parts before conducting the efficiency analysis.

The results of successfully implementing DFM/DFA were important reductions in manufacturing cost. This was partially offset by unintended consequences as well as posing some interesting dilemmas of DFM:

- The need for early manufacturing input (especially when using the supply chain) required propriety methodology to shield new products information from the competition.
- The emphasis on fewer parts and changing their geometry resulted in part integration and more complex but fewer parts.
- The financial relationship of the design project with the supply chain was altered. The tooling cost of all parts might be the same, but the profile is changed: more tooling costs for some parts, but lower number of different parts.
- While assembly time was reduced, it may have resulted in increased cost of fabricating the individual complex parts as well as their tooling costs.
- Disassembly of new products became more difficult as the parts became more complex and joined mostly by nonscrew fastening methods such as snap fits.
- Special tools were required for repairs and/or disassembly of complex parts that are snap fits.
- Cost of repair and replacement of complex parts became more expensive.
- Spare parts inventory and warehouse management were impacted considering the higher cost of these complex parts.

A typical DFM/DFA design optimization process is shown in Fig. 1.1, the mounting of a printed circuit board (PCB) in a product. It shows the new design (the one to the right of the figure) as having a better design efficiency through the reduction of the number of parts by eliminating spacers, screws, and nuts. The new parts are

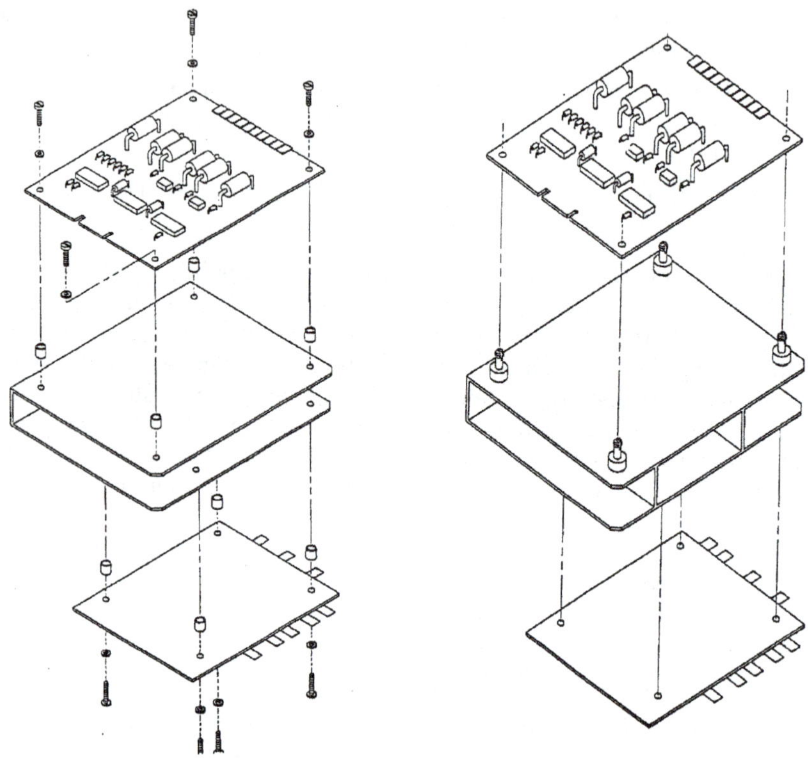

FIGURE 1.1 PCB mounting design optimization.

fewer in numbers but geometrically complex and more expensive than the parts they replaced.

The need for manufacturing input early into the design cycle, especially when the part originated in the supply chain, resulted in the need for early supplier involvement (ESI). ESI creates an additional layer of security and intellectual property concerns for new products. Techniques had to be developed to ensure new product propriety while the suppliers bid on the product early in the design cycle.

Production capability needs could influence DFM/DFA input in terms of handling product size and aspect ratio, especially in the PCB panel fabrication process. The PCB panels are set up in standardized sizes, and DFM input should indicate the desired PCB geometry in the product. Maintaining the ratio of the product PCBs to the panel geometry will result in substantial set-up and materials cost reduction for the PCB fabrications process.

A typical progression of reducing the number of parts as well as assembly time per product is shown for popular printers in the 1980s in Fig. 1.2. The data is based on analysis performed by Hewlett Packard (HP). It shows that the IBM Proprinter, which was designed with DFM emphasis, far outperformed its competitors in numbers of parts and assembly time. Each successive generation of printer products did manage to reduce both parts and assembly time resulting in lower costs.

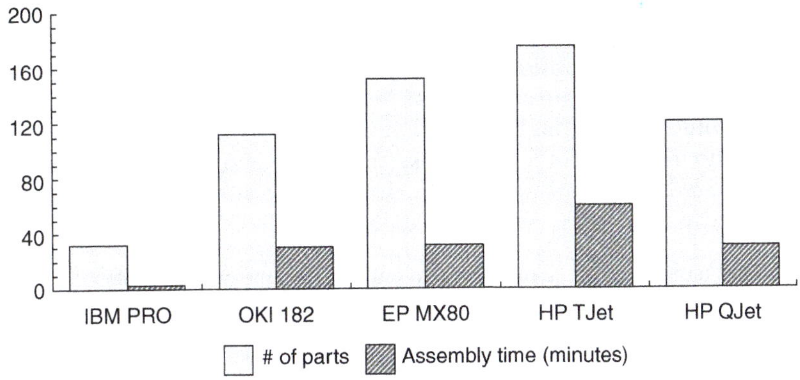

FIGURE 1.2 Printer competitive benchmarks in the 1980s: (IBM Proprinter; OKI data 182; Epson MX80; HP ThinkJet; and HP QuietJet).

1.1.2 Reducing Variability and Optimizing the Design

A popular technique to reduce costs by optimizing the design was the use of DoE techniques. DoE is a tool to vary the design outcome or reduce the variability of the individual parts of the product. DoE resolves design problems through fast design optimization using sample product production without using complex numerical and/or simulation techniques. Before the wide adoption of DoE in the 1980s, the science was the privy of statisticians with advanced degrees. Many companies hired statisticians with a master's degree or Ph.D. to help educate their engineering and product design staff. However, these new employees did not have the product and the process understanding to be effective beyond teaching the DoE concepts to engineers and helping them with interpreting the resulting analysis. Eventually, they had to be eased out of their positions once they completed tutoring their training programs.

DoE can be used to gain more information about the design by having the project teams decide on the relevant factors that influence a design and by assigning different values to each factor. DoE can help discover how to manipulate a design by understanding the effect of each factor and whether the design outcome or the variability can be improved, as shown in Fig. 1.3.

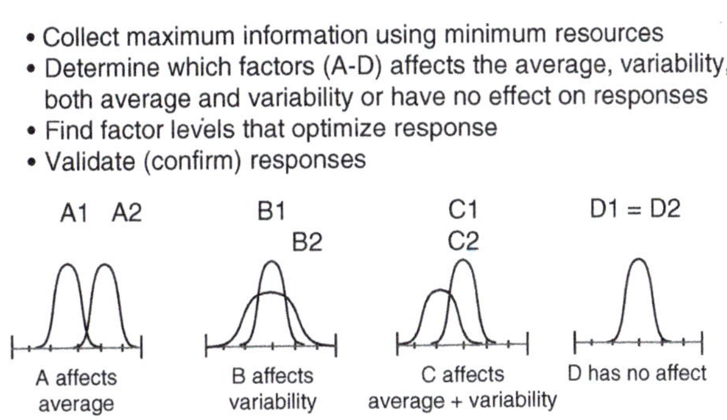

FIGURE 1.3 Objective of a typical DoE with four factors.

The Taguchi method, which became popular in the 1980s, is a technique to simplify the DoE analysis. Developed by Genichi Taguchi of the Nippon Telephone Company, this was a simplification of the classical DoE process through a step-by-step methodology and the use of special tools and techniques to render the design and analysis much more straightforward. It was advocated by Don Clausing, a manager at Xerox who translated Taguchi's books and showed how the Xerox copier design could be optimized using DoE techniques. The major contributions of Taguchi to simplify DoE were as follows:

1. Visualizing the DoE. Taguchi introduced the visual tools of linear graphs and triangular tables to clarify the experiment design stage and to show the relationship between factors and their interactions.
2. Any interaction that was greater than a two-way interaction was considered to be small, therefore it could be ignored. Another factor can thus be assigned to the interaction column, ignoring the confounding (explained below) of the factor with the interaction. That allowed for the following:
   - Reducing the number of experiments based on the number of factors. For example, a five-factor experiment at two levels would require 2^5 or 32 experiments, and is called full factorial. The five-factor DoE can be performed in 16 experiments by ignoring four- and three-way interactions and assigning the fifth factor to the four-way interaction column. This is called a "half fraction factorial."
   - Three factors at three levels would require 3^3 or 27 full factorial experiments. If all interactions are ignored, then up to four factors could be analyzed in a $2^3 = 9$ experiments DoE. The nine experiments are called "screening or saturated design," since all of the columns are being assigned to factors.
   - Seven factors at two levels can be performed in a full factorial 2^7 or 128 experiments. By assuming that all interactions (two and three way) are small and not statistically significant, DoE analysis can be performed using only eight saturated design experiments.
   - If interactions are assumed to be not significant, then factors can be assigned to the interaction columns in a half-fraction or saturated design DoE. If the assumption is incorrect, and interactions are indeed significant as found by subsequent statistical analysis, then they could adversely interfere with the analysis of the DoE and the computed contribution of each factor. This is called "confounding" of the factor with the interactions. It can be resolved by running more experiments such as the remaining half fraction or the full set of full factorial experiments.

In strategically choosing the eight experiments with seven factors at two levels, the effect of each factor could be solved with eight equations using a simultaneous equations solution or Cramer's rule. This screening or saturated design of eight experiments can be seen in Fig. 1.4. Instead of using 128 experiments, full factorial design to study seven factors, only eight experiments (shown as filled in squares) are needed for analysis of how each factor effects the design, assuming all interactions are nonsignificant.

A good DoE methodology for seven factors at two levels is to run the screening DoE with eight experiments. The results can be analyzed statistically to determine which

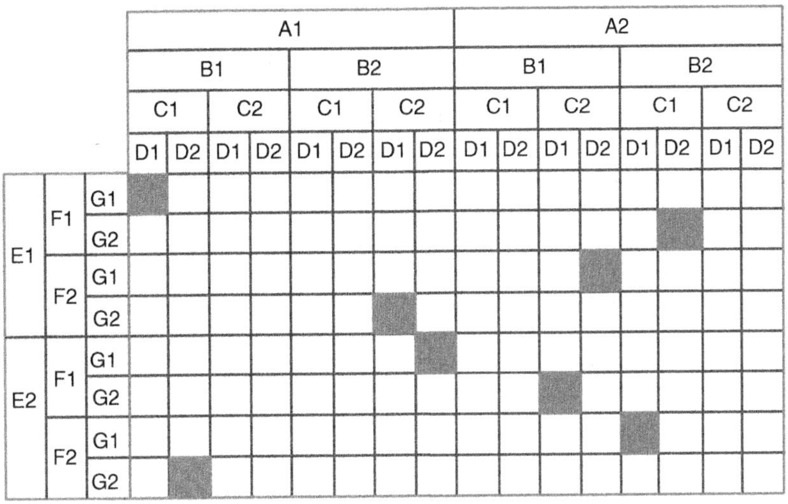

FIGURE 1.4 DoE with seven factors at two levels: saturated versus full factorial experiments.

factors are significant and which factors are not. If only three factors are shown to be significant, then a full factorial eight experiments set (2^3) can be run again with no confounding. By using the two sets of eight experiments for an analysis of seven factors in contrast to a single 128 full factorial experiments, a reasonably good analysis of the seven factors and their influence on the design can be calculated.

The same methodology can be applied to three-level factors. A four factor at three-levels DoE requires 81 (3^4) experiments in full factorial mode with calculating all interactions. This could be reduced by performing two sets of three-level DoEs with nine experiments each. The first set would be a saturated design 9 (3^2) experiments DoE, to identify the two most important out of the four factors. The second set of nine experiments can be used to analyze the two factors and their interactions.

3. Simplification of DoE analysis. Taguchi introduced three tools to augment and simplify the DoE analysis:
   - The percentage contribution $p\%$ is a mathematical expression to determine whether a factor is significant (it does in fact affect the design being studied). It proved to be more useful to engineers than the statistical F test, which was traditionally used for significance determination. The Taguchi rule was that if $p\% < 5\%$, then the factor is not significant, while a value of $p\% > 5\%$ indicates that the factor is significant, and its relative importance is expressed as a percentage of its effects on the design. The $p\%$ was based on the analysis of variance (ANOVA) techniques by taking the sum of the squares of each factor SS_F and subtracting a term equal to the variance of the error multiplied by the degrees of freedom (DoF) of that factor. The new term is called "modified sum of the squares" for each factor (SS_F'). The total modified sum of the squares (SS_T') was made equal to the original sum of the squares (SS_T) by adding back the subtractions made to each SS_F'. The $p\%$ contribution is the percentage representation of the SS_T of the factors in the experiment.

- The S/N ratio is a Taguchi developed term based on his electrical engineering background. S/N is a formulation to determine the effect of each factor relative to the variability of the design being studied. This formed the basis for the variability reduction studies to reduce new product and process manufacturing costs.

  S/N was derived from repeating the DoE several times and then taking each experiment and generating a new expression (S/N) based on the variance of the repetition of each experiment line. This transformation from multiple repetitions to a single S/N number was based on different formulas, depending on whether the design was to be manipulated to a target, maximum or minimum value. Taguchi provided for a transformation formula for each case, making the most positive level of a significant factor the desired level for reducing variability.
- In order to reconcile the average and variability analysis results of the DoE experiments, Taguchi developed another concept to compare the results of the average versus the variability effects of the selected factors. He called it the quality loss function (QLF). QLF is defined as an additional monetary burden created by shipping products that do not exactly meet the design specifications. A QLF value of zero implies the specifications of the shipped products are exactly equal to design nominal with zero variability. QLF gave a simple guideline on how to balance the elimination of quality problems (potential defects for the customers), either by having the factory test to a tighter specification of the new products, versus having the product recalled or repaired in the field at a much higher cost.

An example of the choice offered by the QLF concept is the plating specifications on a new part. Let's assume that the plating thickness on a PCB was specified to be gold plated for better conductivity at .001 in or 1 mil minimum. A plating process normally produces parts in a normal distribution, with certain variability as measured by the standard deviation σ (sigma). What should the engineer specify? If the variability is large, then the engineer should specify a nominal of plating gold as a number that is much higher than 1 mil. This will ensure that most of the plated PCBs will meet the 1-mil spec. However, this strategy will cost much more in plating time and in gold material consumption. If the nominal of the specification is reduced to a number closer to 1 mil to reduce gold use, then many rejects will occur, resulting in the PCBs having to be re-plated. This is the balance dilemma that QLF is trying to resolve.

Engineers did not take to QLF as readily as they did to DoE. One of the concerns was the monetary estimate of the quality loss. Another was the conscious knowledge of shipping potentially defective products, which did not suit the PM philosophy of having to fix any problems that they became aware of. "Zero defects" was the slogan of the day then, and QLF, with its balance of least evils, was not a discussion that would occur at a global American company. The immediate answer was to have zero defects. Eventually that led to the new American-based concept of 6σ.

Perhaps an example of that difference in global companies' mentality of balancing defects in the design or manufacturing versus being susceptible to repair and recalls in the field is the recent Toyota dilemma of uncontrolled acceleration. Toyota determined the cause was due to poor design of the attachments of mats under the driver seat in their cars. Most cars had a "kill" switch that turned off the engines when the ignition key is withdrawn, not so in Toyota cars. In thinking that uncontrolled acceleration would never happen in its cars, and not wanting to incur the extra costs of a kill switch, Toyota proved vulnerable to potential problems. It turns out these balancing defects in factory versus the field as prescribed in QLF did not work in the Toyota case.

These Taguchi techniques were further promulgated through the auto and electronics industries through the American Supplier Institute (ASI), which was set up by the Big Three auto companies to train their suppliers. They held annual meetings in Detroit in the 1980s, where engineers from various industries and companies presented their DoE experiments and met Mr. Taguchi himself. Some companies went as far as decreeing that every engineer should perform one DoE or quality improvement project per year. They set up training programs so that every engineer received DoE training. This focus on DoE was eventually transferred to 6σ, where a Six Sigma black-belt expert is required to complete a quality-improvement project, usually a DoE, to reach that rank.

1.1.3 Design for Quality Tools: Six Sigma and Process Capability *Cp* and *Cpk*

The project management emphasis began to be augmented from strictly managing the engineering design function and producing a single working prototype into more new product realization responsibilities. The PM is now tasked with the additional responsibilities of focusing on reducing manufacturing costs and improving manufacturing quality as well as getting involved in negotiating new product specifications with marketing and customers.

Quality and cost were historically considered to be the responsibility of the manufacturing part of the high-technology company. Technology was the driving factor for these companies, resulting in high product price. The manufacturing cost was considered to be minimal, and, in many cases, it was about 1 to 3 percent of the selling price of the technology product. This led to little emphasis being placed on optimizing quality or cost of high-technology products.

As global competition increased for technology products, the need to focus on product quality and manufacturing costs at the design stage increased significantly. While tools such as DFM focused on reducing cost, and DoE focused on variability reductions, they did not address the need of reducing cost through quality means. Having a higher design quality meant a reduction in factory rejects, rework, and lower field recalls and returns. In addition, higher design quality resulted in much lower technical resources needed in manufacturing by reducing the amount of testing required to weed out defective components in production operations of high-technology products. In some companies, testing costs were equivalent to one-third of the manufacturing labor costs, due to the higher skills and salaries of the test technicians. Customer satisfaction is increased with higher-quality designs with less latent defects in the products and therefore better quality performance in the field. All of these design-quality benefits resulted in real tangible overall lower cost for new high-technology products.

The premise of Six Sigma is simple. A quality level is established by contrasting the manufacturing variability (which is the responsibility of the manufacturing function), with the product specification (which is the responsibility of the design function). The PM's job is to negotiate a quality level, which is either a multiple of sigma (i.e., 4.5 σ), or process capability (*Cp*, capability of the process), between the two functions. Manufacturing's goal is to have the largest possible specification width in order to easily make the product with the smallest number of defects out of specifications. The design engineering goal is to set specifications as tight as possible to reduce the number of variability in the performance of the new product. At the minimum case, the techniques of 6σ would encourage design and manufacturing engineers to communicate about quality and cost early in the design phase. Some of the formulas for 6σ are as follows:

$$Cp = \frac{USL - LSL}{6\sigma} = \frac{\pm SL}{\pm 3\sigma}$$

$$Cpk = \min of \ \frac{USL - \mu}{3\sigma} \quad or \quad \frac{\mu - LSL}{3\sigma}$$

where
Cp = Process capability, negotiated by the PM
Cpk = Process capability as influenced by the average of the manufacturing process deviation from specifications nominal ($\mu - N$)
SL (USL, LSL) = Specifications limit (upper, lower), determined by the design team and tempered by manufacturing input, using Cp or Cpk
μ = Manufacturing process average [as contrasted by the specifications nominal (N) set by the design engineer]
σ = Standard deviation of the process

If the specification limits were set at ±6σ, then $Cp = 2$. If the specifications limits were set at 4.5 σ, then the $Cp = 1.5$, which is alternately called 4.5 σ design. The Cpk is a more exact definition of design quality, which is changed when the manufacturing process average is not equal to the specification nominal (N). Cpk makes the process capability (or the 6σ) concept sensitive to the fact that products can be shipped where some of their performance is not equal to the specification nominal, yet they fall within the acceptable range of specifications interval (±SL). This is the case addressed by the Taguchi QLF term, balancing the process average deviation or reducing the variability of manufactured products.

Specifying a Six Sigma design for a new product implies that the specification of the product and its parts are much wider than the variability due to manufacturing. Therefore, even if there is a substantial shift in the manufacturing process—whether by changing the process average or increasing the variability—this shift does not significantly impact the performance of the product. The quality of the product is maintained, with less defects and customer complaints. A typical 6σ design or process capability $Cp = 2$ is shown in Fig. 1.5.

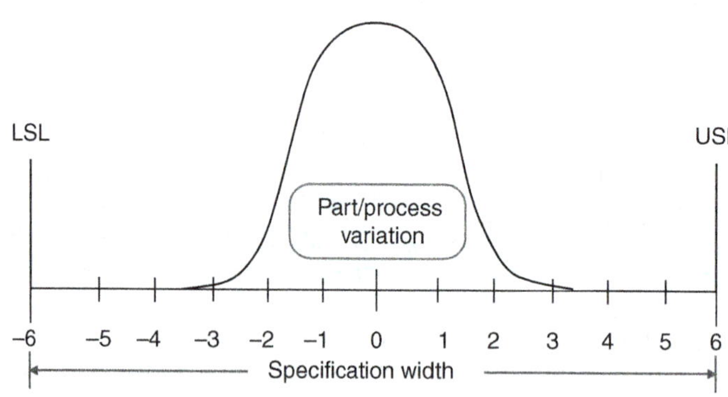

FIGURE 1.5 Typical 6σ design or process capability, Cp = 2.0.

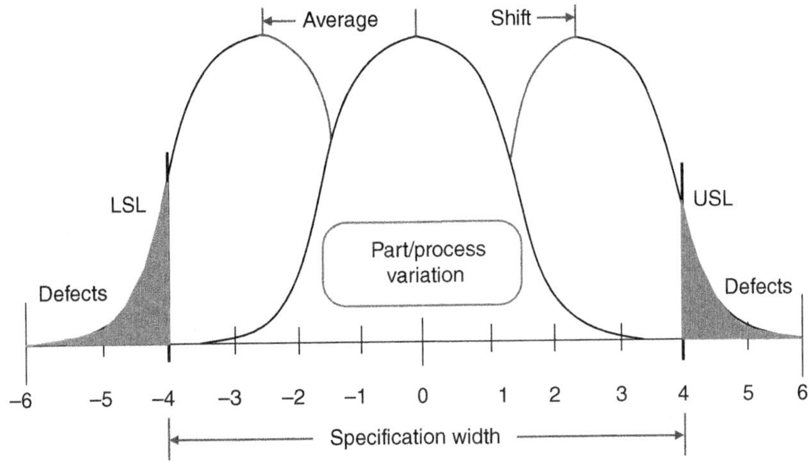

FIGURE 1.6 Typical 4σ design with mean shift or process capability $Cp = 1.33$.

If a product design goal was much lower than 6σ—for example, 4σ or $Cp = 1.33$—then a small shift in either manufacturing process average or variability will result in a much larger defect rate in manufacturing and therefore much higher quality cost, as shown in Fig. 1.6. In this figure, it can be illustrated that a shift in the manufacturing process average (either to the left/smaller value or right/larger value) will quickly result in a much higher defect rate, hence higher manufacturing costs. This is because the normal distribution is shaped like a bell curve. A shift of the process average to the right significantly increases the defects in the product due to more products with process average being closer to the USL. At the other end, there is slightly less defects on the process average being farther away from the LSL. To express this condition mathematically, *Cpk* is measured at both ends of the specification limits, with the right side *Cpk*< left side *Cpk* if the average shifts to the right. The resulting *Cpk* is the minimum of the two *Cpk* values. A lower *Cpk* means lower design quality.

The potential defect rate is based on the interaction of the specifications limits (expressed as multiples of σ) versus the manufacturing variability distribution with an average of μ and standard deviation σ. Any manufacturing distribution can be transformed into a standard normal distribution (SND) by a process called the *z* transformation:

$$z = \frac{x-\mu}{\sigma} = 3^* \; Cp = 3^* \; Cpk \text{ (with no average shift from specification nominal)}$$

The number of defects is then determined from the SND (sometimes called the *z* distribution), which returns *f*(z), for defects due to manufactured products being outside of specifications. The defects from both sides of the specifications limits should be added. The SND is given in Table 1.3.

There are many variations of 6σ quality, represented by two of the major companies that are the premier practitioners: Motorola and GE. Motorola popularized the 6σ concepts through its Motorola University training center in the 1980s. The *Cp* parameters of average (μ) and standard deviation (σ) could be obtained from control charts, the traditional method of monitoring quality through manufacturing. American Telephone

Z	f(Z)	Z	f(Z)	Z	f(Z)	Z	f(Z)	Z	f(Z)	Z	f(Z)
0	0.5										
0.01	0.50399	1.01	0.84375	2.01	0.97778	3.01	0.99869	4.01	0.99996963	5.01	0.99999972742
0.02	0.50798	1.02	0.84614	2.02	0.97831	3.02	0.99874	4.02	0.99997089	5.02	0.99999974123
0.03	0.51197	1.03	0.84849	2.03	0.97882	3.03	0.99878	4.03	0.99997210	5.03	0.99999975436
0.04	0.51595	1.04	0.85083	2.04	0.97932	3.04	0.99882	4.04	0.99997326	5.04	0.99999976685
0.05	0.51994	1.05	0.85314	2.05	0.97982	3.05	0.99886	4.05	0.99997438	5.05	0.99999977873
0.06	0.52392	1.06	0.85543	2.06	0.98030	3.06	0.99889	4.06	0.99997545	5.06	0.99999979002
0.07	0.52790	1.07	0.85769	2.07	0.98077	3.07	0.99893	4.07	0.99997648	5.07	0.99999980076
0.08	0.53188	1.08	0.85993	2.08	0.98124	3.08	0.99896	4.08	0.99997747	5.08	0.99999981096
0.09	0.53586	1.09	0.86214	2.09	0.98169	3.09	0.99900	4.09	0.99997842	5.09	0.99999982066
0.1	0.53983	**1.1**	0.86433	**2.1**	0.98214	**3.1**	0.99903	**4.1**	0.99997933	**5.1**	0.99999982988
0.11	0.54380	1.11	0.86650	2.11	0.98257	3.11	0.99906	4.11	0.99998021	5.11	0.99999983864
0.12	0.54776	1.12	0.86864	2.12	0.98300	3.12	0.99910	4.12	0.99998105	5.12	0.99999984696
0.13	0.55172	1.13	0.87076	2.13	0.98341	3.13	0.99913	4.13	0.99998185	5.13	0.99999985487
0.14	0.55567	1.14	0.87286	2.14	0.98382	3.14	0.99916	4.14	0.99998262	5.14	0.99999986238
0.15	0.55962	1.15	0.87493	2.15	0.98422	3.15	0.99918	4.15	0.99998337	5.15	0.99999986952
0.16	0.56356	1.16	0.87698	2.16	0.98461	3.16	0.99921	4.16	0.99998408	5.16	0.99999987630
0.17	0.56749	1.17	0.87900	2.17	0.98500	3.17	0.99924	4.17	0.99998476	5.17	0.99999988274
0.18	0.57142	1.18	0.88100	2.18	0.98537	3.18	0.99926	4.18	0.99998542	5.18	0.99999988885
0.19	0.57535	1.19	0.88298	2.19	0.98574	3.19	0.99929	4.19	0.99998604	5.19	0.99999989465
0.2	0.57926	**1.2**	0.88493	**2.2**	0.98610	**3.2**	0.99931	**4.2**	0.99998665	**5.2**	0.99999990017
0.21	0.58317	1.21	0.88686	2.21	0.98645	3.21	0.99934	4.21	0.99998722	5.21	0.99999990540
0.22	0.58706	1.22	0.88877	2.22	0.98679	3.22	0.99936	4.22	0.99998778	5.22	0.99999991036
0.23	0.59095	1.23	0.89065	2.23	0.98713	3.23	0.99938	4.23	0.99998831	5.23	0.99999991508
0.24	0.59483	1.24	0.89251	2.24	0.98745	3.24	0.99940	4.24	0.99998882	5.24	0.99999991955
0.25	0.59871	1.25	0.89435	2.25	0.98778	3.25	0.99942	4.25	0.99998930	5.25	0.99999992380
0.26	0.60257	1.26	0.89617	2.26	0.98809	3.26	0.99944	4.26	0.99998977	5.26	0.99999992783
0.27	0.60642	1.27	0.89796	2.27	0.98840	3.27	0.99946	4.27	0.99999022	5.27	0.99999993165
0.28	0.61026	1.28	0.89973	2.28	0.98870	3.28	0.99948	4.28	0.99999065	5.28	0.99999993528
0.29	0.61409	1.29	0.90147	2.29	0.98899	3.29	0.99950	4.29	0.99999106	5.29	0.99999993872
0.3	0.61791	**1.3**	0.90320	**2.3**	0.98928	**3.3**	0.99952	**4.3**	0.99999145	**5.3**	0.99999994198
0.31	0.62172	1.31	0.90490	2.31	0.98956	3.31	0.99953	4.31	0.99999183	5.31	0.99999994507
0.32	0.62552	1.32	0.90658	2.32	0.98983	3.32	0.99955	4.32	0.99999219	5.32	0.99999994801
0.33	0.62930	1.33	0.90824	2.33	0.99010	3.33	0.99957	4.33	0.99999254	5.33	0.99999995079
0.34	0.63307	1.34	0.90988	2.34	0.99036	3.34	0.99958	4.34	0.99999287	5.34	0.99999995343
0.35	0.63683	1.35	0.91149	2.35	0.99061	3.35	0.99960	4.35	0.99999319	5.35	0.99999995593
0.36	0.64058	1.36	0.91308	2.36	0.99086	3.36	0.99961	4.36	0.99999349	5.36	0.99999995830
0.37	0.64431	1.37	0.91466	2.37	0.99111	3.37	0.99962	4.37	0.99999378	5.37	0.99999996054
0.38	0.64803	1.38	0.91621	2.38	0.99134	3.38	0.99964	4.38	0.99999406	5.38	0.99999996267
0.39	0.65173	1.39	0.91774	2.39	0.99158	3.39	0.99965	4.39	0.99999433	5.39	0.99999996469

TABLE 1.3 Standard Normal Distribution

Z	f(Z)	Z	f(Z)	Z	f(Z)	Z	f(Z)	Z	f(Z)	Z	f(Z)
0.4	0.65542	**1.4**	0.91924	**2.4**	0.99180	**3.4**	0.99966	**4.4**	0.99999458	**5.4**	0.99999996660
0.41	0.65910	1.41	0.92073	2.41	0.99202	3.41	0.99968	4.41	0.99999483	5.41	0.99999996842
0.42	0.66276	1.42	0.92220	2.42	0.99224	3.42	0.99969	4.42	0.99999506	5.42	0.99999997013
0.43	0.66640	1.43	0.92364	2.43	0.99245	3.43	0.99970	4.43	0.99999528	5.43	0.99999997176
0.44	0.67003	1.44	0.92507	2.44	0.99266	3.44	0.99971	4.44	0.99999550	5.44	0.99999997330
0.45	0.67364	1.45	0.92647	2.45	0.99286	3.45	0.99972	4.45	0.99999570	5.45	0.99999997476
0.46	0.67724	1.46	0.92785	2.46	0.99305	3.46	0.99973	4.46	0.99999590	5.46	0.99999997614
0.47	0.68082	1.47	0.92922	2.47	0.99324	3.47	0.99974	4.47	0.99999609	5.47	0.99999997744
0.48	0.68439	1.48	0.93056	2.48	0.99343	3.48	0.99975	4.48	0.99999626	5.48	0.99999997868
0.49	0.68793	1.49	0.93189	2.49	0.99361	3.49	0.99976	4.49	0.99999644	5.49	0.99999997985
0.5	0.69146	**1.5**	0.93319	**2.5**	0.99379	**3.5**	0.99977	**4.5**	0.99999660	**5.5**	0.99999998096
0.51	0.69497	1.51	0.93448	2.51	0.99396	3.51	0.999776	4.51	0.99999676	5.51	0.99999998201
0.52	0.69847	1.52	0.93574	2.52	0.99413	3.52	0.999784	4.52	0.99999691	5.52	0.99999998301
0.53	0.70194	1.53	0.93699	2.53	0.99430	3.53	0.999792	4.53	0.99999705	5.53	0.99999998395
0.54	0.70540	1.54	0.93822	2.54	0.99446	3.54	0.999800	4.54	0.99999718	5.54	0.99999998484
0.55	0.70884	1.55	0.93943	2.55	0.99461	3.55	0.999807	4.55	0.99999732	5.55	0.99999998568
0.56	0.71226	1.56	0.94062	2.56	0.99477	3.56	0.999815	4.56	0.99999744	5.56	0.99999998648
0.57	0.71566	1.57	0.94179	2.57	0.99492	3.57	0.999821	4.57	0.99999756	5.57	0.99999998723
0.58	0.71904	1.58	0.94295	2.58	0.99506	3.58	0.999828	4.58	0.99999767	5.58	0.99999998794
0.59	0.72240	1.59	0.94408	2.59	0.99520	3.59	0.999835	4.59	0.99999778	5.59	0.99999998862
0.6	0.72575	**1.6**	0.94520	**2.6**	0.99534	**3.6**	0.999841	**4.6**	0.99999789	**5.6**	0.99999998925
0.61	0.72907	1.61	0.94630	2.61	0.99547	3.61	0.999847	4.61	0.99999798	5.61	0.99999998986
0.62	0.73237	1.62	0.94738	2.62	0.99560	3.62	0.999853	4.62	0.99999808	5.62	0.99999999043
0.63	0.73565	1.63	0.94845	2.63	0.99573	3.63	0.999858	4.63	0.99999817	5.63	0.99999999096
0.64	0.73891	1.64	0.94950	2.64	0.99585	3.64	0.999864	4.64	0.99999826	5.64	0.99999999147
0.65	0.74215	1.65	0.95053	2.65	0.99598	3.65	0.999869	4.65	0.99999834	5.65	0.99999999196
0.66	0.74537	1.66	0.95154	2.66	0.99609	3.66	0.999874	4.66	0.99999842	5.66	0.99999999241
0.67	0.74857	1.67	0.95254	2.67	0.99621	3.67	0.999879	4.67	0.99999849	5.67	0.99999999284
0.68	0.75175	1.68	0.95352	2.68	0.99632	3.68	0.999883	4.68	0.99999856	5.68	0.99999999325
0.69	0.75490	1.69	0.95449	2.69	0.99643	3.69	0.999888	4.69	0.99999863	5.69	0.99999999363
0.7	0.75804	**1.7**	0.95543	**2.7**	0.99653	**3.7**	0.999892	**4.7**	0.99999870	**5.7**	0.99999999399
0.71	0.76115	1.71	0.95637	2.71	0.99664	3.71	0.999896	4.71	0.99999876	5.71	0.99999999433
0.72	0.76424	1.72	0.95728	2.72	0.99674	3.72	0.999900	4.72	0.99999882	5.72	0.99999999466
0.73	0.76730	1.73	0.95818	2.73	0.99683	3.73	0.999904	4.73	0.99999888	5.73	0.99999999496
0.74	0.77035	1.74	0.95907	2.74	0.99693	3.74	0.999908	4.74	0.99999893	5.74	0.99999999525
0.75	0.77337	1.75	0.95994	2.75	0.99702	3.75	0.999912	4.75	0.99999898	5.75	0.99999999552
0.76	0.77637	1.76	0.96080	2.76	0.99711	3.76	0.999915	4.76	0.99999903	5.76	0.99999999578
0.77	0.77935	1.77	0.96164	2.77	0.99720	3.77	0.999918	4.77	0.99999908	5.77	0.99999999602
0.78	0.78230	1.78	0.96246	2.78	0.99728	3.78	0.999922	4.78	0.99999912	5.78	0.99999999625
0.79	0.78524	1.79	0.96327	2.79	0.99736	3.79	0.999925	4.79	0.99999917	5.79	0.99999999647

Table 1.3 *(Continued)*

Z	f(Z)	Z	f(Z)	Z	f(Z)	Z	f(Z)	Z	f(Z)	Z	f(Z)
0.8	0.78814	**1.8**	0.96407	**2.8**	0.99744	**3.8**	0.999928	**4.8**	0.99999921	**5.8**	0.99999999667
0.81	0.79103	1.81	0.96485	2.81	0.99752	3.81	0.999930	4.81	0.99999924	5.81	0.99999999687
0.82	0.79389	1.82	0.96562	2.82	0.99760	3.82	0.999933	4.82	0.99999928	5.82	0.99999999705
0.83	0.79673	1.83	0.96638	2.83	0.99767	3.83	0.999936	4.83	0.99999932	5.83	0.99999999722
0.84	0.79955	1.84	0.96712	2.84	0.99774	3.84	0.999938	4.84	0.99999935	5.84	0.99999999738
0.85	0.80234	1.85	0.96784	2.85	0.99781	3.85	0.999941	4.85	0.99999938	5.85	0.99999999753
0.86	0.80511	1.86	0.96856	2.86	0.99788	3.86	0.999943	4.86	0.99999941	5.86	0.99999999768
0.87	0.80785	1.87	0.96926	2.87	0.99795	3.87	0.999946	4.87	0.99999944	5.87	0.99999999781
0.88	0.81057	1.88	0.96995	2.88	0.99801	3.88	0.999948	4.88	0.99999947	5.88	0.99999999794
0.89	0.81327	1.89	0.97062	2.89	0.99807	3.89	0.999950	4.89	0.99999950	5.89	0.99999999806
0.9	0.81594	**1.9**	0.97128	**2.9**	0.99813	**3.9**	0.999952	**4.9**	0.99999952	**5.9**	0.99999999818
0.91	0.81859	1.91	0.97193	2.91	0.99819	3.91	0.999954	4.91	0.99999954	5.91	0.99999999828
0.92	0.82121	1.92	0.97257	2.92	0.99825	3.92	0.999956	4.92	0.99999957	5.92	0.99999999838
0.93	0.82381	1.93	0.97320	2.93	0.99831	3.93	0.999958	4.93	0.99999959	5.93	0.99999999848
0.94	0.82639	1.94	0.97381	2.94	0.99836	3.94	0.999959	4.94	0.99999961	5.94	0.99999999857
0.95	0.82894	1.95	0.97441	2.95	0.99841	3.95	0.999961	4.95	0.99999963	5.95	0.99999999865
0.96	0.83147	1.96	0.97500	2.96	0.99846	3.96	0.999963	4.96	0.99999965	5.96	0.99999999873
0.97	0.83398	1.97	0.97558	2.97	0.99851	3.97	0.999964	4.97	0.99999966	5.97	0.99999999881
0.98	0.83646	1.98	0.97615	2.98	0.99856	3.98	0.999966	4.98	0.99999968	5.98	0.99999999888
0.99	0.83891	1.99	0.97670	2.99	0.99861	3.99	0.999967	4.99	0.99999970	5.99	0.99999999895
1	0.84134	**2**	0.97725	**3**	0.99865	**4**	0.999968	**5**	0.99999971	**6**	0.99999999901
–0.01	0.49601	–1.01	0.15625	–2.01	0.02222	–3.01	0.00131	–4.01	0.00003037	–5.01	0.00000027258
–0.02	0.49202	–1.02	0.15386	–2.02	0.02169	–3.02	0.00126	–4.02	0.00002911	–5.02	0.00000025877
–0.03	0.48803	–1.03	0.15151	–2.03	0.02118	–3.03	0.00122	–4.03	0.00002790	–5.03	0.00000024564
–0.04	0.48405	–1.04	0.14917	–2.04	0.02068	–3.04	0.00118	–4.04	0.00002674	–5.04	0.00000023315
–0.05	0.48006	–1.05	0.14686	–2.05	0.02018	–3.05	0.00114	–4.05	0.00002562	–5.05	0.00000022127
–0.06	0.47608	–1.06	0.14457	–2.06	0.01970	–3.06	0.00111	–4.06	0.00002455	–5.06	0.00000020998
–0.07	0.47210	–1.07	0.14231	–2.07	0.01923	–3.07	0.00107	–4.07	0.00002352	–5.07	0.00000019924
–0.08	0.46812	–1.08	0.14007	–2.08	0.01876	–3.08	0.00104	–4.08	0.00002253	–5.08	0.00000018904
–0.09	0.46414	–1.09	0.13786	–2.09	0.01831	–3.09	0.00100	–4.09	0.00002158	–5.09	0.00000017934
–0.1	0.46017	**–1.1**	0.13567	**–2.1**	0.01786	**–3.1**	0.00097	**–4.1**	0.00002067	**–5.1**	0.00000017012
–0.11	0.45620	–1.11	0.13350	–2.11	0.01743	–3.11	0.00094	–4.11	0.00001979	–5.11	0.00000016136
–0.12	0.45224	–1.12	0.13136	–2.12	0.01700	–3.12	0.00090	–4.12	0.00001895	–5.12	0.00000015304
–0.13	0.44828	–1.13	0.12924	–2.13	0.01659	–3.13	0.00087	–4.13	0.00001815	–5.13	0.00000014513
–0.14	0.44433	–1.14	0.12714	–2.14	0.01618	–3.14	0.00084	–4.14	0.00001738	–5.14	0.00000013762
–0.15	0.44038	–1.15	0.12507	–2.15	0.01578	–3.15	0.00082	–4.15	0.00001663	–5.15	0.00000013048
–0.16	0.43644	–1.16	0.12302	–2.16	0.01539	–3.16	0.00079	–4.16	0.00001592	–5.16	0.00000012370
–0.17	0.43251	–1.17	0.12100	–2.17	0.01500	–3.17	0.00076	–4.17	0.00001524	–5.17	0.00000011726
–0.18	0.42858	–1.18	0.11900	–2.18	0.01463	–3.18	0.00074	–4.18	0.00001458	–5.18	0.00000011115
–0.19	0.42465	–1.19	0.11702	–2.19	0.01426	–3.19	0.00071	–4.19	0.00001396	–5.19	0.00000010535

TABLE 1.3 Standard Normal Distribution (*Continued*)

Z	f(Z)	Z	f(Z)	Z	f(Z)	Z	f(Z)	Z	f(Z)	Z	f(Z)
–0.2	0.42074	**–1.2**	0.11507	**–2.2**	0.01390	**–3.2**	0.00069	**–4.2**	0.00001335	**–5.2**	0.00000009983
–0.21	0.41683	–1.21	0.11314	–2.21	0.01355	–3.21	0.00066	–4.21	0.00001278	–5.21	0.00000009460
–0.22	0.41294	–1.22	0.11123	–2.22	0.01321	–3.22	0.00064	–4.22	0.00001222	–5.22	0.00000008964
–0.23	0.40905	–1.23	0.10935	–2.23	0.01287	–3.23	0.00062	–4.23	0.00001169	–5.23	0.00000008492
–0.24	0.40517	–1.24	0.10749	–2.24	0.01255	–3.24	0.00060	–4.24	0.00001118	–5.24	0.00000008045
–0.25	0.40129	–1.25	0.10565	–2.25	0.01222	–3.25	0.00058	–4.25	0.00001070	–5.25	0.00000007620
–0.26	0.39743	–1.26	0.10383	–2.26	0.01191	–3.26	0.00056	–4.26	0.00001023	–5.26	0.00000007217
–0.27	0.39358	–1.27	0.10204	–2.27	0.01160	–3.27	0.00054	–4.27	0.00000978	–5.27	0.00000006835
–0.28	0.38974	–1.28	0.10027	–2.28	0.01130	–3.28	0.00052	–4.28	0.00000935	–5.28	0.00000006472
–0.29	0.38591	–1.29	0.09853	–2.29	0.01101	–3.29	0.00050	–4.29	0.00000894	–5.29	0.00000006128
–0.3	0.38209	**–1.3**	0.09680	**–2.3**	0.01072	**–3.3**	0.00048	**–4.3**	0.00000855	**–5.3**	0.00000005802
–0.31	0.37828	–1.31	0.09510	–2.31	0.01044	–3.31	0.00047	–4.31	0.00000817	–5.31	0.00000005493
–0.32	0.37448	–1.32	0.09342	–2.32	0.01017	–3.32	0.00045	–4.32	0.00000781	–5.32	0.00000005199
–0.33	0.37070	–1.33	0.09176	–2.33	0.00990	–3.33	0.00043	–4.33	0.00000746	–5.33	0.00000004921
–0.34	0.36693	–1.34	0.09012	–2.34	0.00964	–3.34	0.00042	–4.34	0.00000713	–5.34	0.00000004657
–0.35	0.36317	–1.35	0.08851	–2.35	0.00939	–3.35	0.00040	–4.35	0.00000681	–5.35	0.00000004407
–0.36	0.35942	–1.36	0.08692	–2.36	0.00914	–3.36	0.00039	–4.36	0.00000651	–5.36	0.00000004170
–0.37	0.35569	–1.37	0.08534	–2.37	0.00889	–3.37	0.00038	–4.37	0.00000622	–5.37	0.00000003946
–0.38	0.35197	–1.38	0.08379	–2.38	0.00866	–3.38	0.00036	–4.38	0.00000594	–5.38	0.00000003733
–0.39	0.34827	–1.39	0.08226	–2.39	0.00842	–3.39	0.00035	–4.39	0.00000567	–5.39	0.00000003531
–0.4	0.34458	**–1.4**	0.08076	**–2.4**	0.00820	**–3.4**	0.00034	**–4.4**	0.00000542	**–5.4**	0.00000003340
–0.41	0.34090	–1.41	0.07927	–2.41	0.00798	–3.41	0.00032	–4.41	0.00000517	–5.41	0.00000003158
–0.42	0.33724	–1.42	0.07780	–2.42	0.00776	–3.42	0.00031	–4.42	0.00000494	–5.42	0.00000002987
–0.43	0.33360	–1.43	0.07636	–2.43	0.00755	–3.43	0.00030	–4.43	0.00000472	–5.43	0.00000002824
–0.44	0.32997	–1.44	0.07493	–2.44	0.00734	–3.44	0.00029	–4.44	0.00000450	–5.44	0.00000002670
–0.45	0.32636	–1.45	0.07353	–2.45	0.00714	–3.45	0.00028	–4.45	0.00000430	–5.45	0.00000002524
–0.46	0.32276	–1.46	0.07215	–2.46	0.00695	–3.46	0.00027	–4.46	0.00000410	–5.46	0.00000002386
–0.47	0.31918	–1.47	0.07078	–2.47	0.00676	–3.47	0.00026	–4.47	0.00000391	–5.47	0.00000002256
–0.48	0.31561	–1.48	0.06944	–2.48	0.00657	–3.48	0.00025	–4.48	0.00000374	–5.48	0.00000002132
–0.49	0.31207	–1.49	0.06811	–2.49	0.00639	–3.49	0.00024	–4.49	0.00000356	–5.49	0.00000002015
–0.5	0.30854	**–1.5**	0.06681	**–2.5**	0.00621	**–3.5**	0.00023	**–4.5**	0.00000340	**–5.5**	0.00000001904
–0.51	0.30503	–1.51	0.06552	–2.51	0.00604	–3.51	0.000224	–4.51	0.00000324	–5.51	0.00000001799
–0.52	0.30153	–1.52	0.06426	–2.52	0.00587	–3.52	0.000216	–4.52	0.00000309	–5.52	0.00000001699
–0.53	0.29806	–1.53	0.06301	–2.53	0.00570	–3.53	0.000208	–4.53	0.00000295	–5.53	0.00000001605
–0.54	0.29460	–1.54	0.06178	–2.54	0.00554	–3.54	0.000200	–4.54	0.00000282	–5.54	0.00000001516
–0.55	0.29116	–1.55	0.06057	–2.55	0.00539	–3.55	0.000193	–4.55	0.00000268	–5.55	0.00000001432
–0.56	0.28774	–1.56	0.05938	–2.56	0.00523	–3.56	0.000185	–4.56	0.00000256	–5.56	0.00000001352
–0.57	0.28434	–1.57	0.05821	–2.57	0.00508	–3.57	0.000179	–4.57	0.00000244	–5.57	0.00000001277
–0.58	0.28096	–1.58	0.05705	–2.58	0.00494	–3.58	0.000172	–4.58	0.00000233	–5.58	0.00000001206

TABLE 1.3 *(Continued)*

Z	f(Z)	Z	f(Z)	Z	f(Z)	Z	f(Z)	Z	f(Z)	Z	f(Z)
−0.59	0.27760	−1.59	0.05592	−2.59	0.00480	−3.59	0.000165	−4.59	0.00000222	−5.59	0.00000001138
−0.6	0.27425	**−1.6**	0.05480	**−2.6**	0.00466	**−3.6**	0.000159	**−4.6**	0.00000211	**−5.6**	0.00000001075
−0.61	0.27093	−1.61	0.05370	−2.61	0.00453	−3.61	0.000153	−4.61	0.00000202	−5.61	0.00000001014
−0.62	0.26763	−1.62	0.05262	−2.62	0.00440	−3.62	0.000147	−4.62	0.00000192	−5.62	0.0000000957
−0.63	0.26435	−1.63	0.05155	−2.63	0.00427	−3.63	0.000142	−4.63	0.00000183	−5.63	0.0000000904
−0.64	0.26109	−1.64	0.05050	−2.64	0.00415	−3.64	0.000136	−4.64	0.00000174	−5.64	0.0000000853
−0.65	0.25785	−1.65	0.04947	−2.65	0.00402	−3.65	0.000131	−4.65	0.00000166	−5.65	0.0000000804
−0.66	0.25463	−1.66	0.04846	−2.66	0.00391	−3.66	0.000126	−4.66	0.00000158	−5.66	0.0000000759
−0.67	0.25143	−1.67	0.04746	−2.67	0.00379	−3.67	0.000121	−4.67	0.00000158	−5.67	0.0000000716
−0.68	0.24825	−1.68	0.04648	−2.68	0.00368	−3.68	0.000117	−4.68	0.00000144	−5.68	0.0000000675
−0.69	0.24510	−1.69	0.04551	−2.69	0.00357	−3.69	0.000112	−4.69	0.00000137	−5.69	0.0000000637
−0.7	0.24196	**−1.7**	0.04457	**−2.7**	0.00347	**−3.7**	0.000108	**−4.7**	0.00000130	**−5.7**	0.0000000601
−0.71	0.23885	−1.71	0.04363	−2.71	0.00336	−3.71	0.000104	−4.71	0.00000124	−5.71	0.0000000567
−0.72	0.23576	−1.72	0.04272	−2.72	0.00326	−3.72	0.000100	−4.72	0.00000118	−5.72	0.0000000534
−0.73	0.23270	−1.73	0.04182	−2.73	0.00317	−3.73	0.000096	−4.73	0.00000112	−5.73	0.0000000504
−0.74	0.22965	−1.74	0.04093	−2.74	0.00307	−3.74	0.000092	−4.74	0.00000107	−5.74	0.0000000475
−0.75	0.22663	−1.75	0.04006	−2.75	0.00298	−3.75	0.000088	−4.75	0.00000102	−5.75	0.0000000448
−0.76	0.22363	−1.76	0.03920	−2.76	0.00289	−3.76	0.000085	−4.76	0.00000097	−5.76	0.0000000422
−0.77	0.22065	−1.77	0.03836	−2.77	0.00280	−3.77	0.000082	−4.77	0.00000092	−5.77	0.0000000398
−0.78	0.21770	−1.78	0.03754	−2.78	0.00272	−3.78	0.000078	−4.78	0.00000088	−5.78	0.0000000375
−0.79	0.21476	−1.79	0.03673	−2.79	0.00264	−3.79	0.000075	−4.79	0.00000083	−5.79	0.0000000353
−0.8	0.21186	**−1.8**	0.03593	**−2.8**	0.00256	**−3.8**	0.000072	**−4.8**	0.00000079	**−5.8**	0.0000000333
−0.81	0.20897	−1.81	0.03515	−2.81	0.00248	−3.81	0.000070	−4.81	0.00000076	−5.81	0.0000000313
−0.82	0.20611	−1.82	0.03438	−2.82	0.00240	−3.82	0.000067	−4.82	0.00000072	−5.82	0.0000000295
−0.83	0.20327	−1.83	0.03362	−2.83	0.00233	−3.83	0.000064	−4.83	0.00000068	−5.83	0.0000000278
−0.84	0.20045	−1.84	0.03288	−2.84	0.00226	−3.84	0.000062	−4.84	0.00000065	−5.84	0.0000000262
−0.85	0.19766	−1.85	0.03216	−2.85	0.00219	−3.85	0.000059	−4.85	0.00000062	−5.85	0.0000000247
−0.86	0.19489	−1.86	0.03144	−2.86	0.00212	−3.86	0.000057	−4.86	0.00000059	−5.86	0.0000000232
−0.87	0.19215	−1.87	0.03074	−2.87	0.00205	−3.87	0.000054	−4.87	0.00000056	−5.87	0.0000000219
−0.88	0.18943	−1.88	0.03005	−2.88	0.00199	−3.88	0.000052	−4.88	0.00000053	−5.88	0.0000000206
−0.89	0.18673	−1.89	0.02938	−2.89	0.00193	−3.89	0.000050	−4.89	0.00000050	−5.89	0.0000000194
−0.9	0.18406	**−1.9**	0.02872	**−2.9**	0.00187	**−3.9**	0.000048	**−4.9**	0.00000048	**−5.9**	0.0000000182
−0.91	0.18141	−1.91	0.02807	−2.91	0.00181	−3.91	0.000046	−4.91	0.00000046	−5.91	0.0000000172
−0.92	0.17879	−1.92	0.02743	−2.92	0.00175	−3.92	0.000044	−4.92	0.00000043	−5.92	0.0000000162
−0.93	0.17619	−1.93	0.02680	−2.93	0.00169	−3.93	0.000042	−4.93	0.00000041	−5.93	0.0000000152
−0.94	0.17361	−1.94	0.02619	−2.94	0.00164	−3.94	0.000041	−4.94	0.00000039	−5.94	0.0000000143
−0.95	0.17106	−1.95	0.02559	−2.95	0.00159	−3.95	0.000039	−4.95	0.00000037	−5.95	0.0000000135

Table 1.3 Standard Normal Distribution (*Continued*)

Z	f(Z)	Z	f(Z)	Z	f(Z)	Z	f(Z)	Z	f(Z)	Z	f(Z)
−0.96	0.16853	−1.96	0.02500	−2.96	0.00154	−3.96	0.000037	−4.96	0.00000035	−5.96	0.00000000127
−0.97	0.16602	−1.97	0.02442	−2.97	0.00149	−3.97	0.000036	−4.97	0.00000034	−5.97	0.00000000119
−0.98	0.16354	−1.98	0.02385	−2.98	0.00144	−3.98	0.000034	−4.98	0.00000032	−5.98	0.00000000112
−0.99	0.16109	−1.99	0.02330	−2.99	0.00139	−3.99	0.000033	−4.99	0.00000030	−5.99	0.00000000105
−1	0.15866	**−2**	0.02275	**−3**	0.00135	**−4**	0.000032	**−5**	0.00000029	**−6**	0.00000000099

TABLE 1.3 *(Continued)*

and Telegraph (ATT) developed control charts prior to World War II. It used sampling theory techniques to allow for accepting parts from the supply chain or through manufacturing, as long as they were within a confidence interval X double bar ±3*s* (s is the sample standard deviation) and X double bar is the grand average of historical data of sample averages. This made the *Cp* calculation much easier if the data from the manufacturing control charts were used.

The result of connecting control charts with *Cp* posed a dilemma. Control charts allowed the manufacturing process average sample value to be acceptable within X double bar ±3s (sample). The Motorola approach was to assume that the process average μ can shift around ±1.5σ to account for the acceptability of average variation in the control charts. The *Cpk* term uses the actual process average shift in the calculations and does not assume a range of probable average shift.

These two different interpretations of 6σ led to confusion when calculating the defect rate for a new product. For a Six Sigma design ($Cp = 2$), the defect rate will be 3.4 parts per million (ppm), given the assumption that the process average μ inherently shifts by ±1.5σ. The shift causes the *z* value in Table 1.3 to be reduced from $z = -6$ to $z = -4.5$, making $f(z) = 0.0000034$ or 3.4 ppm. When this average shift is taken into account in the *Cpk* formula, it could be more accurately described as $Cpk = 1.5$ or a 4.5 sigma design. A true 6σ design with no average shift should have a defect rate of 0.002 ppm or two parts per billion as opposed to 3.4 ppm in the *Cp* definition. The other side of the specifications curve defects is not added up since it is very small ($Z = -7.5\sigma$).

Some industries have adopted a design quality target of 4σ, which roughly translates to 64 ppm or approximately one part defect in 10,000 defect opportunities. This is based on using Table 1.3 for $z = -4$, resulting in $f(z), = 0.000032$. Adding up the two sides of the specification limits defects results in 0.000064 or 64 ppm.

There are many interpretation issues with the 6σ design, which makes it difficult to use this technique in complex new high-technology products. The PM has to decide on how to proceed when tasked with implementing a 6σ project:

- If the goal is to develop and design a 6σ product, then does the entire product have to meet 6σ, or do individual components of the product have to meet 6σ or both?
- If the product is composed of hundreds of parts, each part or subassembly could meet 6σ, but adding up all of the defects from all of the parts makes it difficult for the total product of parts and assemblies to meet the 6σ requirements.

- One resolution of this issue is to calculate a product defect rate that is the sum of all of the defects of the individual parts, added up to the product level. It can then be converted back to a *Cpk*, as can be seen in the example in the next section.
- Another approach is to have the product design σ level assume the level of the lowest quality part in the product. In this approach, the lowest-quality part can be identified then targeted for improvement.

Many of the measurements of defects are not based on the variability of individual parts, but on counting the defects found in a particular number of parts or operations in manufacturing, called defect attributes. The equation for finding defects, as illustrated above in the *Cpk* formulae, depends on the part manufacturing variability against the specifications by calculating the value of z, then making the defects equal to $2*f(z)$ from Table 1.3. Many projects count the number of defects in manufacturing the product, based on opportunities for defects, and then convert the defect rate back to a *Cpk*. The process is the reverse of finding defects from the SND Table 1.3. By equating half of the defects in the SND as $f(z)$, the z value can be extracted, making $Cpk = Z/3$. This implies that the defects are occurring against a certain complex set of specifications, and the *Cpk* calculations are derived indirectly, assuming there is no average shift versus specifications nominal.

For example, a defect rate of 4/1000 or 0.004 for a part in manufacturing is assumed to be two sided. A one-sided defect rate (from each ±SL) is 0.002, which can be set equal to $f(z)$ in Table 1.3 and the value of z is extracted. In this example, a value of $f(z) = 0.002$ in Table 1.3 returns a value of $z = 2.88$, resulting in a $Cpk = z/3$ or 0.96. This mathematical process assumes that variability distribution of the part is normal and that there is no average shift versus the nominal (N) of the part design specifications. There were no actual calculations of the variability of the part due to manufacturing process variations.

An assembly might have many different opportunities for defects. Some PMs can increase the amount of opportunities by double counting some operations in order to reduce the defect rate. For example, in the electronics industry, assemblies such as PCBs have many operations and components imbedded in the assembly as well as joining operations.

There are five major parts of the PCB assembly: (1) the raw (unloaded) PCB, (2) the components loaded into the PCB, (3) the placement of solder dots on the PCB surface to prepare the components for soldering onto the PCB, (4) the placement operations to locate the components on the PCB, and (5) the joining operation of the components using a conveyer oven to reflow the solder and attach the components. Any of these operations are opportunities for defects, and the defect number per PCB can be interpreted differently. In the case of PCB defect rates, the relevant industry association for connecting electronics industries (IPC) has published a standard to resolve this issue, specifically counting the opportunities of defect generation in PCBs for end product and in-process calculations. They are IPC 7912A/9261A end item and in-process DPMO (defects per million opportunities). The goal is to level out the different claims of quality in terms of σ level capability for different PCB manufacturing and supplier facilities.

The PM has to be careful when tasked with a project designated as 6σ. They should keep in mind that 6σ is a management tool to encourage the communication between design and manufacturing, with the ultimate goal of designing and producing high-quality products, with maximum performance in the field and minimum manufacturing cost. Understanding these issues and agreeing beforehand how to handle them will make for a successful 6σ project.

1.2 The 1990s

During this decade, high-technology companies focused on collecting these various tools and techniques developed earlier into a cohesive set to allow successful integration into CE. The tools mentioned in the 1980s were then grouped as to their relative and proper position and contribution to successful project management.

The CE culture in Fig. 1.7 was proposed to be built on a solid quality ethic that encouraged the use of TQM tools into problem solving, decision-making, and the future impact of these decisions. CE encouraged the grouping or co-locating of all parts of the project development team together, from the core engineering groups to the support organizations such as quality, marketing, manufacturing, finance, and accounting. The idea was to encourage communication and speedy problem solving by having meetings that included all members of the extended project team. The leadership of the team was to be clear, with the PM designated to be the technical leader and the product champion. The PM would advocate for the project to be completed on time, within budget, and meeting all of the quality, cost, and customer satisfaction goals.

The four major components of CE for the PM to focus on were the following:

1. Robust design → Higher quality using the tools of 6σ, DoE
2. Low cost → Lower product cost through DFM/ESI using the tools of design efficiency (BDI/GE Hitachi) and activity-based costing (ABC)
3. Improving development program and new product release → Time to market (TTM) using the tools of reducing ECOs and "Half the Time" to develop new products
4. Meeting expectations → Customer satisfaction through QFD

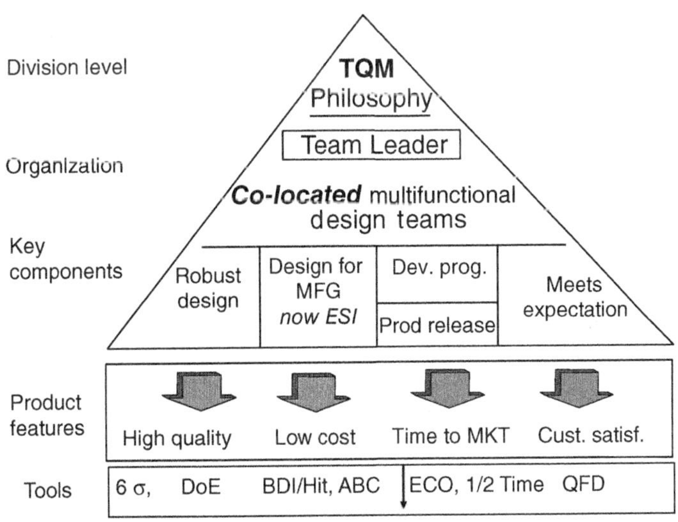

FIGURE 1.7 Summary of the concurrent engineering culture.

1.2.1 Robust Design of the High-Technology Product

This is a measure to introduce the tools of design for quality into project development. Robust design of new products leverages the benefits of 6σ and DoE for reducing variability, either by experimentation or by negotiating a design quality level with manufacturing through *Cp* or *Cpk* targets. Conducting extensive DoE experiments during the design and early prototype stages of production is also encouraged to reduce variability. Six Sigma tools encourage design engineers to incorporate wide enough specification limits for new parts and products to match the capabilities of manufacturing processes.

1.2.2 Low Costs for New Products

These strategies are encouraged by the use of tools such as DFM, including ESI, to establish very early in the design phase a goal of high design efficiency through ease of assembly and manufacture. The PM is also encouraged to focus early on the cost estimate of the product through ABC, which is a tool to estimate product cost early in the design cycle, to be updated at each design review step. (ABC will be covered in Chap. 11 in this book.) It is similar to DFM in that it uses common features of the parts in the product as cost drivers to calculate the cost of every manufacturing step leading to a reasonable estimate of the cost of the part. For example, in a PCB, the number as well as the cost of parts in the PCB is used to establish the total expected cost, including all assembly operations. Usually, ABC is also called the cost model of the product and is used by the PM to track the product cost as development continues from early design to later stages.

1.2.3 Time to Market

This was the third and most important concern of CE for project managers. Many early CE projects had the goal of developing products in half the time. If the development time was cut in half, then the development cost (in terms of engineering staff assigned to the project not sitting around and getting paid while waiting for others to complete their assignments) would be significantly decreased. This reduction in TTM took on many facets:

1. Reducing the iterations of design engineering changes (ECOs)
2. Design products in "half the time"
3. Role of technology in new products → first to market
4. Project management metrics

1.2.3.1 Reducing the Iterations of Design Engineering Changes (ECOs)

Many studies were performed in the 1990s to compare global companies in Japan or the United States in terms of development cycle, and the impact of shorter development on product introduction cycles. Figure 1.8 and Table 1.4 illustrate the major difference in project management between the two countries and the subsequent impact of these differences.

As illustrated in Fig. 1.8, this comparison of the profile of engineering changes on the manufacturing release of a product clearly shows the impact of neglecting the project management process. In the American company example, the ECOs continue

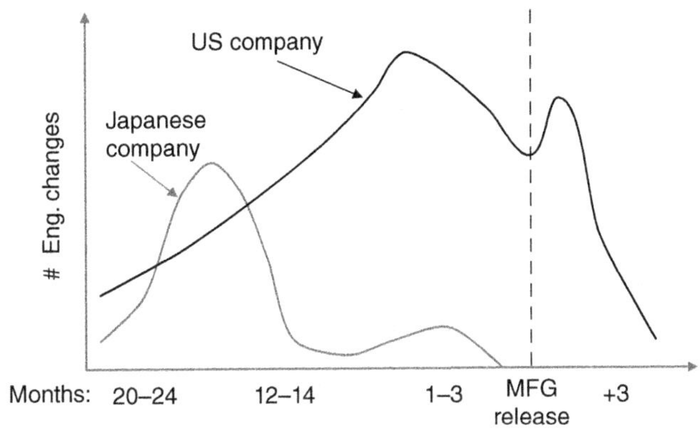

Figure 1.8 Auto industry example in comparing engineering changes.

~~Japan~~	Japan	U.S.	
Design time/model	47	60	Months
Design effort/model	1.7	3.0	M Man H
R&D investment	5	8	$ Billion
Average replacement	4.2	9.2	Years
Annual production/model	120	230	000 cars
Models in production	72	36	

Table 1.4 Japanese/U.S. Auto Product Design Lifecycle Comparisons

to increase as the project nears its completion, only to be reduced within three months of release, but then increase sharply after release due to poor focus on manufacturability that can be offset by tools such as DFM/ESI.

The Japanese counterpart profile is the continued reduction of ECOs, resulting in the release to manufacturing without major transition problems. This strategy of minimum ECOs at release shows the clear intention of the PM to deliver a defect-free design to manufacturing as well as minimize the cost through faster product introduction and reduced redesigns. The resulting effect was the making of a much more effective auto industry, as shown in Table 1.4, which was published by *The Economist* on April 14, 1990. Table 1.4 summarizes data from the eight Japanese and three U.S. auto companies.

The first three items in Table 1.4 show the effect of reduced ECOs, in terms of design project duration (months), design effort (in million man/hours) and total engineering investment (in billions of dollars). The results of this imbalance in project management can be seen in the last three items. The reduced cost of introducing new models is reflected back in quicker replacement of older models and introduction of newer ones that can quickly adopt to new technology and customer demands. More new models can be also offered, which can target more niche markets while sustaining a lower sales volume per model. For the Japanese auto companies, selling fewer cars per model can increase market share by having the ability to satisfy a more diverse marketplace. For the U.S. model, the larger cost of new car development results in fewer new models, forcing the auto companies to develop cars that can satisfy as many customers as possible. They continued to lose niche and specialty corners of the market to their competition.

1.2.3.2 Design Products in Half the Time

One of concurrent engineering main themes is the goal to develop products in "half the time." This challenge was applied in many industries, including automobile, electronics, communications, and computer. The focus became the design/development process itself and how to make design engineers more productive. The global technology companies felt that the manufacturing cost of their products was greatly improved by the quality and cost tools mentioned previously. They were now looking for similar improvements in project time and engineering productivity, thereby creating many more new products out of the same development budgets. Many companies showed their concerns for these issues by creating the new position of "productivity manager" and hiring experts to manage and focus on this effort. The author was one of these productivity managers at the Hewlett Packard Corporation in the 1990s.

Studies in TTM across many global companies indicated the same results as shown in the auto industries. The keys to reducing TTM were to focus on reducing ECOs through better project management of the project design cycle and properly identifying the new product and customer expectation through a much more detailed concept stage of the product. The focus was also to make sure that the detailed design phase first iteration was as complete and accurate in meeting the project specifications as possible. This could be accomplished by using more frequent design reviews, use of tools that solicit inputs from other parts of the organization such as marketing (QFD) and manufacturing (DFM/DFA/ESI, 6σ, and DoE), and the use of advanced design and analysis tools such as:

- CAD and CAD/CAM interfaces to make fast prototypes
- Analysis software such as finite element analysis (FEA) to reduce the need for design iterations
- Rapid prototyping methodologies such as stereo lithography and 3D printing to validate the design and make sure those different parts of the product are in full geometric synchrony to each other

Many studies have been conducted to measure the effect of the focus of PMs on the concept stage and first iteration in the detailed design stage in properly identifying customer needs and manufacturing capability requirements, as shown in Fig. 1.9. It shows a typical outcome of spending more time on the initial concept design phase as a percentage of total development time. The longer time for the concept stage was needed to reduce future ECOs through communications within different parts of the global company and the use of analytical tools and/or expert engineers' input through design reviews to validate the product design. The initial concept design stage will be longer than the traditional stage, while the total project development cycle might be shorter, by as much as 40 percent, through all of its phases of detailed design, design revisions, and data dissemination to manufacturing.

Figure 1.10 shows the effects of collaborative design in a similar pattern of reducing the overall design cycle. Communications through design reviews and use of analytical tools increase the validation of the product meeting its intended goals. While a non-verified design could produce first prototypes faster, it might lead to a much longer total design time because of ECOs. A longer verified initial design might lead to only one redesign.

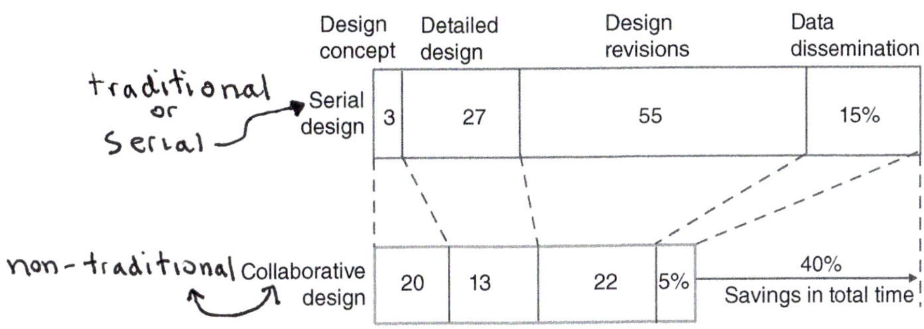

FIGURE 1.9 Time to market profile from industry surveys.

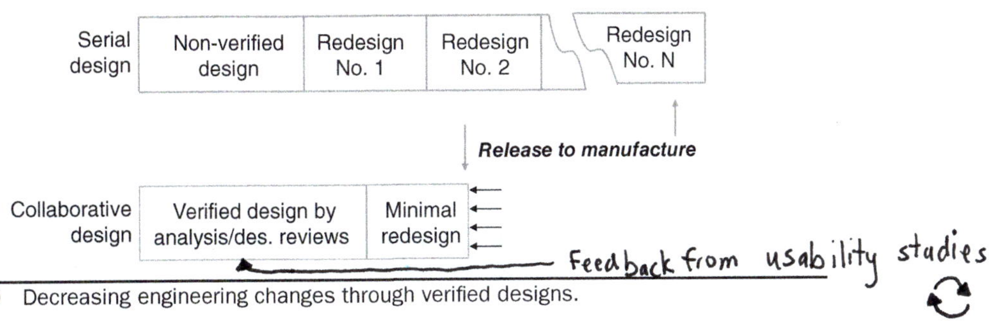

FIGURE 1.10 Decreasing engineering changes through verified designs.

PMs were instructed to plan on only one minimal redesign of the new product in their proposals of new product schedules and budgets.

A typical study for shortening TTM was to record in great detail the amount of time spent for each project activity and determine the percentage of time spent by engineers on design and analysis, as opposed to other nontechnical activities. Figure 1.11 shows a typical study of an electronic product that took 72 weeks in development, with four major ECOs accounting for most of the time. Project activities were recorded to find out

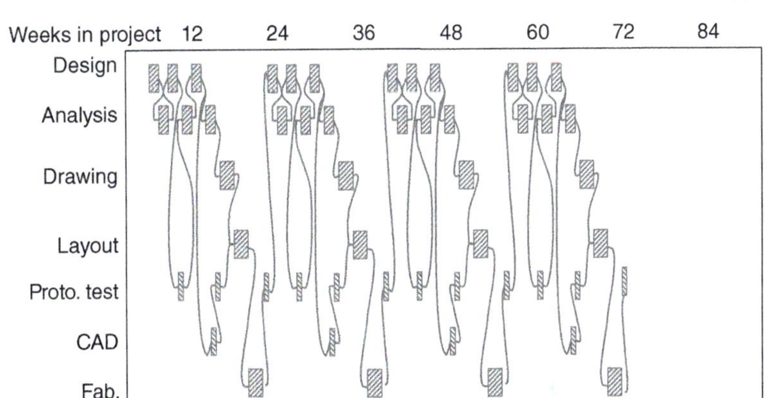

FIGURE 1.11 Reducing the number of revisions to speed up design time.

project time distribution and the impact of multiple ECOs as well as their causes. It showed that the design and analysis time, which is the primary duty of engineers, is a small portion of the project timeline, since the engineers must wait for the ECOs to be completed, redesigned, manufactured, and tested. It is speculated that the engineers use this slack time to get updated on technical skills and paperwork for the project. Up to 25 percent of the ECOs were found to be caused by poor communications.

The typical reaction of engineers to this chart was to demand faster tools, such as CAE/CAD, to speed up the turnaround time for each revision, while management was interested in assigning engineers to multiple projects to better utilize their slack time. The successful resolution to TTM is to focus on the following:

- Increase communications through more meeting facilities and informal project meetings
- The use of more sophisticated design analysis tools and faster prototyping methods
- More frequent informal as well as formal design reviews

1.2.3.3 Role of Technology in New Products

This is important in terms of the need to keep engineering development projects on track and understand the life cycle of new high-technology products. It is well understood in global high-technology companies that the first new technology product to market captures 70 percent of market share. In addition, the obsolescence date of a technology product is not set by the company, but by market forces such as competition and technology development, as shown in Fig. 1.12. The PM has to be conscious of the impact of technology on their project plans in the following areas:

- Keeping project on track is critical
- The first high-technology product to market gets 70 percent of market share
- To control project development time, use only well-defined technology or introduce successive products that incorporated technology advances in incremental steps
- If new technology could force a major ECO, scrap the current project and go for a new product

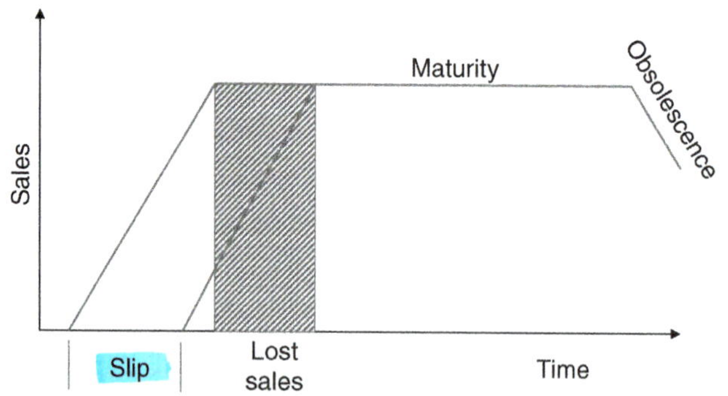

Figure 1.12 Consequences of project schedule slip.

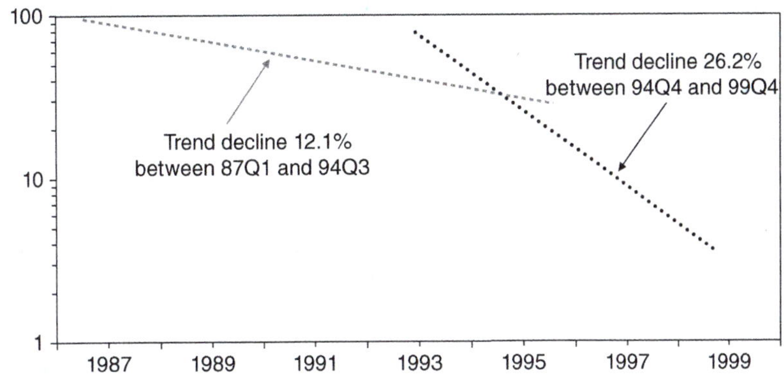

FIGURE 1.13 Relative price declines in computers in the 1990s.

- A one-month slip in development is one less month of sales
- End of product life is fixed, determined by market forces, not the company

Moore's law, first introduced in an article in *Electronics Magazine* in 1965 by Gordon Moore, the co-founder of Intel Corporation, best illustrates the impact of technology on a project. Summarized briefly, it indicates that, "Every 18 months technology capability doubles while price halves." Fig. 1.13 best illustrates the impact of Moore's law in the 1990s.

The PM has to be conscious of this trend shown by Moore's law by incorporating it into their plans in the following areas:

- Project inception phase and how much technology to incorporate into the new product
- Proper balance of technology adoption versus project development time
- Adopting new technology in incremental new product introductions
- Develop a sound price versus performance strategy. Either add more performance in a new product but keep the same price or keep performance the same, but offer the new product at a lower price
- Understand the impact of technology on product development tools, cost, components, assemblies, capital equipment purchases, and return on investments (ROI) for new products
- Understand the impact of technology on product life cycle and obsolescence
- Make a decision on development in house or acquire new technologies for the new product.

All of these issues of technology will be dealt with in the following chapters of this book.

1.2.3.4 Project Management Metrics

As interest increased in the evaluation of engineering productivity, various companies attempted to quantify and measure engineering-effectiveness metrics, to the extent of using the data for individual engineer's performance evaluation. This idea was borrowed from the TQM philosophy of measuring a parameter before improvements are made. Some these metrics took the following forms:

- Time spent by engineers on various activities (technical versus nontechnical)
- Cost impact of engineering activities
- Performance measures of engineers and projects
- Time spent by engineers on various activities was derived from the PM's concerns that engineers spend the majority of their time on non-technical activities. An example is shown in Fig. 1.14, taken from surveys conducted by the author.

Typically, the PM would be concerned with the two largest components of time spent by engineers, communication, and documentation. Many solutions can be implemented to facilitate communications, as well as efficiency in documentation, such as:

- More meeting rooms or spaces to hold impromptu meetings
- More frequent (daily) update meetings scheduled at set times (in the mornings)
- Allowing only stand-up meetings (no chairs in meeting rooms) to reduce meeting time
- Making sure that meeting agendas are published well in advance as well as meeting summaries and conclusions
- Limiting time allotted for verbal progress reports by use of rolling clocks, lights, and buzzers

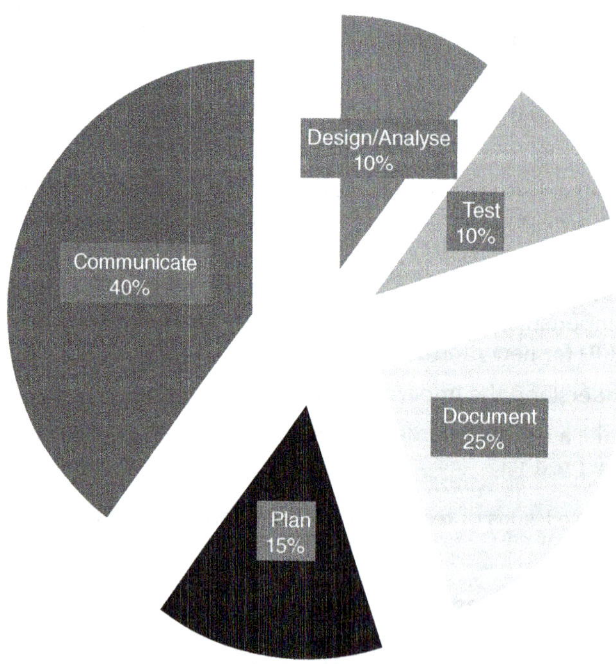

FIGURE 1.14 Distribution of a typical engineer's time in a project.

- Making sure that templates are available for all documentation for easy authoring and reading
- Encourage the use of non-engineers for generating or auditing some documents
- Use of sophisticated commercial software for project communications and documentation

To optimize an engineer's time in the traditional engineering topics (plan, design, analysis, and test), several innovative techniques can be used to improve the engineering workspace. Cleanliness and reducing noise of the work environment can be implemented as follows:

- Encourage the use of noise-reducing partitions around the engineers' workplaces to increase privacy without reducing team communications
- Use of sound-absorbing floors, carpets, walls, and ceilings
- Use of noise-cancelling technology in the workplace
- Encourage the use of quiet time in the workplace, such as two hours each morning and afternoon when engineers are not allowed to make or take phone calls or hold conversations in the aisles
- Use of false or raised floors to route communication and power cables to reduce clutter
- Use of ergonomically correct furniture such as stools and adjustable height desks

Another metric is to investigate the division of the engineer's time between working on original or innovative product ideas versus design reuse or variant design. Design reuse or adaptive design is adapting already-existing designs from older-generation products incorporated into the new products with minor modifications. Variant designs are those originating from other products incorporated into the new product. The principle of design reuse is to reduce the number of original parts in the product since their design time is longer than using older designs. Figure 1.15 (design effort distribution) is

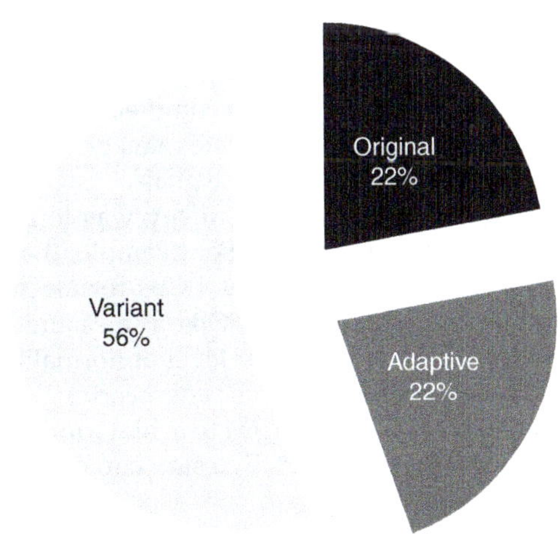

FIGURE 1.15 Design effort distribution.

Design time/CAD drawing	30–50	Hours
Design cost/CAD drawing	40–50	K$
ECO/CAD drawing	1–2	Hours
$ cost of ECO/drawing	1.5	K$

Table 1.5 Mechanical CAD Drawings' Statistics

an illustration of studies in maximizing engineering time. Keeping track of these activities will help the PM set a goal of original versus reused parts into the new product. Design reuse and variant designs have these additional benefits:

- Reduce design effort for new products
- Reduce the cost of parts maintenance and inventory management
- Reduce the cost of tooling and capital equipment needs

Cost impact of engineering activities. In order to illustrate the cost impact of various engineering activities to reduce the negative aspects of redesigns and ECOs, studies were undertaken to quantify the cost of the various types of original design versus engineering change activities, as shown in Table 1.5.

The values in Table 1.5 are dependent on other factors that impact the numbers such as:

- Design/time per CAD drawing might have to be different depending on the complexity of the drawing. A geometry-based system of complexity might have to be implemented with different expectations for each level of complexity.
- Design cost/CAD drawing might include engineer's time, CAD control and archiving.
- Cost of ECOs/CAD drawing might include average cost of rework and scrap of parts.

Performance measures of engineers and projects. Many high-technology companies began to track the progress of their engineering projects toward achieving more rigorous goals of quality, cost, and timeliness in completing specified project tasks on schedule. Goals were set for project and corporate managers and enforced by pinning their performance rewards on adherence to these goals.

A concern in implementing such a strategy was to normalize the effect of these goals as technology advances increased. For example, the number of ECOs had to be normalized per assembly as in ECOs per PCBs for electronic products. Technology enhancements resulted in more complex PCBs, as measured in component counts, density, and performance. Therefore, another level of normalization had to be included in the ECOs' metric. These trends in electronic engineering design efficiency over time can be tracked and used to set goals for the PM to achieve for the next product design cycle. An example is given in Fig. 1.16 for PCB design complexity over time and in Fig. 1.17 for design quality over time.

It can be seen from Fig. 1.17 that a straight metric of complexity and quality such as the number of iterations (ECOs) per PCB is not adequate to describe the increased

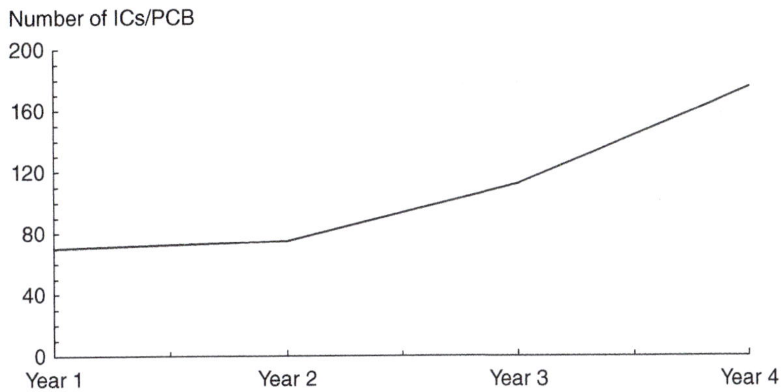

Figure 1.16 PCB design complexity over time.

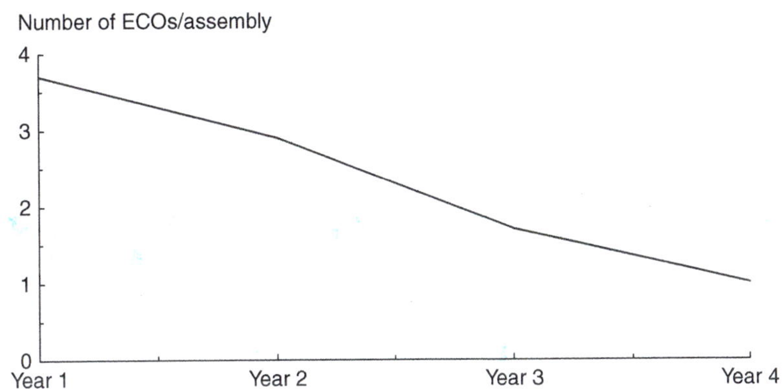

Figure 1.17 Design quality over time.

complexity of the PCB over time. The progress of less ECOs/assembly is rendered more significant once the design complexity, as measured by the number of ICs/PCB, is taken into effect.

For Fig. 1.17, the goals of one ECO per assembly took four years to accomplish, even with intensive focus on working to eliminate the reasons for the ECOs and enhancing project communications. Metrics for measuring and improving the PM's progress in completing projects on time, with high quality and low cost can be summarized as follows:

- Set out a process to measure important parameters of project activities for all engineering disciplines, including electrical, mechanical, and software components
- Implement a system to measure design complexity based on number of parts or components
- Parameters to be measured include design time normalized by design complexity
- Measure design activity (original design and redesigns) by cost to illustrate cost of quality

- Set multiyear goals to achieve significantly higher design quality (one ECO per assembly)
- Tie-in those design quality and cost goals to the PM and the project engineers' performance evaluations and rewards

1.2.4 Meeting Expectations and Customer Satisfaction through QFD

This is the last leg of the CE formula for success for the PM. A project could perform all recommended activities and successful use of the tools discussed above, yet can still be a financial failure. This occurs when the new product does not meet customer expectations in terms of product availability, performance, features, adherence to specifications, and adoption of relevant technology.

In new product introduction, risks occur when the interfaces between the PM and the pre- and post-design activities are not well understood. Improper attention to these risks can cause severe problems such as ECOs, high product cost, and low product quality as well as the inability to complete the project on time or adequately satisfy customer demands. A combination of these problems can result in the scrapping of the project.

Technical risk occurs when the issues of engineering and manufacturing interface of quality and cost, as well as the engineering design productivity issues, are not handled adequately by the PM. The tools and methodologies of concurrent engineering have been developed in response to technical risks. These risks can be mitigated using tools and concepts such as 6σ, DoE, DFM, ESI, and ABC as well as the methodologies of reducing ECOs and TTM, as previously explained.

The risk of not meeting customer expectations is called "market risk." In this case, the new product specifications do not meet customer requirements. There is also an additional layer of regulatory and industry-specific requirements for design specifications. There are two types of engineering project management and, subsequently, two requirements for success in mitigating market risk:

1. Design to Customer or DTC (specifications-driven projects). In this case, marketing risk is much smaller, since the specifications are negotiated or handed over from the customer. The PM's focus on these projects is to make sure that the specifications are well understood and to communicate to the customer a sense of the price of technology adoption versus the project cost and schedule. Cost control and managing the project schedule while maintaining the product performance to negotiated specifications is critical to the success of the project. In summary, the issues of DTC projects are as follows:
   - Bid driven, used in military [Department of Defense (DoD)] products, construction projects, and subcontracting.
   - DoD projects evolved from cost +, whereby military contractors charged the design of new products with their overhead directly to their customers. This technique was abandoned due to excess prices for overhead-driven parts such as an airplane toilet seat for $600 (which will be explained in a Chap. 11). Currently, defense companies bid for contracts while absorbing the initial development cost. An example is the F-35 Joint Strike Fighter where two defense contractor consortia designed and competed for the contract through a fly-off competition. The consortia shared the large initial development costs, but were assured they would eventually share in the manufacturing of the planes. Attributes of these DTC projects are:

- Contractors have to be prequalified for proven design experience, staffing, quality, and industry standards compliance prior to bidding on contracts to facilitate the selection process.
- Control of expenses, good task definitions, attention to schedules, and accurate cost estimating are critical for DTC projects, with small price differences separating the top bids.
- Most of DTC projects are schedule-driven, which encourages use of proven technology.
- In most cases, the project delivery schedule is fixed with penalties for delivery delays.
- Sometimes the contractor is encouraged to complete the project before the deadline with significant rewards for each time period of early delivery. This might lead to risk taking and adoption of new technology to speed up the schedule. An example is the completion of the I-10 overpass reconstruction in Los Angeles after the 1994 Northridge earthquake. The successful bidder had to analyze the complex interrelation of project cost and early delivery benefits as well as adopt a new technology of rapid set cement to successfully bid for the contract. The Los Angeles region was estimated to suffer economic losses of up to $1 million a day due to the freeway's closure. Contractor C.C. Myers was awarded the job to completely rebuild two bridges with a significant bonus of $200,000 a day for early completion.
- Technology development issues and costs are part of the contract negotiations process.
- Companies cannot analyze potential profit based on traditional return on investment (ROI).

"Pull-to-Market"

2. Design to Market or DTM (market-driven projects), where there has to be a careful market requirement analysis, including the competitive position of the new product, in price and performance, as well as the degree of technology adoption and impact on the company's current product portfolio. The PM is expected to complete an in-depth economic analysis (sometimes called the business plan) to convince management that the project is in line with the company expectations of maintaining its financial position and return on assets. This business plan entails the following:
   - Analyze total potential market of the new product
   - Estimate new product market share based on customer feedback and the competition
   - Estimate product lifecycle and price fluctuation before obsolescence
   - Predict the impact of new product on the existing product portfolio sales
   - Set a sale price commensurate with technology adoption and market and product competition analysis
   - Estimate all development costs, including engineers and supporting staff salaries, tooling costs, outside engineering services, and prototype costs
   - Estimate profit based on selling price minus expected material and manufacturing costs

Most of this will be included in the Product Performance specification/DIR/Marketing Requirements/ etc.

- Calculate ROI, making sure it is in line with the company return on investment strategy
- Review the ROI results at each project milestone meeting

This project economic analysis will be further explored in Chap. 5. In summary, the issues of DTM with increased market and technical risks are as follows:

- Market-analysis and technology-driven strategies are needed for successful projects
- Estimate risks in technology adoption and development costs of new technology. These technology risks can be mitigated by:
  - Buying other companies to adopt their technology (Cisco Model in the 2000s)
  - Adopting new technology in stepwise fashion for each new product cycle
  - Use product options to allow greater participation of a wider range of customers
  - Always use standards-driven technology for new products. Use propriety technology when competitive analysis shows that the technology is a real product differentiator
  - New technology development for better new products can be risked for greater return

A major problem for high-technology companies is that engineers are good at designing new products for use by other engineers or scientists, but are not as successful in designing new products for standard commercial applications. For DTM products, there is a greater need to estimate customer needs through marketing inputs, surveys, and focus groups as well as competitive analysis to determine the proper level of specifications and technology adoption in new products.

There are many examples to illustrate this point of engineers needing customer input to set the proper specifications. New product opportunities can arise when technology offers automation to resolve errors in manual procedures or the imposing of new government regulations. The PM can then set the proper product complexity commensurate with the product's intended use, based on customer inputs:

1. When deaths in operating rooms increased because the wrong gas was given to patients during anesthesia, a medical technology company realized a product opportunity by inventing a CO_2 monitor. The monitor can be directly plugged into the patient's airway and produces an alarm when the patient undergoing surgery is not receiving an adequate level of oxygen. The dilemma is what accuracy level to use in the CO_2 specifications? A technical PM might inherently want to develop a product with ±1 percent accuracy, while an anesthesiologist might be satisfied with ±10 percent accuracy.
2. When the European Union (EU) established the Restriction of Hazardous Substances (RoHS) standard for lead-free electronic products in 2006, a market opportunity developed for *lead sniffers*. Customers could be electronic manufacturers or customs agents who needed to test their incoming components and products. The dilemma for the PM was how to set the specification for these disparate groups: whether to design a universal complex product with adjustable

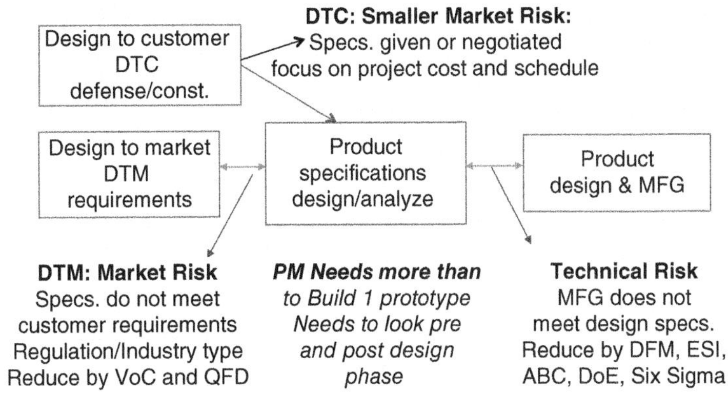

Figure 1.18 Summary of market versus technical risk for new products.

accuracy levels for manual as well as automatic use, or a family of low-cost niche products with varying accuracies and ease of use.

A summary of the risks involved in developing new products is shown in Fig. 1.18. Most of the issues outlined in this section regarding meeting customer expectations are shown in this figure. The need for PMs to better understand customer needs encouraged the use of QFD. This tool was developed in Japan and brought to wide use in U.S. companies in the 1990s through an article in the *Harvard Business Review* in 1988 by Don Clausing, the same gentleman who translated Taguchi's DoE books.

QFD was developed as a structured tool to better understand customer needs in developing the next generation of products. It is based on customer feedback on the current product portfolio as well as an analysis of the competitive environment, in order to design the new product as a best match to those customer needs. It had many other notations such as the "House of Quality" and "Matrix of Matrices." It was widely adopted in the 1990s by many global technology companies and credited with numerous successful product launches. Its main attraction is to focus the PM and design teams on customer needs, competitive analysis, and better communication between the designers and potential users of new products.

QFD starts with customer input, whether through marketing, surveys, or focus groups on their opinion as to the current level of satisfaction with the current-generation products as well as comparison with the competition. The customer needs are then collected into synergy groups and matched to the new product design attributes in matrix format. The needs are ranked as to the relative importance placed by the customer.

The relationship of each of the customer's needs versus each of the product attributes is then defined at four levels (strong, medium, small, and no relationship) jointly by the project team and either the customers or their representation in the marketing department. This relationship is tempered by two additional influences:

1. The customer perception of the position of the company versus its competition in best meeting each customer need. If there is a marked difference in the negative direction, then the output of the QFD exercise is boosted higher for a particular need to compensate for this shortcoming.

2. The interaction of various elements of the relationship. This interaction is shown as a triangular shape above the matrix, hence the name "House of Quality." For example, in a car door, customer needs versus design attributes could take the following scenarios:
    - Customer needs/comments on the ease of opening the door influence design of the door latches as well as door hinges
    - Customer concerns about safety could influence the design of the locking mechanisms, but that design could interact with the ease of opening the door
    - Customer wishes about a rich sound when closing the car door could influence the tightness of the fit of the door to the frame

A QFD exercise would also ask the customer to rank these needs, based on their relative merit, as well as rank the company's attention to those needs versus the competition. The result of the QFD exercise would be to guide the design team as to which specification of each design attribute should be significantly improved and, therefore, require major design development or improvements. Other specifications could remain the same as the existing product families, and no further design improvements needed.

The QFD exercise can be seen as a good methodology for DTM or market-driven products and less so for DTC projects. Many defense contractors have adopted hierarchical techniques similar to QFD, where the negotiations with the customer are structured to make sure that customer intent for making the product are well understood. Therefore, prospective bids from the company will better match the customer needs. For both DTC and DTM products, the use of QFD has to be tempered with the level of new technology adoption in the product:

1. If the new product is a follow-on product or a new product generation, the proper input from the customers (via QFD) based on the previous product portfolio, as well as their evaluation of the competition, should help in setting the desired performance and options level in the new product.
2. If the product is totally revolutionary, using new technology in an original way to greatly increase the productivity of the customer, then the customer input is not as critical. The perceived benefits of the new technology will vastly outperform existing products. The customer experience in using older products might not be relevant in guiding the specifications or the design of the revolutionary or disruptive technology product. An example would be that the users of slide rules cannot convey their experiences into advising on an electronic calculator design criteria.

QFD is a great tool for developing products that are evolutionary or next generation, but not for revolutionary or DTC (specification-driven) projects. The advantages of QFD can be summarized as follows:

- Excellent tool for increasing communications between engineering and marketing functions
- Best used for evolutionary and next-generation new products

- Encourages the PM and design team to focus on customer needs for new products
- Customer needs are collected through surveys, marketing assessment, and competitive analysis
- Customer needs are ranked based on customer input
- Customer needs rankings are adjusted by the perception of the company competitors' products
- Customer needs can be classified into several groupings to match product design requirements
- Customer needs are sometimes translated from surveys through automatic software means
- The QFD process matches customer needs with new design requirements
- Interaction between design requirements should be taken into account
- The result of the QFD analysis is where to focus the new product design improvements
- QFD process could influence which specifications/designs need to be improved in the new product and which specifications/designs can be maintained as in older products

The QFD process is not without its shortcomings. It is not as beneficial for revolutionary products since current customers have no basis to compare the new product with existing products and technologies. In addition, it is not useful in DTC specifications projects, since the customer already fixed the specifications. Many industries that are suited for DTC, such as defense contractors, use a variation of QFD in order to better understand customer needs and to clarify some of the specifications, as will be shown in Chap. 4. In addition, the QFD process does not handle the issue of customer choices of product specification through options or product price easily. The idea is that once the PM and the design team go through the QFD process, the new product will have customer satisfaction at a reasonable cost, assuming that the design team is using all of the tools that make them competitive with world-class companies. The QFD shortcomings are as follows:

- Not very useful in revolutionary or DTC products
- Does not handle the issues of product cost versus design requirements or specifications
- QFD assumes the specifications are for a generalized product. It does not handle options and niche products in the higher or lower end of the market

The QFD process will be examined in detail in Chap. 3.

1.3 The 2000s and Beyond

The turn of the century brought the advent of global communications, facilitated by the Internet. Instead of design teams meeting on the Internet via satellite hookups at $5000/hour, new product design issues can now be resolved by using Internet communication that was made easy and simple by purchasing a $50 camera to attach to a laptop computer worldwide. These inexpensive, rapid, and constant communications in turn enabled companies to develop new products, designed by distributed teams, which could be manufactured anywhere in the world. Outside expertise in design, analysis, local, and global standards can be easily purchased and leveraged when needed through the Internet. Increased use of standards by competing design and analysis tool suppliers enable engineering teams to communicate effectively, even when using different design and analysis tools.

The CE tools such as 6σ, DoE, DFM, ESI, ABC, and QFD were made into standards of design and manufacturing, enabling global companies to communicate and resolve issues of design and manufacturing in the same framework and with a common language. This standardization allowed for increased use of the supply chain for design and manufacturing. Manufacturing became a commodity, with a high level of quality and optimum level of cost. The supply chain leverages its size into lower-cost materials than their customer companies. Global companies began to seek methods of outsourcing design functions, while keeping their core competency.

New product design became a collaborative process, overseen by the PM with team members as well as outside technical resources and a global supply chain, to create new products that will meet a global set of environmental, safety, and regulatory standards. The PM decision-making process expanded into a wide span of control, over many scattered resources, some under their direct control, while others report to different functional areas inside and outside of the project and company boundaries. At the same time, non-core members of the design team have to also navigate the different tasks they must accomplish through different types of projects. The challenges to the PM in this period are summarized in Table 1.6.

An example is that of a quality engineer in a global technology company who might be assigned to multiple projects and also needs to support indirect tasks such as quality audits, parts obsolescence, and parts purchasing reviews. This dilemma of *serving multiple objectives* is challenging for the quality engineer and the PM. Constant communications

1. Distributed design resources
• Successful companies operate as global enterprises
• Design teams can be in multi-sites using different tools
• Outside expertise can be leveraged when needed
2. Using the supply chain for design and manufacturing
• Retain core competency and outsource other functions
• Manufacturing is a global commodity
• With increasing standards, part or all of the design process can be easily outsourced
3. Competitive environment
• Educated, demanding customers with increased expectations
• Diverse and rapid changing technology
• Varying global regulatory and environmental concerns

TABLE 1.6 Challenges of the Collaborative Project Manager

Management Organizational layers	Many →	Few
Project Manager Span of control	Narrow →	Wide, Intermixed
Team members skill set	Specialized →	Interdisciplinary
Customer base	Local, narrow →	Global
Supply chain	Local, manufacturing only →	Global design and MFG
Market segments	Few and large →	Many and niche
Project team	Colocated →	Distributed worldwide

TABLE 1.7 Changing Role of the Project Manager in a Global Economy

and balancing of priorities are essential for all involved to ensure smooth development of new products. The PMs of global products also had to change their method of operations and leveraging of distributed resources, as shown in Table 1.7.

1.4 Conclusions

The historical developments of the tools and concepts of the high-technology project management were outlined for the purpose of understanding how they were developed and what role they played in creating more competitive global products. The main responsibilities of the PM were amplified from developing the new product to a more universal product champion responsible for the total collaborative new product lifecycle: from product inception to the product sunset, either by obsolescence or by replacement with a newer technology product. The new major concerns of a PM became as follows:

1. Keep the project on track. Use a project-development process with phases and gates. Keep track of the schedule through good planning, risk assessment, and cost reports. Use Gantt and Pert tools and charts (to be discussed in Chap. 7) to specify design activities start and end times. For a technology product, first to market captures 70 percent market share, and the obsolescence time of the product is determined by the marketplace and not the company. If the product release to market is late, it could have a severe impact on company profits and product ROI.
2. Control engineering development costs. Use cost, scheduling, and risk-tracking tools to maintain original project cost targets. Most important, stay on schedule by reducing ECO revisions and rework, resulting in lower engineers' salaries and expenses charged to the project. Initiate a good initial analysis of market scope and competitive position to reduce design revisions. Manage the project team using a suitable project organization. Reduce technology risk by using proven technology or incremental technology advancements in each product cycle. Reduce interpersonal conflict within the team and increase communications

while keeping other functions such as manufacturing, accounting, and quality involved. Keep a core team, while other team members are engaged in an as-needed basis. Leverage the use of outside knowledge consulting resources and the supply chain.

3. Keep product manufacturing costs low, based on good marketing estimates and proper balance of tooling versus expected sales volume. Using the supply chain to lower development and manufacturing cost. Study the impact of profit or loss due to declining sales of existing products. Most important is to ensure product is completely designed before start of production; otherwise, warranty and repair costs will be very high. Keep an accurate estimate of product manufacturing cost and profit by using tools such as ABC costing. Keep quality high and cost low by using tools such as DFM, ESI, 6σ, and DoE.
4. Make sure that specifications are commensurate with the marketplace, competition, and technological innovation as well as customer expectations. Use tools such as QFD for follow-on evolutionary or next-generation products. Use self-directed teams for revolutionary products. Adjust new product specifications by continuously monitoring the competition and customers.

When a PM is faced with a delay in the project schedule due to a problem in one or more of the four issues outlined above, what is the best course to take to get the project back on track? Many studies were made to compare the different scenarios a PM could implement to keep the project on schedule. They could take one of the following forms, which are graphically outlined in Fig. 1.19. In this figure, the four alternatives outlined below are shown with their impact on the lifecycle profitability, giving a scenario of 20 percent annual growth in sales, 12 percent price erosion per year, five-year lifecycle:

1. Accept the project delay. This is the least desirable option since the first technology product released captures 70 percent market. In addition, product obsolescence is determined by the competition and technology advances, not by the company developing the product. The negative impact of late product release on future profits is the most severe.

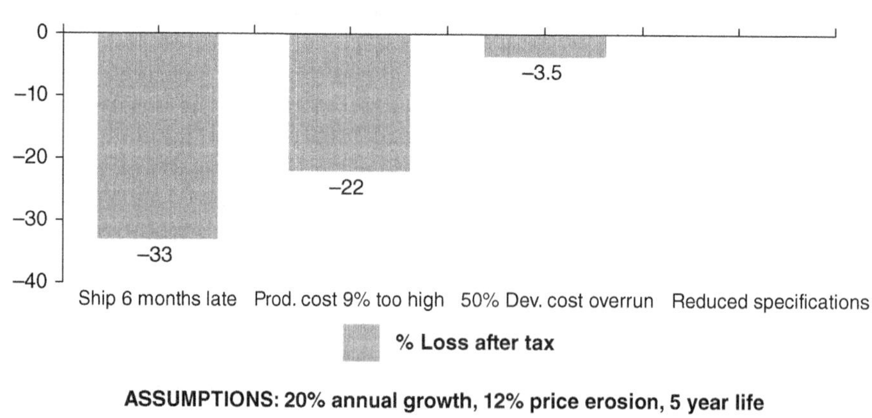

FIGURE 1.19 Sensitivity of project manager decisions due to project schedule delays.

2. The PM could release the product on schedule, even if the design is not complete, with poor quality. Products would have to be reworked in the factory or at the customer site, leading to greater product cost. This option is the second least desirable, since the monetary impact is augmented by the loss of customer satisfaction and its impact on future product sales.
3. The PM could seek more resources in terms of more engineers added to the project. The request has to be delicately handled; otherwise, the additional engineers will trip over one another. This is the most desirable scenario to reduce schedule slips.
4. Finally, the PM can review the new product specifications through a process such as QFD, then determine if the specifications can be altered to reduce the engineering design effort.

References and Bibliography

Akao, Y. (ed.), *Quality Function Deployment: Integrating Customer Requirements into Product Design*, Productivity Press, New York, 1991.

Advanced Product Quality Planning and Control Plan (APQP), Automotive Industries Action Group (AIAG), Southfield, MI, 1995.

Atkinson, A., *Optimum Experimental Designs, with SAS*, Oxford University Press, Oxford, 2007.

Bailey, R., *Design of Comparative Experiments*, Cambridge University Press, Cambridge, 2008.

Bajaria, H., "Six Sigma-Moving Beyond the Hype, "*Annual Quality Congress Proceedings*, ASQ, 1999.

Boothroyd, G., P. Dewhurst, and W. Knight, *Product Design for Manufacture and Assembly*, 2nd ed., Marcel Dekker, New York, 2002.

Box, G., J. Hunter, and W. Hunter, *Statistics for Experimenters: Design, Innovation, and Discovery*, 2nd Ed., New York: Wiley-Interscience, 2005.

Clausing, D., and Simpson, H., "Quality By Design," *Quality Progress*, pp. 41–44, January 1990.

Gilliland, M., "Is Forecasting a Waste of Time?" *Supply Chain Management Review*, July/August 2002.

Goos, P., and B. Jones, *Optimal Design of Experiments: A Case Study*, Wiley, New York, 2011.

Hahn, G., et al., "The Evolution of Six Sigma," *Quality Engineering*, 12(3): 317–326, Issue 3, 2000.

Harry, M., *The Nature of Six Sigma Quality*, Motorola University, Rolling Meadows, IL,1988.

Harry, M., and R. Schroeder, *Six Sigma*, Doubleday, New York, 2000.

Hauser, J., and D. Clausing, "The House of Quality," *Harvard Business Review*, 3:63–73, May–June, 1988.

Henchy, L., "American Manufacturing Fights Back," *Business Solutions*, p.10, Feb. 22, 1988.

Hwang, J., "Isle of Man," *Environment Friendly Electronics: Lead-Free Technology*, Electrochemical Publications Ltd., 2001.

Iversen, W., "The Six Sigma Shootout," *Assembly Magazine*, pp. 20–24, June 1993.

Kemtovicz, R., *New Product Development: Design and Analysis*, Wiley, New York, 1992.

Kinkelmann, K., and O. Kempthorne, *Design and Analysis of Experiments*, 2nd ed., Wiley, New York, 2008.

Liker, J., *The Toyota Way: 14 Management Principles from the World's Greatest Manufacturer*, McGraw-Hill, New York, 2003.

Mason, R., R. Gunst, and J. Hess, *Statistical Design and Analysis of Experiments with Applications to Engineering and Science*, Wiley, New York, 1989.

Miyakawa, S., and T. Ohashi, "The Hitachi Assembly Evaluation Method (AEM)," *Proceedings International Conference on Product Design for Assembly*, Newport, RI, April 15–17, 1986.

Ohno, T., *Toyota Production System: Beyond Large-Scale Production*, Productivity Press, New York, 1988.

Ohno, T., *Just-In-Time Time for Today and Tomorrow*, Productivity Press, New York, 1988.

Pilkington, A., "Manufacturing Strategy Regained: Evidence for the Demise of Best-Practice," *California Management Review*, 41(1): 31–42, 1998.

Production Part Approval Process (PPAP), Automotive Industries Action Group (AIAG), Southfield, MI, 1995.

Ross, P. J., "The Role of Taguchi Methods and Design of Experiments in QFD," *Quality Progress*, XXI(6):41–47, June 1988.

Shina, S., *Green Electronics Design and Manufacturing*, McGraw-Hill Professional, New York, 2008.

Shina, S., *Six Sigma for Electronics Design and Manufacturing*, McGraw-Hill Professional, New York, 2002.

Shina, S., and A. Saigal, "Manufacturing Costs for Electronic Products," Vol. 3 of the *Encyclopedia of Materials*, pp. 2727–2735, Elsevier Press, Waltham, MA, November 2001.

Shina, S., *Successful Implementation of Concurrent Engineering Products and Processes*, Wiley, New York, 1993.

Shina, S., *Concurrent Engineering and Design for Manufacture of Electronic Products*, Kluwer Academic Publishers, Norwell, MA, 1991.

Smith, B., "Six Sigma Quality, a Must not a Myth," *Machine Design*, pp. 13–15, February 12, 1993,

Smith, P., and D. Reinertsen, *Developing Products in Half the Time* 2nd ed., John Wiley and Sons, New York, 1998.

Srihari, K., "A DFM Framework for Surface Mount PCB Assembly," *International Conference on Technology Management, Design for Competitiveness*, pp. 207–217, Denver, CO, 1993.

Sullivan, L. P., "Quality Function Deployment," *Quality Progress*, XIX(6):39–50, June 1986.

Ulrich, K., and S. Eppinger, *Product Design and Development*, 2nd ed., McGraw-Hill, New York, 2005.

Womack, J., and T. Jones, *Lean Thinking: Banish Waste and Create Wealth in Your Corporation, Revised and Updated*, Harper Business, New York, 1991.

Womack, J., T. Jones, D. Roos, and D. Carpener, *The Machine That Changed the World: The Story of Lean Production*, Free Press, New York, 1990.

Discussion Topics

1. If you followed Taguchi's principles, and you wanted to investigate a design with five factors at two levels, how many experiments do you need? How many would you need if you used five factors at two levels in classical statistics with full factorial experiments? Some of the participants can try their own numbers such as three factors at two levels; two factors at two levels; two factors at three levels; three factors at three levels, etc. For more information consult the following websites: http://en.wikipedia.org/wiki/Fractional_factorial_design and https://controls.engin.umich.edu/wiki/index.php/Design_of_experiments_via_taguchi_methods:_orthogonal_arrays

2. If your company policy decided on a $Cp = 1.33$ design, what are the specification limits going to be equal to in terms of number of σ? And what is the expected defect rate? Some of the participants can try their own numbers such as $Cp = 1.67$, $Cp = 1$, etc.
3. According to the example of taking the defect rate and translating it into a Cpk (assuming that the specification limits are two sided and the process average μ = design nominal N), what is the equivalent Cpk for a defect rate of five units per 1000 tested? Some of the participants can try their own numbers such as six defect rate per 1000 tests, etc. Each participant should have a unique reject rate translated into Cpk.
4. List the major concerns of the PM. How would you rank the impact of not meeting these concerns on the project lifecycle of high-technology new products?
5. How would the impact of not meeting these concerns of the PM change in priority if the project were being developed in the following industries? Please provide the reasons for your selection:
   a. A high-technology smartphone manufacturer
   b. A commercial airplane manufacturer
   c. A regulated manufacturer (defense or medical)

   Show your results in matrix form (PM concerns against three products types). Indicate the change of the PM concerns against the product type. Rank them in priority as HI, MED, and LO.
6. Discuss the major difference in the PM role is DTM or DTC. List at least three issues based on that difference.

Problems

1.1. Write a generalized view of the evolution of concurrent engineering and product design, in terms of increasing product performance, reducing cost through DFM/DFA/ESI or increasing quality and variability reduction through DoE and Six Sigma. Give an example of products from your own personal experience, or compare and contrast two different companies such as Motorola versus GE or Dell versus HP in quality and cost. You may also choose any other product family such as printers, TVs, computer displays, TV recorders, cell phones, autos, and computers (desktops, laptops, or tablets), and trace their evolution in performance, cost and quality, using the tools mentioned in Chap. 1.

1.2. Compare and contrast the different tools mentioned in Chap. 1. Please limit yourself to a maximum of three issues or one paragraph for each of the three answers:

a. DFM: BDI versus GE/Hitachi
b. DoE: Classical DoE versus the Taguchi method
c. Design quality: Cp (using the Motorola definition) versus Cpk (using the GE definition)

1.3. Write a generalized view of engineering project cycle times, new model introduction speed, and the increasing globalization of design and manufacturing resources. Discuss the topics of metrics and how they could be used in measuring the technical performance of engineers, managers, and engineering projects. Give one example of a global product and its successive generations from major companies. You may also choose from any product family such as autos, airplanes, printers, TVs, computer displays, TV recorders, cell phones, and computers (desktops, laptops, or tablets), and trace their evolution in performance, cost, and quality through successive generations of new products now and in the past.

1.4. Compare and contrast the different constraints and concerns for the PM in DTM or DTC projects. Mention at least three concerns. Show how these two types of projects differ in the following areas:

a. Priority of the different types of concerns
b. Project risks and where in the project lifecycle they are of most concern
c. How to mitigate these concerns

CHAPTER 2

Product and Project Perspectives and Managing Different Types of Engineering Projects

The new product development cycle for technology companies has been undergoing rapidly accelerating changes due to the increased technology development and competitive pressures. These pressures are rendering new products critical to the financial future of companies and accelerating older product obsolescence. In this chapter, we will discuss the evolution of the new product development process in technology companies and its implementation within the overall new product lifecycle.

In the global high-technology industry, the lifecycle of the typical product is different than that of historical models thanks to the impact of rapid technological change. The investment in new product development is greater than in other industries because the rate of new product development and technological changes is much higher. These development costs tend to be the highest during the start-up phase of companies and lowest as products mature. Costs vary from about 3 percent of revenues in established products to more than 40 percent of revenue in new market segments. In addition, the first product with technology innovation in the global technology market captures 70 percent of potential total sales.

Companies are looking for ways to optimize and preserve their investments. They have targeted product development as the most cost-effective element in lowering their overall new product costs. In this chapter, we will discuss methods and techniques for assuring success in new product strategies by adopting different development methods, depending on the technology and maturity of the market. The project-management structures needed to develop these new products will be analyzed. The topics to be discussed in this chapter are as follows:

2.1 *The Overall Product Lifecycle Model.* In this model, the relative stages of market development of the product use and its technology are presented: start-up, growth, maturity, and its concluding stage, whether in decline or the commodity stage. The role of the product development in each step will be outlined.

2.2 *The Role of Technology in Product Development and Obsolescence.* The technology strategy in new product development is presented in terms of price versus performance curves. Emphasis on the risks of adopting either alternative will be discussed. The types of new product lifecycles and engineering investments are also illustrated, depending on technology adoption and industry type. The role of technology in the obsolescence strategy of global products is shown, with emphasis on product changeover risks.

2.3 *Technology Product Types and the Project Management Structures Needed To Develop Them.* The different types of products—from evolutionary, platforms, and architectures to revolutionary products—are discussed. The project-management structures needed to manage these different product developments are shown, with the benefits and concerns related to each model.

2.4 *Creating an Environment for Successful Project Management.* The creation of a quality culture within the organization and the focus of the enterprise on new products and team approach to project management are key elements for ensuring success in new product creation. The phase review process for project management and the key processes to support a successful project are also discussed.

2.1 The Overall Product Lifecycle Model

Technology products go through many stages in their lifecycle, as shown in Fig. 2.1. The first stage is *start-up* or *market development*. During this stage, emphasis is on the performance of the product. Features such as speed, accuracy, capacity, response time, and other *bells and whistles* dominate the product cycles. At this stage, the benefits of the product to the customer are perceived to be high in increased productivity or personal comfort and satisfaction. The number of competitors is large, since entry into the market is wide open, and a new company can establish a *niche* in the market place for a relatively low investment. Product development during the start-up stage is marked by

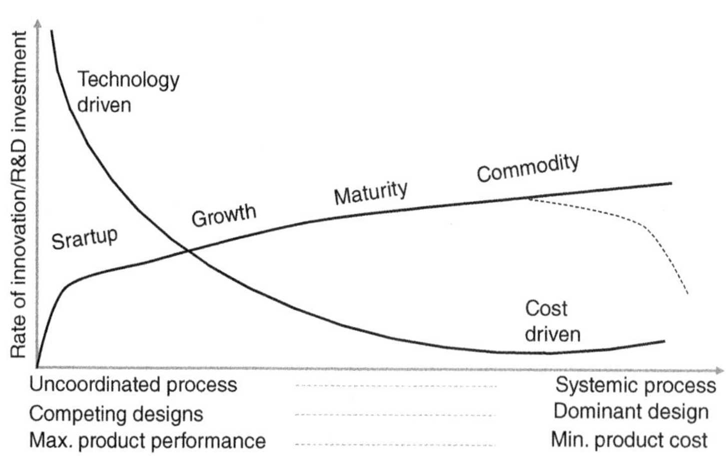

Figure 2.1 Product lifecycle versus technology innovation.

the intense drive to arrive at the market as early as possible, with minimum concern over manufacturing cost or quality. A good indicator of this stage is the number of wire cuts and changes to printed circuit boards (PCBs) in new electronic products. The quality and reliability of the new product in manufacturing is achieved by extensive inspection and tests.

In summary, the attributes of the start-up stage are:

- Technical performance is the key driver of the start-up product
- There is no dominant design for start-up products
- There are few industry standards available
- Companies compete on features such as speed, accuracy, capacity, and response time
- Customers' perceived benefits of buying the product are high
- Price is not affected by the competition and is much higher than that of traditional products
- Market entry by new companies is relatively easy
- The project manager's (PM) focus is on schedule or time to market (TTM); there is less concern with manufacturing cost and quality
- Project development costs are high compared with that of traditional projects (25 to 40 percent of sales revenue)
- Long-term lifecycle costs and cost of ownership are not yet important considerations to the customer
- Job-shop-style specialty manufacturing is the dominant process for making the start-up product
- Manufacturing site is local
- Sales volume is low (in the 10,000s)
- Employee skills are critical and multifunctional
- Product availability might be an issue with lower quality of design and manufacturing hindering the ability to meet increased customer demands

The second stage is *growth*. As the marketplace expands, and general acceptance of the product and its technology is assured, the number of competitors drops, and the rate of market and technology development begins to slow. The driving issues of product development are not the acceptance of the technology or the particular use of the product, but the differentiating aspects of the products offered from different companies. There is increased customer confidence in the stability of the technology and future product evolution and offerings. In addition, companies will emphasize their long-term commitment to this business segment. The long-term cost of ownership of the product such as reliability, field support, and the growth of ancillary products and services supporting the product and its technology are emphasized.

Product development during the growth stage is characterized by the focus on introducing the manufacturing guidelines of capabilities and constraints to the new product and beginning to concentrate on manufacturing as a strategic weapon to emphasize low cost and high quality. Working with suppliers is increased, with parts of the manufacturing or design efforts subcontracted to the supply chain.

In summary, the attributes of the growth stage are:

- Technical performance and option differentiators begin to slow down
- There are fewer different design alternatives in this stage
- Some industry standards become available
- Companies continue to compete on features such as speed, accuracy capacity, and response time
- Customers' perceived benefits remain high
- Competition is beginning to affect price and thus is trending lower
- Market entry is moderate
- Many companies with the wrong technology or design methodology go out of business
- PM focus is on schedule, with increasing concern with manufacturing cost and quality
- Project-development costs are moderate compared to traditional projects (10 to 25 percent of company revenue)
- Long-term lifecycle costs and cost of ownership become important considerations to the customer
- Batch-shop-style specialty manufacturing is the dominant process for making the growth product
- Manufacturing sites are national
- Sales volume is medium (in the 100,000s)
- Employee skills are increased with cross training
- Product availability becomes a differentiating attribute of companies

The third stage is *maturity*. This phase is characterized by the emergence of a dominant technology or technique for the product design and the development of industry standards that encourage a dominant design and uniform customer perception in using the product. At the same time, the relative growth of the market is slowed, being only proportional to the growth of the population or the customer base, as the product saturates the market. The number of manufacturers continues to decrease, as they either go out of business or get bought out by the larger companies. The competitive emphasis in this stage is on price and quality, as the dominant technology does not allow too much variation on the basic design of the product. Examples of competing technologies that were decided by the customer are video-recording formats and the corresponding players. First there was VHS-versus-Betamax formats, with Betamax having the better technology but losing on the lower cost of VHS. Then it was Blu-ray versus DVD. The two formats are similar, and the players and discs look the same, and even the disc menus are similar. Blu-ray brings three major improvements over DVD: better image quality, better sound quality, and more special features. All three are made possible by the larger storage capacity of Blu-ray, which is capable of storing 50GB of information on a single Blu-ray disc, compared with DVDs, which can hold about 8GB. In this case, the large distribution channels can decide on the winning technology. Wal-Mart, the world's largest retailer, phased out HD DVDs by June 2008 and made Blu-ray the technology winner.

Product development in the maturity phase is focused on continued improvement in the manufacturing process. There is a stronger emphasis on quality through the tools discussed in Chap. 1, including DFM/DFA/ESI, DoE, 6σ, ABC, and QFD. The suppliers for the product are involved early and often, and the design process is more robust through the use of analysis and simulation tools. There is an emphasis in speeding up the new product introductions to leverage technology adoption as early as possible.

In summary, the attributes of the maturity stage are:

- Most common product in the global technology industry
- Market growth begins to slow as new products saturate the market
- Technical performance and option differentiators will slow down
- A dominant design emerges and is used by most competitors
- Dominant design is encouraged by major distribution channels or material availability from major suppliers
- Common industry standards are adopted
- Companies compete on cost, quality, and delivery
- Customers' perceived benefits are medium
- Price and perceived quality are prime competitive factors
- Only few companies compete because of consolidation and buyouts
- Many barriers exist to companies seeking market entry, including dominant design and standards competency
- PM focus is on project lifecycle, with extensive use of concurrent engineering (CE) tools
- Project development costs are similar to that of traditional projects (3 to 10 percent of company revenue)
- Long-term lifecycle costs and cost of ownership become important considerations
- Assembly-line manufacturing is the dominant process for making mature products
- Manufacturing sites are global
- Sales volume is high (in the 1,000,000s)
- Workforce reduction is achieved by automation and outsourcing
- Product cost and quality become the differentiating attribute of companies

The last stage can take one of two forms: either the product will decline as the need for the product is overwhelmed by new technology or the product will develop into a *commodity*. This process of technology turnover is sometimes called "disruptive technology" and can result in quickly obsolescing commodity products. An example is the music player business, where eight-track cassette players were replaced by CDs, which were then replaced by Internet music downloads stored on personal electronic devices.

The difference between a disruptive and a revolutionary product is that the revolutionary product is creating an entirely new market where one did not exist before, while the disruptive product is accelerating the demise of a commodity product by replacing it with alternative technology. A tablet computer and a smartphone are revolutionary products in that they created their own customer demand.

In the commodity stage, the number of product manufacturers will decline to a select few large companies, and entry into this market will become expensive and risky in terms of taking market share from established competitors. The emergence of standards of use, manufacture, interconnection, and quality will make price the only competitive leverage. The products will essentially be interchangeable from one manufacturer to another, with high customer expectations of quality and reliability. The revenue per unit decreases rapidly as competition is solely on price, and manufacturing techniques become the major factor in ensuring the long-term survival of the company. Follow-on products will be evolutionary, with a market leader establishing a careful trend, which locks on his customer base and provides a definite upgrade path for every new generation of products.

The product development emphasis in the commodity stage is on reducing manufacturing cost while maintaining the high quality expected by the customer. There is a much higher level of automation, as manufacturing knowledge and the stability of the design are increased. Few companies can enter into a market at the commodity stage, as the costs are prohibitively high for recruiting the expertise required to design the product to established standards and customer expectations as well as name recognition. In addition, the costs to manufacture the product at elevated levels of production, automation, and quality are high. The project development process is detailed and systematic to ensure maximum profit returns from minimum engineering investments. In some cases, companies begin to share resources, especially in manufacturing, to round out their product set and maintain their presence in the marketplace. A single factory in Mexico can turn out many competitors' TVs in the same line with different companies' labels.

The global high-technology industry has followed many other industries into this pattern. The automobile industry is a prime example. In the early part of the last century, there were hundreds of auto manufacturers, and any of the competing technologies could have become dominant: electric, steam, or internal-combustion engines. The computer industry has gone through these stages for various products. The mainframes have declined and were replaced by the personal computers, then with laptops, minibooks, and tablets, all of which have become commodities, with exchangeable software programs and plug-in components. The sales of laptops are declining, being rapidly replaced by personal tablet devices.

In summary, the attributes of the commodity stage are:

- Price is the main differentiator, since most companies achieve high levels of cost, quality, and delivery
- Technical performance and option differentiators are at a minimum between competing suppliers
- A single design dominates, with few component and system suppliers available
- Mature industry standards
- Companies compete mostly on price, since customer perception is that all products are the same

- Few survivor companies remain, and entry to market is prohibitive
- Project managers focus on project lifecycle, with extensive use of design and manufacturing outsourcing
- Project development costs are below those of traditional projects (1 to 3 percent of company revenue)
- Long-term lifecycle costs and cost of ownership become important considerations, with some companies offering extensive warranties
- Commodity companies share resources such as the same plants for competing products and supply chain
- Flow-line manufacturing is the dominant process for making the mature products
- Manufacturing sites are globally distributed
- Sales volume is high (> 10,000,000s)
- Workforce size and skills are reduced by heavy automation
- Commodity products could become obsolete by disruptive technologies
- Companies can afford saturated advertising to maintain name recognition

Table 2.1 summarizes the design attributes trend of technology-based products at each stage of innovation, and Table 2.2 shows the impact of these stages on the manufacturing activities of technology products. It is clear from these two tables that the transition from start-up to commodity products indicates a maturity of skills and processes. Most global high-technology companies keep a mix of products in their product portfolio to keep their profits margins stable, by continually filling the new product pipeline with products at different stages of the lifecycle continuum.

The customer perception of the role of technology adoption and inclusion of new features and options in new products is further discussed in the following Chap. 3, which deals with customer-satisfaction concepts such as the voice of the customer (VoC), the Kano diagram, and the quality function deployment (QFD).

	Start-Up	Growth	Maturity	Commodity
Product Variety	Great variety	Standardization	Dominant design	Mature standards
Sales Volume	Low	Increasing	Very high	10's millions sold
Industry Structure	Many competitors	Consolidation and buyouts	Few companies Barrier to entry	Survivors Sharing resources
Competition Point	Product performance	Product delivery	Quality and options	Price
Critical Processes	Time to market	New products cycle	Focus on CE tools	Cost control

Table 2.1 Attributes of Product Lifecycle Design Stages

	Start-Up	Growth	Maturity	Commodity
Manufacturing Process	Specialized job shop	Batch shop	Assembly line	Flow line
Manufacturing Volume	10,000s	100,000s	1,000,000s	10,000,000s
Manufacturing Location(s)	Local	National	Global	Distributed
Labor Emphasis	Employee skills	Employee training	Labor reduction	Lowest skills

TABLE 2.2 Attributes of the Product Lifecycle Manufacturing Stages

2.2 The Role of Technology in Product Development and Obsolescence

The technology changes in the global high-technology industry stem from many different sources: faster and larger capacity of electronic components such as memories and microprocessors; new techniques in information storage, retrieval and display; and new production processes, materials, and plastics innovation. In addition, global competition is forcing the turnover of products at a much higher rate. New products quickly become old products as global manufacturers rush to produce similar products through reverse engineering.

Traditionally, most revenues from new products occur while it is early in the product lifecycle. If the product offers unique features that are not duplicated by the competition, the revenue from this product will be even greater. The first technology product to market captures 70 percent market share. As the product matures, and there are many competitors in the market offering similar capability, price erosion will ensue as more companies compete for scarce customers.

Typical price performance curves have been developed for technology products, and they follow the basic Moore's law. They show the price coming down for the same performance/capability over time, as competitive pressures and technological advances take hold. At the same time, new performance curves tend to form at the top price and then continue to come down to the lower price over time. The creation of a new performance curve has to be significant, such as 30 to 50 percent improvement in speed, measurement capability, or response time. Customers would not support a new pricing structure for the higher performance product unless there is a measured improvement to their productivity through the product use.

The application of technology trends in product development can proceed along two different paths, depending on the new product strategy for the company. They are called price performance curves and are shown in Fig. 2.2. They are:

1. Emphasis on more performance with constant price. The advantage of this scenario is the creation of a new market demand for the new technology. It also provides customer upgrade and growth paths into the product, as the customer becomes more sophisticated in fulfilling his or her needs. The pricing mechanism is easier to set for maintaining the current level of profitability, since the new products are priced similar to older ones. The risk of this strategy is that

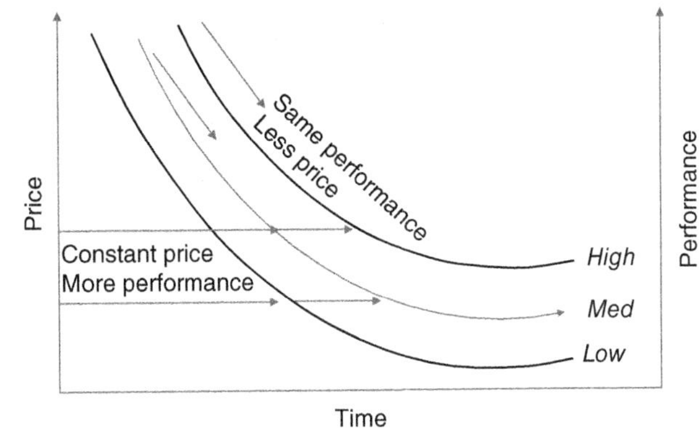

FIGURE 2.2 Price performance curves.

competitors can easily upset it by offering the same or increased performance at a lower price. In addition, this strategy requires a well-developed market and sophisticated customers who can understand the nuances of the price performance curves.

2. Emphasis on providing existing capability or same performance with less price. This same performance for a lower price strategy can be either a defensive or an offensive move. It has the advantage of protecting the low-end customer base and could broaden the appeal of the product to a larger constituency. In addition, it leverages off the market-development cost associated with new products.

One of the drivers of this strategy is to hopefully increase product sales volume because of lower product selling price. However, if the sales volume does not increase substantially to offset the lower sales price, profitability will be decreased. This could, in the long-term, erode the financial position of the company.

Another risk to this strategy is the possible existence of hidden costs that can crop up when volume increases. The manufacturing process has to be well developed in documentation and employee competency, and the product material and labor costs should be well defined.

Another concern when developing a lower-price strategy for the same performance is that lowering the price may not generate more customer sales at the expense of the competition. Surveys have shown that the effects of flexibility, perceived quality, and quick delivery are becoming more important than price in the customer purchasing decision matrix, especially in mature or commodity products. Table 2.3 is a summary of issues with the price performance strategies.

Another important role of technology in product development and strategy is the effect of technological change in making components and processes obsolete, therefore shortening the life span of technology products. A case study in point is the obsolescence of electronic components by a major company. The average obsolescence rate of all parts was 7 percent per year. Semiconductor memories had the highest obsolescence rate at 15 percent per year. This obsolescence can be mitigated by making lifetime buys, which is a dangerous proposition, considering there might be shelf-time issues in the storage of these components, which can alter some of their physical attributes.

More Performance with Constant Price	Same Performance with Less Price
Advantages	
Create new market demand	create low-end customers
Provides customer growth path	Leverage current technology and market conditions
Easier to set price and manufacturing cost	Pressure competitors in maintaining profits
Concerns	
False customer expectations long term	May erode long-term profits
May require market development effort	Hidden costs may cause profit surprises
Danger of false performance upgrade	May fall behind in technology adoption

Table 2.3 Issues with the Price Performance Strategies

These technological factors and competitive pressures determine the obsolescence timeline for the product. Many manufacturers hold on to products too long, only to discover that the negative effects far outweigh the easy profit they think they can generate. Customers may feel that these companies are lagging in the technological field and will not purchase these products. The manufacturing department will have to incur the extra expense of procuring and maintaining outdated parts and processes. In addition, there is a loss of the professional staff morale, as engineers struggle with maintaining old technologies.

Another issue of being late in the design phase is the effect of releasing a new product on the current sales of existing products. If the new release plan does not dovetail perfectly with the overall production schedule, the company could be vulnerable to a dry period, where customers convert their orders from old products to new ones as soon as the new products are announced. This could create a serious gap in the revenue stream and production schedules if the new products have late shipment due to production problems. The production turnover plan has to be smooth to carry the company financially during the product changeover.

The classical case of this type of dilemma concerns the Osborne computer in the early 1970s. The company was a small start-up, with a unique idea (at that time) of having a personal computer in a briefcase. The Osborne I was successful, but when plans were announced for the Osborne II, customer orders shifted to the newer model. The company was unable to ship the Osborne II on time and eventually went out of business. Figure 2.3 further illustrates this point for successive generations of microprocessors in personal computers. Every release of a new product generation in the product family displaces the older product and has to be available to the market in large volume to replace the older-generation product.

Another important issue to consider is the product lifecycle in certain industries. The need for technological change and therefore new products is much slower in industrial products than in high-technology products. Industrial products tend to have a much longer lifecycle than information-technology products. Companies purchase them for use in production lines or warehouse activities, and they represent an investment that has to pay for itself before a new replacement product is purchased. If the rate of innovations and technology turnover in industrial products is slow, then the need for a large engineering staff to design new products is small. This design engineering need level, usually identified as the research and development (R&D) budget, should be commensurate with the product lifecycle and technology improvement over time, as

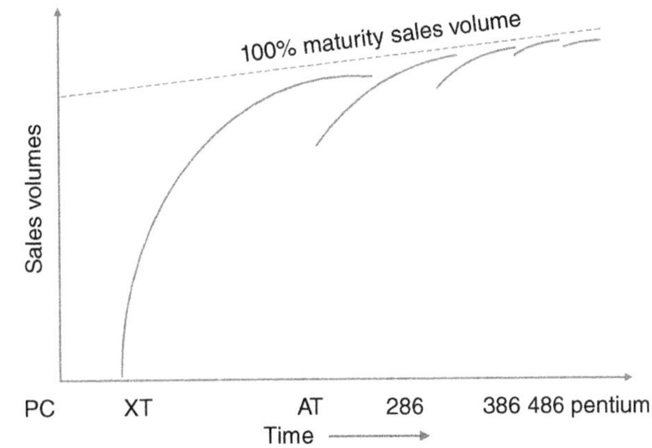

FIGURE 2.3 Faster time to market in a product family.

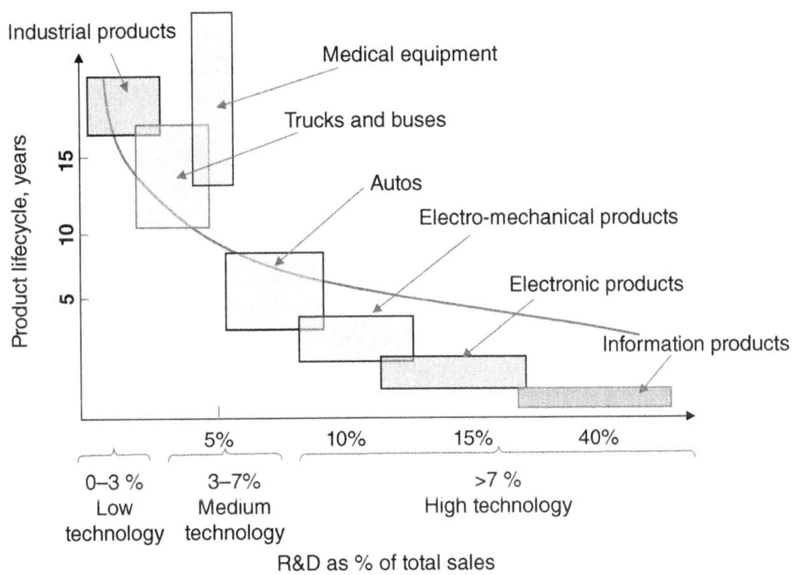

FIGURE 2.4 R&D investments versus industry-specific product lifecycle.

shown in Fig. 2.4. The lifecycle is shown in years of new product development, and the R&D investment is shown as a percent of sales, a common number available from companies' balance sheets.

The following observations can be made from Fig. 2.4:

- Industrial products technology adoption is small, and the customers want to gain the maximum return on investment (ROI). Some products are designed to last the same cycle as the industrial structures they are in (i.e., elevators).
- Trucks and busses have long lifecycles due to regular maintenance and less customer focus on style changes.

- Medical products have variable lifecycles due to issues of product regulations and customer training investments for doctors and nurses.
- Automobiles have yearly cosmetic model and style changes. The lifecycle is that of major chassis or engine or component system changes.
- Electro-mechanical products have wear-out issues, which limits the lifecycle of products.
- Electronic products have extensive development costs that need to be amortized before introducing the next-generation products.
- Information technology products have the lowest price, greatest volume, and maximum technology adoption requiring the shortest lifecycle and hence the largest R&D costs.

2.3 Technology Product Types and the Project Management Models Needed to Develop Them

The PM for a new technology product has to be aware of the traditional competitive position of the company and the proposed technology adoption in the new product. These factors determine the size and cost of the development project. In addition, the project management model is also influenced by whether the product is design to customer (DTC) or design to market (DTM).

The PM should explore the concurrency of the proposed project tactical plans with the strategic direction of the company. They should then propose a project plan with the proper tactical elements and obtain the consensus and signoff by the management team. Internal processes may have to be identified or developed to implement these tactical proposals and specific tools selected for managing the project.

Technology adoption is the biggest dilemma for the PM. A structure is needed to balance the amount of technology adoption in the new product versus the risk of not completing the project on time. In larger companies, research labs are established to develop new technologies without a firm timetable. Once these labs develop a new technology, an advance development function begins to implement this technology into possible commercial uses. A painstaking process for advance development is used to characterize the technology, packaging it into a useful form for the customer, and developing the manufacturing process to make new products with low cost, high quality, and reliability. Only then can a stream of products emanate for this new technology in various applications for different customer markets.

2.3.1 Types of Products That Can Be Created with New Technology Adoption

These types are shown in Fig. 2.5 and are comprised of the following:

1. Revolutionary (breakthrough) products. These are technology-based products, which are driven by basic research in the corporate research laboratories in new materials or technologies. They are brought over to the development organization through the process of advance development. This advance function is where the product feasibility is determined at the design and manufacturing stages in a process that usually occurs three to five years before actual production. This phase is outside of the tight new product development schedule (12 to 18 months),

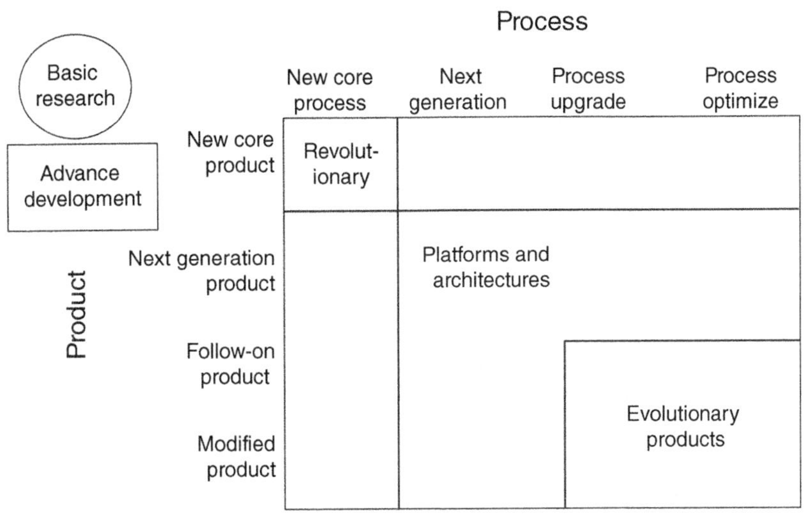

Figure 2.5 The product and process development types.

but has a definite set of requirements, budgets, and schedule goals to be met. The products are then scheduled for development through the new project management process.

The revolutionary product is usually either a strikeout or a homerun in terms of profit potential for the company. An established company with several profitable products (cash cows) might develop an opportunity to use its technology in a new application, with a potentially much larger ROI than the existing product portfolio. A product opportunity might also arise for a company to develop a revolutionary product for its customer base. In these cases, the revolutionary product will be a new core product, with a new core process to manufacture it. Eventually this revolutionary product will mature based on the product lifecycle chart shown in Fig. 2.1.

Companies create R&D labs for basic research. These labs do not have specific goals or timelines to explore new technologies. They are expensive to manage, since most are populated by advanced degree engineers. For example, AT&T labs developed the transistor (Shotckey, 1951) and had the most Ph.D.s in the United States at its New Jersey site. Hewlett Packard developed the original calculator (later spawning its computer business) and laser printing in its research labs, which became the core competency of the company.

Many companies develop new technologies through their research labs, but do not develop the necessary product or manufacturing strategy to take those technologies into the marketplace, allowing other companies to take the lead. Xerox developed the mouse, and AMP developed video recording, but never turned these technologies into successful products in their portfolios.

Advance development is a function dedicated toward a commercialization goal of new technology. It usually takes a medium time-horizon of three to five years to focus on specific issues of developing the product packaging or manufacturing process for the technology. An example would be the transition from light emitting diode (LED) technology invention used in early calculators to LED screens in cell phones to touch screens on phones and tablet computers. Another example is the advent of component miniaturization starting with the SONY portable mini radios in the 1980s to cell phones. The revolutionary smaller product size necessitated smaller components to be manufactured, new interconnections, and packaging technologies to be developed as well as new manufacturing materials and methods to join these components.

Recent trends and global competition have reduced the number and size of corporate research labs, shifting the research burden in the U.S. to government-supported research universities.

2. Follow-on products. These are new products within the existing product families, which are introduced depending on the price/performance characteristic of the marketplace. They are driven by changes in customer expectation, competition, technology, market development, and product opportunity. These products are defined internally by a combination of inputs from sales, product management, and the internal marketing managers. They normally go through a regular development schedule with minimum interface to advance development or research labs. These follow-on products are sometimes called "platform" and "architectures," referring to the base technology used in the product. They can be one of the following:
   a. Next generation product. In this case, an update to the product with the latest technology level might be required to keep the product price/performance curves on track, as was discussed in the previous section.
   b. Product family follow-on. In this case, a new product opportunity is identified at a certain point in the price/performance curves. It could be a low-cost or a high-performance version of the same product technology.
3. Modified products. These are sometimes called "enhancements" or "evolutionary products." These enhancements on existing products are made essentially for cost reduction or specific customer inputs or options. They are simple in nature and require little engineering effort. In most cases, they require only one functional area (manufacturing) to complete the product development cycle, without input from other functional areas.

Engineers handle modified products in a single function. They sometimes require a thorough knowledge of the manufacturing process as well as the technology of the product and the material choices involved. In addition, there is the advantage of shielding the development engineers from current production problems. It has been estimated in some U.S. companies that as much as 50 percent of the development engineer's time is spent on handling production problems. This could adversely affect the new product schedule.

The manufacturing process for the three types of products—revolutionary, platforms, and evolutionary—all experience a similar evolution. From a possible new core process for revolutionary products to next-generation process as these products mature into product platforms and further into evolutionary products.

Revolutionary Products	Follow-on Products
Complete development process	Mostly development phase required
Research, advance development, development	Cash cow
Long total development cycle (3 to 5 years)	Short development cycle (12 to 18 months)
Large impact on profits (home run)	Competitive positioning, market share protection
Technology and innovation driven	Cost and market driven
Thrives in uncoordinated processes (skunk works)	Requires a systemic process
Many competing designs	Dominant or proven design
Migh development risks	Risks are traditional in product development

TABLE 2.4 Comparing the Two Major Product Types

The manufacturing process continues to mature thorough enhancements, upgrades, and optimization.

Of the three types of products, the revolutionary products have the largest impact on the long-term financial health of the company. For Hewlett Packard (HP) Corporation in the 1970s, the handheld calculator business was one-third of the company's sales and profits, which disappeared quickly as the competition reacted. In the 1980s, the laser printer, which was developed by three engineers, made more money for the company than the thousands of engineers in HP's computer business. In the 1990s, it was the inkjet printer technology that kept HP in the vanguard of profitability. In summary, the attributes of the two different products are compared in Table 2.4.

2.3.2 Project Management Structures Needed to Support Product Creation

These structures are dependent on the product development types shown in Fig. 2.5. After World War II, technology companies tended to organize in a top-down command structure, similar to military hierarchy. The company might be split into different business units, each unit focusing on a different market or customer segment. Most of the technical managers were functional managers (FMs), with a central focus on the traditional functions: engineering, marketing, manufacturing, quality, and finance in design to market (DTM) companies. The marketing department focused on customer service, specifications, product management, and market forecast.

In design-to-customer (DTC) companies, the marketing function is replaced with program management, with a focus on customer relations and task scheduling. Further subdivision of the FM engineering function was made into individual disciplines such as mechanical, electrical, and software engineering. The subdivision in the operations function included manufacturing, materials purchasing and control, and manufacturing engineering. Other functions in the organization, such as finance and quality, were supportive of the different types of programs. Figure 2.6 shows a typical organization for DTC products. Different types of project-management structures are developed to meet the needs of new product creation.

In each type of organization, DTC or DTM, program or section managers supervise the new product creation process. These managers might be in charge of several products in a business section, product families, or one large new product, program, or system.

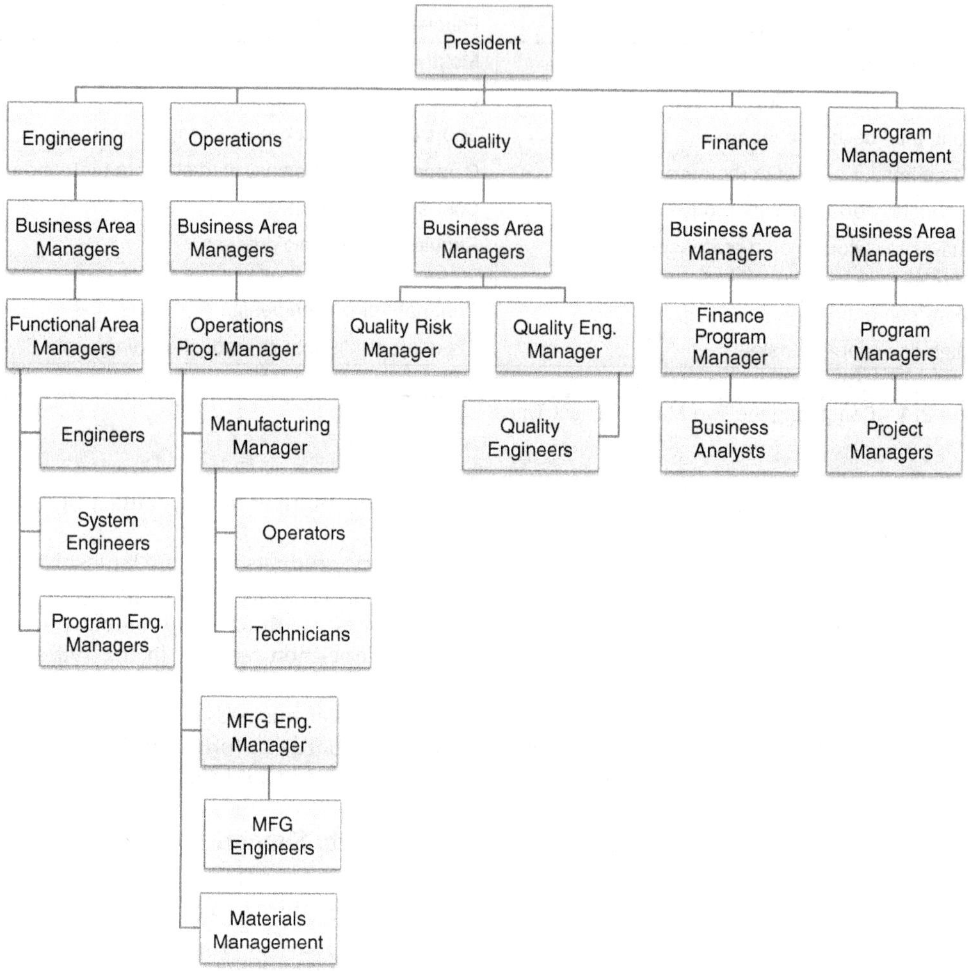

FIGURE 2.6 Typical organization for make to customer products.

Individual project managers supervise either a complete smaller product creation team or a component of a larger product or system. The relationship of the project manager to the product design team as well as to the rest of the organization can be classified as in Fig. 2.7, which shows the structures needed to support new product creation.

1. Liaison project management structure is used to coordinate all activities in managing new projects through a separate function. In a DTC organization, the liaison project managers report to a department called "program management." Similarly, in a DTM organization, the project managers report to a product management function in the marketing department. Liaison PMs have a skill set developed either through marketing/sales experience or through function management in engineering or operations. Some of the liaison PMs tend to be nontechnical (or technically obsolete), which might cause potential problems

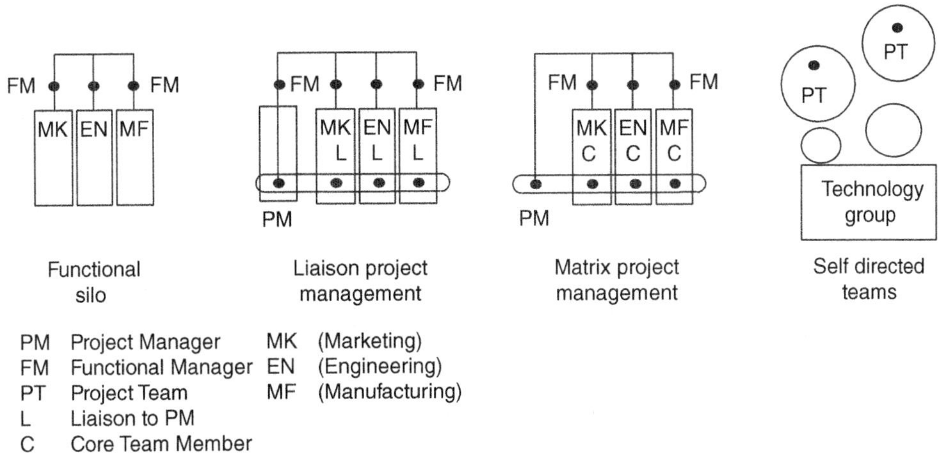

FIGURE 2.7 Project management structures needed to support new product creation.

when managing or directing technical design engineers. In global companies, this problem is made more severe with the geographical distribution of the company resources. PMs are forced to develop their personal skills through remote communication links to manage disparate resources and resolve conflicts among distributed project team members. Some of these conflicts are based on technical issues, which are difficult for the liaison manager to resolve quickly.

In a top-down management structure, a central engineering department plays an important role in developing new products. The department is broken down into sub-departments, which are organized along the traditional disciplines: electrical, mechanical, software, and support, with a functional manager (FM) for each discipline. The liaison manager's focus is on managing the schedule and the engineering and material resources. Engineers are borrowed from a central engineering pool, with a dual reporting scheme. An engineer would report to the FM for technical guidance and at the same time to the PM for task scheduling. All communications, negotiations, and directions originate from the PM and FMs. Engineers are considered another commodity such as capital and equipment.

The liaison structure is clearly designed to maximize resources and control for new products. The liaison organization is focused on the process, not the product. The work is negotiated by the managers, and instructions are then handed down to the engineers. Liaison management creates an environment that discourages engineers' creativity and leaves the managers with all decision-making powers. It is a system that discourages direct engineer-to-engineer communication, since interdisciplinary decisions have to be negotiated between the PM and the FM. The system is supposed to control engineers' activities by micro-managing their time. It tends to shield the engineers from interfacing with many other activities in the company such as manufacturing, marketing, purchasing, and production. As a result, their designs tend to reflect their lack of understanding how to maximize these resources for their product's benefit. In addition, longer communication lines produce misunderstandings and delays in conveying critical information, making for a poor and untimely decision-making process.

Liaison management has been successful in large companies with vertical integration, diverse geographic resources, and corporate research laboratories to investigate and develop new technologies. Success depends on project managers who negotiate resources, schedules, and product performance. Since the focus of the project managers is the product development schedule, their success with the engineers is based on personal rather than technical leadership skills.

In the liaison structure, the individual project engineers might receive direction from multiple sources. The PM and FM might provide conflicting tasks, which will need to be clarified at the manager's level. Additional direction might also come from the other discipline managers, such as finance or quality, which should provide a balance between the needs of the PM and FMs. Unless a team member is dedicated 100 percent to a particular project (core team member), these conflicting directions can be confusing and frustrating. A non-core team member might be assigned to multiple projects/programs and also has to support indirect nonproject specific tasks. For example, a quality engineer might have to attend obsolescence component meetings or component purchasing reviews. This dilemma of *serving multiple masters* is a challenge for engineers and management. Constant communication and balancing of priorities are essential for all involved to ensure success.

The liaison organization and its long communication lines and process focus tend to isolate the company's technical resources and its management from its customers and market trends. Many companies have been evaluating their liaison project management and plan to move into other structures better suited to manage particular product development. Other companies are satisfied with this liaison model to manage large multidisciplinary projects.

In summary, liaison project-management emphasis is on negotiating the project requirements by managers of the functions required to meet the project goals. The project is schedule-driven, and the decisions are made by managers and acted through instructions to project design teams. Communication loops are long, and engineers and support members of the design team might be assigned to multiple projects. Constant communications are needed to resolve technical problems and set priorities for the design team.

2. Matrix project management is a modified version of liaison project management. It is a structure that attempts to correct the perceived problems with liaison management, while creating some of its own potential problems.

   Matrix management is best suited for quickly reducing communication loops and resolving technical problems. There is a core project team, which manages all project activities. It is supported by sub-teams of the different functions involved. The team members are the individual engineers performing the scheduled tasks. The matrix project manager is the driver of the project, making sure the schedule is on time and within budget. The PM has a strong technical skill set. Decisions made by the PM are based on technical and business considerations of the project. Trade-offs are discussed collectively and decided upon to maintain the overall goals. These trade-offs include project scope, resources, and schedules. The differences between the two types of project management are shown in Table 2.5.

Liaison Project Management	Matrix Project Management
Project manager (PM) focus as scheduler	Project manager as product champion
Problems resolved at PM and FMs level	Problems solved directly by PM + team
Team members have dual reporting	Core team members report directly to PM
Some team members assigned multiple projects	Core team members assigned to one project
Tasks negotiated by managers	PM + team have common vision
Long communication loops between members	Direct communication loops in the team
Managers assume and delegate responsibility	Members and PM agree on commitment
Members are shielded from other functions	Members exposed to the total organization

Table 2.5 Project Management Styles Contrast Between Liaison and Matrix Project Managers

Although one might assume from Table 2.5 that there are many advantages to matrix project management, careful analysis and mitigation of potential problems have to be planned in order for matrix management to be successful. Matrix project management is a temporary structure made to accommodate the design of a new product in the midst of many other projects in various stages of completion in the company. Once a project is completed, the engineers are then released from the project and reassigned to new projects. There is inherent conflict in that the engineers might want to jump prematurely from their current project to take on the next more exciting and up-to-date technology project. No engineer wants to be the one left in the project to correct design problems and risk making his or her skill set obsolete.

Another problem in the matrix management model is the disparity in the skill set between the PM and the engineers reporting to the PM. The PM might be a specialist in one discipline, while directly managing engineers in another discipline. For example, mechanical engineers might object to a software PM completing their technical evaluation form. In addition, there is inherent conflict between the PM's priority in completing the project and other duties and desires of the project team. An example would be that the PM might object to other non-project functions that the engineers in the design team might want to do, such as attending conferences or training sessions to update their skills. Matrix management does not encourage skill specialization, since that renders the skilled engineers more in demand by other projects, reducing their benefit to the project at hand.

It can be seen that there remains a strong need for FMs in matrix project management to allow the engineers to have a base to deal with technical discipline issues as well as advocacy. It is customary that while an engineer is assigned to a project team, the PM is the sole manager that will assign tasks and make all personnel decisions such as performance evaluations and time off for travel and training. In summary, the concerns in matrix management that have to be considered by the PM are as follows:

- The performance evaluations process has to be well-defined to allow the PM to administer them objectively, even to an engineer with a skill set unfamiliar to the PM.

- The PM should consult with FMs to scope discipline based tasks correctly.
- Scoping of design team members' contribution has to be identified accurately to allow team members to leave the current project and join new ones in a timely fashion.
- Specialized skill-set engineers should be assigned to support roles for design teams and not as core team members. They can join the team for a period of time when needed. Alternatively, special skill sets can be purchased as needed.

3. Functional silo project management is used for modified products, where the technology adoption or feature enhancements are small. There is no need for multidisciplinary resources in the project, and it can be managed by one functional area alone, most likely manufacturing/operations. Communications loops are short, and issues of cost, performance, and price are well understood. Silo project examples might include projects dealing with a special housing or environmental packaging for the product, or a special option requested by a small minority of customers.

   The advantages of functional silo project management are resource optimization and a focus on the specialized expertise of the single function. The structure allows for a narrow span of control and a good opportunity for an individual engineer's accountability and performance evaluation. Conflicts based on discipline issues, organizational boundaries, and task selections are much less acute.

   The disadvantages of silo project management are that the projects are task-oriented and do not offer as much satisfaction to the engineers as do broad-based multidisciplinary projects. In addition, there is a risk of technical obsolescence since there is no constant examination of the design alternatives using a combination of multi disciplines to solve technical problems.

4. Self-directed team project management is used where there is an opportunity to develop and apply new technology or a new core product in the company portfolio. It is epitomized by the term "skunkworks," which is an official alias for Lockheed Martin's development of the U2 and SR71 spy planes. It is widely used to describe a project team within an organization that was given a high degree of autonomy in task and schedule operations. It is usually unhampered by the company's bureaucracy and standard sets of operating and planning procedures.

   Self-directed team project management should be used when certain conditions make it imperative for a company to explore new products with a high level of technology adoption or using their technology in a new market application. It is the best project management structure for revolutionary product creation. The risks are high in terms of development, but the project has the potential to create a new core product family or a new business market for the company. In many cases, the integration of the self-directed team back into the company organization (liaison or matrix) is difficult because of management and team culture clash, and eventually some companies might spin off the new revolutionary product team into a separate business.

   Advantages of the self-directed team include greater team member commitment and motivation for the project, since they are setting their own goals and tasks. The lack of bureaucracy encourages new ideas and innovations both in technology

Project Type	Product Type	Advantages	Drawbacks
Functional Silo	Advance development Modified products	Resource optimization, expertise control, accountability, evaluations	Task focus, lack of breadth, skill obsolescence
Liaison	Follow-on products Large systems	Resource sharing, process focus	Long communication loops
Matrix	Platforms and architecture	Project focus, team commitments accountability, PM independence	Staffing, multiple disciplines difficult to manage, scoping
Self-Directed	Revolutionary products Basic research	Team sets own goals and operations	Integration back into company Excessive risk taking

TABLE 2.6 Project Management Organization Advantages and Drawbacks

development and project management. This in turn encourages risk-taking to fully realize the maximum potential rewards. The disadvantages are too much risk-taking endangering the project goals, and the subsequent difficulty in integrating the project team back in to the company's structures. These concepts are summarized in Table 2.6.

2.4 Creating an Environment for Successful Project Management

Successful project management is the correct mix of technology advancement, market conditions, consumer trends, and competitive factors. Planning is the key to developing a coherent product introduction stream that anticipates these factors. Otherwise, new product creation becomes a reactive process, with the subsequent risk in developing a product too late to capture a significant presence in the market or lowering the existing product prices to protect market position.

The worse scenario is the scheduling of a new product with unrealistic expectations or timeline. It results in tremendous strains on the organization in general and the development project team in particular. In addition, marketing plans, which are set in motion based on the false schedule, will be undermined for existing as well as new products.

Product creation should be a team-centered activity. The balance of the mix between technology and market input can be determined best by practitioners of both crafts. The inputs from marketing and sales organizations, the research laboratories, the advance development group, and the current development organization should be evaluated and a collective decision reached for the timing of the next product roll-out. A risk benefit analysis should be made of the trade-offs between rushing a new technology or idea to market versus properly investigating the development and manufacturing problems.

Properly performing the traditional marketing functions in their entirety is critical. These activities include:

- Technology adoption strategy
- Total market analysis
- Competitive analysis

- Product planning and positioning on the price performance curve
- Product business plans and ROI
- Marketing communications and input from the customers

To create a successful new product creation process in a company, the PM should adopt the best management model to match the product development they manage, according to the previous sections in this chapter. In global competition, the smaller, more efficient organizations are those that are nimble and fast reacting to the market. They are successful because of their focus on quality and cost of their own products and the team approach to project management. Each of these topics will be discussed in detail as follows:

2.4.1 Create a *total quality* culture within new products development projects
2.4.2 Develop *product focus* organizations within the company
2.4.3 Emphasize the team-focus approach to project management
2.4.4 Implement a phase review process for project management control
2.4.5 Key processes to enhance the project management process

2.4.1 Create a Total Quality Culture within New Product Development Projects

The implementation of total quality is a critical element of success for project management. It is the base from which all other ideas, procedures, methodologies, and tools of concurrent engineering and successful project management can be developed, nurtured, and successfully implemented.

Although the focus of a total quality culture is the control and enhancement of quality, it has evolved into highly successful methodologies for many different aspects of successful management and operation of companies. Total quality infuses the whole organization with a common set of terminology and procedures to perform the following important tasks:

1. Problem identification and resolution. The organization is trained to spot problems (quality or otherwise), identify them properly, and suggest methods for improvement. Alternatives are studied and weighed carefully, decisions properly made, and adverse consequences evaluated. Management is kept informed and provides guidance, encouragement, and resources for successful completion of the tasks.
2. Team process. Total quality is synonymous with the team process. It encourages working in groups, helping team members reconcile individual-versus-group goals, set team objectives and expectations, make collective decisions, and learn to operate with less management direction.
3. Continuous improvements. This is the idea of not being satisfied with the status quo, *not doing things the same old way*, and constantly seeking better performance from team members, resources, and processes. Part of continuous improvement is the challenge to realize that a limit has been reached with the current condition and to seek other alternatives and original ideas to improve. Expectations should be set correctly, and improvements can be achieved in large steps only when using new technologies or methodologies. Total quality allows for many small steps that, when accumulated over time, lead to large steps.

One of the inhibitors of total quality at the engineering level is the engineer's view that it is for manufacturing and lower-skilled personnel in the company or that it is *fourth-grade stuff*. In addition, engineers by nature have been trained in universities to compete instead of collaborate: grades and exams stress individual contribution rather than teamwork. These cultural inhibitors have to be dealt with by treating them as procedural as well as training issues. Emphasizing quality and teamwork in performance evaluations sends a strong message that the company is serious about implementing a total quality program.

The formation of total quality teams that work together for completion of successful projects at all the engineering and marketing functional levels is a good precursor and training ground toward a successful project management culture in the company.

2.4.2 Develop Product Focus Organizations within the Company

The company or the business units of major corporations are comprised of a collection of functions required to manage a business. These include the traditional three: marketing (or program management), development (R&D), and manufacturing (operations). In the following text, each functional unit is analyzed and issues outlined for successful support to the project management process:

1. Marketing/product management or program management in DTM companies

   The marketing department is normally responsible for formulating new product plans and strategies to profitably penetrate existing markets or open new ones. The department also implements programs to support the field sales force. They perform market research and analysis to provide R&D with information on new market opportunities.

   Marketing identifies trends and forecasts general business activities, providing long- and short-term forecasts. Detailed product sales forecasts are also provided to establish component inventory and production schedules. The department evaluates and reports on the competition, providing information on trends, market share, and product features and positioning. It directs, in cooperation with the sales force, a continuing and coordinated program for sales and promotion, including brochures, specification sheets, etc. It recommends new product pricing and performance levels and implements the introduction of new products into the field.

   Product management is the marketing representative on the new project team. The product manager provides the prime interface, or liaison, from the product to the market. The PM provides the primary interface with the technology, according to the schedule, time, and cost goals.

   Development is responsible for the design of the product and, jointly with product management, evaluates the design specifications, making sure they meet the market needs and justify the project investments.

   The formal processes for these important marketing functions should include the following:

   - Marketing research and analysis (market segmentation)
   - Product definition and positioning
   - New product business plans
   - Business development plans

A business development plan could consist of the following:

- Statement of the purpose of the plan
- Specific market objectives to be achieved within the intermediate period (three to five years)
- Description of the market, including potential customers and channels of distribution
- Description of the competition, their technical capabilities and potential, and their current product profile and emphasis
- Description of new products and services necessary for success in the next three to five years, and the plan for development or purchase of such products and service plans
- Financial analysis of costs and returns
- Project risk assessment
- Tactical implementation plan matching the tactical business horizon (12 to 18 months)

2. The R & D department is responsible for the following functions:
    - A communication function with the corporate research laboratories or the general technology base of the company for initiating new product ideas and technologies
    - An advanced development function to transfer new technologies into product and process feasibility; it is beneficial to isolate this function from the tightly scheduled development projects
    - A development project function with cost and schedule control supervised by PMs to create new products
    - A technical service function tasked with supporting the design systems the engineers are using

    Technical problems in production should be handled by the manufacturing engineers. Otherwise, they can be hidden impediments to the timely completion of new product schedules. In some recorded instances, as much as 50 percent of the development engineers' time was consumed by solving technical problems in production. The contrary argument that solving production problems is the responsibility of the engineers who designed the product can be resolved through the participation of manufacturing engineers on the new product development team.

    New product development projects are staffed by a team of multi-disciplined engineers and managed by PMs selected from either the development ranks or from product management in marketing. The FMs are responsible for the technical content and the technology to be used in the development project. They interface directly with the PMs on solving technical problems as well as coordinating the project schedule and the scheduling of tasks for development engineers assigned to the project team.

3. The production department (operations) is responsible for the following functions for products and processes:
    - Current products support, including scheduling, planning, documentation, fabrication, assembly, and test. Production should also maintain a continuous

improvement posture to achieve higher quality, lower cost, and superior performance within the advertised specifications of the product.

- New product introduction, including manufacturability assessment, production scheduling and layout, material acquisition plans, product data transfer, and test development for in-process as well as final test.
- Technical support for automation processes such as computer aided manufacturing (CAM) and product data transfer.
- Process development for fabrication, assembly, test, and automation technology.
- Communication with the other parts of the organization such as new product and project managers, logistics, materials, quality, and development departments to ensure meeting of shipments and targets.

The production department performs all the functions necessary to maintain a viable shipping profile. The interaction and communications with the rest of the operation's organization and its support functions as well as marketing and development should be open and bidirectional.

2.4.3 Emphasize the Team Focus Approach to Project Management

The team concept for project management has been shown to be effective for new product development in all of the four different types of project management discussed above. In small evolutionary/modified new product projects, a product manager can double as a liaison PM. In this case, the new product development can be implemented by having one person responsible for technical and project leadership.

In larger projects, depending on the product development stage, either a liaison or matrix PM has the final authority to resolve, with other team member and FMs, any technical problems that may arise. There is a core team, composed of all the engineers, who work on new development from concept to production to field performance. Members of this core team are the prime experts of the different functions involved. The core team communicates with the PM either directly (matrix management) or through a liaison.

Core team members are identified by job function and contribution to the project tasks. They are skilled in the various disciplines: electrical, mechanical, software, production, logistics, quality, and other functions. In the matrix management model, cross-functional communications occur directly from one team member to another, without having to deal through a FM, while in the liaison management, communications pass through the functional liaisons.

The challenges for a successful PM are as follows:

1. Poor visibility of external and internal dependencies and their potential impact on the project schedule
2. Focus on the technical specification and lack of coupling to financial, logistics, quality, production, and other functional expectations
3. Lack of communication channels up to management and down to individual engineers
4. Clear understanding of management goals and expectations
5. Strong belief in the benefits of planning and execution of project management guidelines and techniques

6. Difficulty in allocating people in a dynamic environment and defining the skill level necessary to successfully complete assigned tasks
7. Unclear definition of project objectives and specifications and reactive market decisions during the development phase

The uncertainties when implementing successful development projects are:

1. For management, it is the uncertainty whether the project is actually on schedule, within the prescribed cost, and with the defined functionality. This feeling of uncertainty intensifies as the project nears completion.
2. For the development engineers, it is the clear understanding of the tasks required as well as the importance of the task in the overall project plan and the amount of risks involved in meeting the technical objectives.
3. For the PM, how to keep schedule, cost, and product performance targets on track. If a problem occurs, when to react and report to management that a *schedule buster* problem has occurred and seek the best solution.

2.4.4 Implement a Phase Review Process for Project Management Control

One of the proven methods for effectively facing challenges and reducing uncertainty of project schedule management is to use a phase (gate) review process for controlling projects. Specific phases (gates) are identified and allowed to proceed to the next phase after successfully completing assigned tasks for each phase. Each phase can be viewed as a stand-alone stage with objectives, deliverables, product cost, quality, serviceability, and manufacturability status and project costs to date. More details on the phase review process will be given in Chap. 8.

The phase review process brings the PM and the core project team with a selected management group together formally at the end of each phase. These milestone meetings include a review of the project status conducted in terms of achieving the phase objectives, analyze recommendations, make appropriate decisions, and commit to the next phase.

The spacing of these milestones is important. They should be scheduled as required in groupings of like tasks, with more and more frequent milestone meetings as the project nears completion. The functional and top management should use the milestone meetings as an opportunity for detailed review of the project and its current direction and fit to the changing overall objectives.

Each functional area should have specific plans, measures, and goals for each phase as part of the overall project plan. These plans should be reviewed at the milestone meetings in addition to the technical review of the project. A sample of the plans could be:

- Manufacturing should have testability, quality, yield, and throughput goals reviewed at each milestone meeting. Responsibilities include manufacturability (DFM) feedback, test strategy and plans, production documentation, technical competence to handle the product after release to production, operator training, review of the manuals, and updates on production equipment installation and production process optimization.
- Product (program) management should review the competition, market surveys, profitability, pricing, obsolescence, and overall scheduling at each milestone meeting.

- Sales should review the forecast, the product introduction plans, the promotional plans, and the feedback from customers and dealers.
- Service should review the mean time between failures and the mean time to repair (MTBF, MTTR) and the service and repair plans, resources, training, and equipment.
- Quality should review the product quality specification according to the quality milestone plans.
- Materials should review part procurements and qualifications. Supply chain status, especially global suppliers, should also be reviewed.
- The controller department should review expenditures to date and remaining funds and timing of major purchases. In addition, the costs and profit calculation should be redone for each phase.

A good project-management plan will contain the schedule details as well as input from all functions necessary for the overall success of the product. A set of goals for the plans should be:

1. Provide concise project definitions with phases and milestones, with specific deliverables at each phase
2. Identify team members, their responsibilities, and keep management and team members informed of progress
3. Force team members and managers to continuously evaluate and re-plan
4. Identify potential problems and help in taking preventative action in time
5. Improve communications and delegation of duties and responsibilities
6. Identify all project risks and plans to mitigate these risks
7. Help the team focus on the activities at hand

2.4.5 Key Processes to Enhance the Project Management Process

These processes can be developed to operate in the project management organizational structure in order to create an environment for continuous improvement and operational excellence through the use of proven methodologies:

1. Phase Review Process

   This should be the primary vehicle for project-management control. Specific phases are identified in the product development process, with each phase being a collection of task completions. Each phase can be viewed as a stand-alone entity with objectives, deliverables, product cost, quality, serviceability and manufacturability status, and project costs to date.

   The PM and core team should carefully plan each phase and milestone, with shorter time between the later milestones to give upper management more information as the project nears completion. The process should be used as the primary vehicle to update the management and project teams with the current status of the project. The phase review process brings the core project team with a selected management group together formally at the end of each phase (milestone) to review the status in terms of achieving objectives, analyze recommendations, make appropriate decisions, and commit to the next phase.

2. Quality Advocacy and the Quality Systems Review

   This procedure is used to ensure that the quality system is effective in achieving total quality and customer satisfaction. The historical focus on regulatory and product quality and reliability issues should be augmented by a quality advocacy at each functional level. The quality function at the company's highest management level should put sufficient emphasis on facilitating an organization-wide adoption of total quality across the entire organization. It should have a process rather than a product focus. The role of this function is a consulting one assigned to assist other functions in integrating quality methods in their day-to-day operation.

   The quality-systems review is an assessment vehicle to evaluate the status of quality in each function and department. It defines the quality vision of how business should be conducted; it sets a common goal of quality and provides awareness of quality systems requirements across the organization. The quality-system review process should be used as a measure of the quality progress to provide opportunity for exchanging ideas and serve to refocus each part of the organization on the basic issues of quality.

3. Manufacturability Assessment

   This process serves to evaluate new products for ease of manufacturing, to ensure a high level of quality, and to maintain lower production costs. The issues of design for manufacturing (DFM) can be outlined in the following areas:

   a. Fabrication/tooling assessment and review. This assessment ensures a high degree of automation and tooling for fabricated parts, including plastic and machined parts, castings, sheet metal, and electronic components, including PCBs. By ensuring that only approved parts, tools, and processes are used, quality and variability reduction can be completely handled by the existing production equipment or the supply chain.
   b. Testability assessment. This operation will ensure that all tests, including in-process, final test, and systems tests can be handled adequately by ensuring proper physical access to the product internals as well as the product output. In addition, the use of appropriate test methodologies, integration of production test and other customer services, self-test procedures, and test algorithms are encouraged. The quality assessment of the new design can also be used to plan the testing strategy of new products.
   c. Product manufacturability assessment (DFM review). This assessment ensures that proper current and future product cost reduction can be applied by reducing the number of distinct parts, assembly motions, part geometries, and increased part-design symmetry. This review can lead to enhancing automatic and robotic assembly if justified by higher sales volume.
   d. Tools to support suggested processes, shown in Table 2.7. Training on these tools should be provided to assist in achieving success in applying these tools.

2.5 Conclusions

This chapter has presented the total new product development cycle for global products in terms of the product lifecycle, the impact of technology, the product development stage, and the project management model needed to manage new product development. In addition, the modern global enterprise organizations and

Tool	Use
Engineering process	Engineering process metrics
Phase review process	Control of project schedule
ABC	Product cost estimating tool
Six Sigma	Design for quality
DoE	Design optimization and variability reduction
ESI	Early supplier involvement
DFM	Design for low cost
MTBF, MTTR	Reliability analysis
QFD	Understanding customer expectations

TABLE 2.7 List of Tools and Their Use in Project Management

their functions as well as current methods for project tracking and control were presented. The importance of the role quality and CE tools were clarified in the business plan and identified for specific project goals.

This chapter serves as the introduction to the product development stage as well as understanding the effects of technology and the marketplace on new products and their specifications. The future success of project managers can be achieved only in the context of efficiently concluding the overall product development plan. Understanding these issues of market strategy, technology impact, and project organization and control for project managers is of great value in developing successful new products.

References and Bibliography

Box, J., "Extending Product Lifetime Prospects and Opportunities," *European Journal of Marketing*, 17: 34–49, 1983.

Brentani, U., and E. Kleinschmidt, "Corporate Culture and Commitment: Impact on Performance of International New Product Development Programs," *Journal of Product Innovation Management*, 21(5):309–333, 2004.

Carbone T., *Critical Success Factors in the Front-End of High Technology Industry New Product Development*, Doctoral Dissertation, University of Alabama in Huntsville, November 2011.

Clark, K., and S. Wheelwright, *Revolutionizing Product Development: Quantum Leaps in Speed, Efficiency, and Quality*, The Free Press, a Division of Macmillan, Inc., New York, 1992.

Cooper, R., "Predevelopment Activities Determine New Product Success," *Industrial Marketing Management*, 17(2):237–248, 1988.

Cross, N., *Engineering Design Methods*, Wiley, Summerset, NJ, 1989.

Day, G., "The product life cycle: Analysis and Applications Issues," *Journal of Marketing* 45: 60–67, 1981.

Gordon, M., C. Musso, E. Rebentisch, and G. Nisheeth, *The Path to Developing Successful New Products*, MIT Sloan Management Review Press, Cambridge, MA, 2009.

Kahn, K., *The PDMA Handbook of New Product Development*, 2nd ed., John Wiley & Sons, New York; Hoboken, NJ, 2005.

Kim, J., and D. Wilemon, "Sources and Assessment of Complexity in NPD Projects," *R&D Management*, 33: 16–30, 2002.

Khurana, A., and S. Rosenthal, "Towards Holistic 'Front Ends' in New Product Development," *Journal of Product Innovation Management*, 15(1):57–75, 1998.

Leonard-Barton, D., "Core Capabilities and Core Rigidities: A Paradox in Managing New Product Development," *Strategic Management Journal*, 13 (S1):111–125, 1992.

Levitt, T., "Exploit the product life cycle," *Harvard Business Review*, 43: 81–94, 1965.

Menz, M., "Functional Top Management Team Members: A Review, Synthesis, and Research Agenda," *Journal of Management*, 38(1):45–80, 2012.

Ullman, D., *The Mechanical Design Process*, 4th ed., McGraw-Hill, New York, 2009.

Westkämper, E., "Live Cycle Management and Assessment, Approaches and Visions Towards Sustainable Manufacturing," *Annals of the CIRP*, 49: 501–522, 2000.

Wheelwright A., and C. Steven, "Operations as Strategy: Lessons from Japan," *Stanford Graduate Business School*, 50(1):1981/1982.

Discussion Topics

1. List the product lifecycle major stages and give examples of products that are commonly used today for each of the different stages (start-up, growth, maturity, and commodity) now and in the past.
2. Think of thresholds that these products had to pass through on each stage.
3. Give examples of price/performance strategies for companies now and in the past and which path they followed (same performance less prices; same price more performance). What strategy would you recommend for the company(ies) you chose. What are the benefits and concerns?

Problems

2.1. Write a generalized view by selecting a product from personal experience or a company that you admire or work for. Analyze the product position in the product process continuum (evolutionary, platform, or revolutionary) and the project management type (silo, liaison, matrix, or self-directed team). Although it might be difficult to get information as to the type of project management for the product you selected, consider the most reasonable management organization that best fits the product. Consider popular products in the electronics or auto industry. You may also choose any other product family such as autos, airplanes, printers, TVs, computer displays, TV recorders, cell phones, or computers (desktops, laptops, or tablets), and trace their evolution in technology, industry standards, commodity status, and whether they are revolutionary, revolutionary, or disruptive products.

2.2. Compare and contrast the different types of industries and how difficult it is to enter the market at different phases in the product process continuum (evolutionary, platform, or revolutionary). List the different phases of technology adoption over time in these industries. Mention disruptive products and services that altered that industry. Please limit yourself to one paragraph in each part.

a. Voice recording
b. Video recording
c. Cell phones

2.3. You have been asked to consult to an oil industry services company that specializes in horizontal-drilling techniques used in fracking for oil and gas. The technology uses sensors in the drill bit to send back status (temperature and pressure) and GPS information back to the surface using calibrated digital pressure waves in the liquid column in the pipe. The electronics in the drill bit get very hot and frequently fail due to poor design and manufacturing specifications. The company compensates for that problem by prescreening the electronics package and heating it in a high-temperature oven. Shipping only the ones that survive and scrapping the ones that fail.

Due to increased demand of the company's products, this method of production has become untenable. Your host manager asked you to be on call for 24 hours due to the busy schedule of the board of directors, and when they gather for their quarterly meeting from their travels around the world, you are to fly to the company, tour the facilities, and then make suggestions for solving the shipping problem to the board. Obviously money is no problem, but a quick solution is required.

You arrive in Houston and stay at the Marriot on the company property. During the morning tour, you notice there is no professional support staff of design or manufacturing engineers. There is little process documentation in the factory, and you learn that the design was subcontracted many years ago, with no documentation about the qualification process of the design. Obviously a redesign or a large technical investigation is warranted. As you prepare for your presentation to the board, what would you recommend to tackle this problem?

Please discuss organizational, technical issues, and use of new tools discussed in this and the previous chapter. What training and management structures, including time horizons, would you like to recommend to the board?

CHAPTER 3

Project Inception: Benchmarking, IP, and VoC

In this chapter, the project-inception phase is examined in terms of developing new product technology and process strategy. Topics include benchmarking the competition in positioning the new product on the technology and price performance curve, ensuring the integrity of the product's intellectual property and understanding customer needs. The use of proven tools and techniques for combining these elements into a structured analysis methodology will be illustrated. These tools will yield concrete direction to project managers (PMs) on setting the optimum level of new product specifications and project deliverables.

The purpose of the project-inception phase is to manage the activities of the new product proposal prior to presentation to the company's upper management, outside investors, or the customers in order to ensure a successful new product proposal. In design to customer (DTC) proposals, the emphasis is on identifying a bid opportunity, making a bid decision, and communicating with the customer prior to the bid in order to better understand customer requirements. Plans are then developed for meeting these requirements and converting them into product specifications that best meet customer expectations. In design to market (DTM) products, the emphasis is on evaluating customer needs and matching these needs to the new product specifications. At the same time, a thorough analysis of the competitive position of the company's product portfolio and the new product position in the price performance curves should be conducted.

Concurrently, the management will be concerned in maintaining the return on investments (ROI) in new product development costs, and whether this proposal will continue the profitability and growth goals of the company. This topic will be discussed in Chap. 5. In addition, the proper level of new product development investment should be commensurate with the product lifecycle mentioned in Chap. 2. A mismatch between the new product development technology adoption, project cost, or resulting product price versus the current customer expectations could result in a cancelled project or failed bid.

Topics to be discussed in this chapter are as follows:

3.1 *Benchmarking of Products and Processes.* The tools of benchmarking will be discussed, with attention given to world-class companies, products, and processes. Different methods of illustrating benchmarking techniques through visual means will be provided as well as understating how the different tools discussed in the previous chapters might be aggregated for a successful new product introduction. Finally, the concerns and pitfalls of benchmarking will be discussed.

3.2 *Intellectual Property Concerns in New Technology Product Inception.* The different issues of safeguarding the new product from potential litigation, patent protection, and infringement as well as examples of infringement cases will be explored.

3.3 *Voice of the Customer* (*VoC*). Several structured methodologies for quantifying the voice of the customer, for DTM and DTC products will be explored and examples given.

3.1 Benchmarking of Products and Processes

Benchmarking is defined as "the best performance by a competitive product, service, or process or other processes performing a similar function." One of the most interesting benchmarking diagrams is used by Motorola to justify the Six Sigma (6σ) concept, published by Motorola University in 1988 and used to compare annual improvements in key performance measures of cost, quality, and services for three types of companies: a typical U.S. company, a world-class global competitive company, and a best-in-class company in any one of the measures. The 6σ benchmark is shown in Fig. 3.1, with a goal of achieving 6σ levels of quality (3.4 parts per million defects) in a reasonable time horizon of around six years.

3.1.1 Attributes of Benchmarking Global Technology Companies

The attributes shown in Fig. 3.1 are in percent improvements per year over an extended period. It shows that the typical U.S. company is improving by 10 percent per year, while the global world-class companies are improving by a factor of 200 percent per year, and the best-in-class companies are improving by a factor of 10 times per year. While these improvements cannot be sustained indefinitely, it clearly showed the path

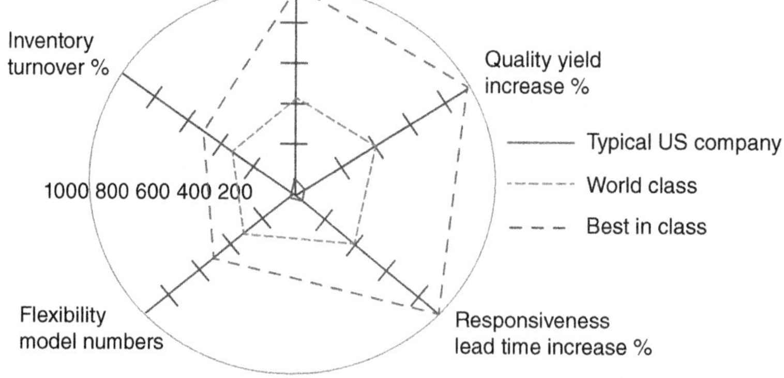

FIGURE 3.1 Six Sigma original competitive benchmarks (1988).

to achieve 6σ levels in quality, cost, and responsiveness to customer needs, flexibility, and materials management. The typical U.S. company was perceived to be unaware of the potential of significant improvement in these key attributes, as shown below:

- The typical U.S. company is somewhat profitable and growing.
- Market prices of their products are declining.
- New competition entering marketplace.
- Typical U.S. company is spending 10 to 40 percent of manufacturing cost in testing, reworking, or scrapping products due to low-quality and high-defect rates.
- Unaware that "best-in-class" is 100 times better in quality.
- Believes zero defects or 6σ are neither realistic nor achievable.
- Has 10 times the number of suppliers required to manage the business.
- Five to 10 percent of the customers are dissatisfied and would not recommend others to purchase products from the company.

As was told to the author by Bill Smith, Motorola's quality manager in the 1980s and the "father of Six Sigma," and verified by Robert Galvin, the CEO of the company, to the author as well, the 6σ goals originated as the best method to maintain the company growth at its current 18 percent per year at that time.

For a company to achieve 6σ levels in these attributes, it is important to find measurable criteria for each attribute. In some cases, a standard is required to explain how to measure each attribute. In the first chapter, it was shown that the defect rate for printed circuit board (PCB) assembly could be interpreted in many different methodologies, leading to different defects rates depending how the opportunities for defects are calculated. To benchmark different manufacturing facilities for PCB assembly, an industry consortium (IPC) standard was created. Table 3.1 shows some suggested metrics that can be used to benchmark company attributes.

As can be seen in Table 3.1, many attributes can be measured in terms of benchmarking the company with its competitors. Each attribute to be measured or improved should be studied individually for its fit into the overall strategy of the company and its desire to compete with other world-class companies.

A qualitative method that a PM can use to encourage internal benchmarking in new product development is to set these attributes of project improvement per quarter (or per year) on a single chart to communicate progress toward achieving these goals. A PM or a team can be tasked with making regular improvements on a single attribute to the product development process. These improvements can be measured concurrently on a tracking or spider chart, as shown in Fig. 3.2.

Many companies will survey their competitors or their industry to benchmark these attributes in managing new products. They can be prioritized as to which attributes have great strengths and which attributes are lagging. Attributes can then be organized into a matrix, showing strengths and weaknesses, versus high and low priority for improvements, as shown in Table 3.2. Attributes in the high-priority and weakness corner of the matrix in Table 3.2 are those that need attention from the PM (warranty costs and field failures).

Attribute	Metric
Cost	Cost of each type of manufacturing operation such as component placement (high and low volume)
Quality	Yield at final test
	Yield at in-process tests
	Defects per unit in each step in manufacturing
	Cp/*Cpk*/Six Sigma level of quality
	Number of DoEs performed per engineer/year
	Number of repairs in systems test
Reliability	Cost of new product warranty
	Number of new product customer returns
	Number of product recalls
	Number of dead on arrival (DOA) units/product
	Number of errors in the users/repairs manuals
	Number of software glitches and updates
	Number of systems resets in the field
	Mean time between failures
	Mean time to repair
Development	Design efficiency score for new assemblies
	Number of ESI meetings per new product
	Number of engineering changes/new design
	Number of software defects per 1000 lines of code
	Number engineering changes after release to MFG
	Time to reach full production volume after release
	Turnaround time for new prototypes (EE, ME)
	Return on investment for new projects
	Return factor for new products
	Number of patents filed per engineer per year
	Mandated training hours per engineer per year
	Engineering conference visits per engineer per year
	Number of team building activities in hours per year
	Number of engineers' turnover/year
Materials Management	Inventory turnover/year
	Purchased part cost reduction/quarter
	Number of unique parts in new designs
	Number of inventory days per new product on hand
	Number of inventory days per products in stock
	Number of meetings with supply chain per quarter

TABLE 3.1 Suggested Metrics That Can Be Used to Benchmark Company Attributes

Attribute	Metric
Customer Service	Number of new products introduced per year
	Number of phone rings to answer a customer query/order
	Time to respond and to complete a service call
	Frequency of updates to the company website
	Time to provide a customer order quote/prepare bid
	Number of visits/exhibits in major conferences/year
	Number of customer referrals/year
	Number of major customer visits/year
	Product delivery time to customers

TABLE 3.1 *(Continued)*

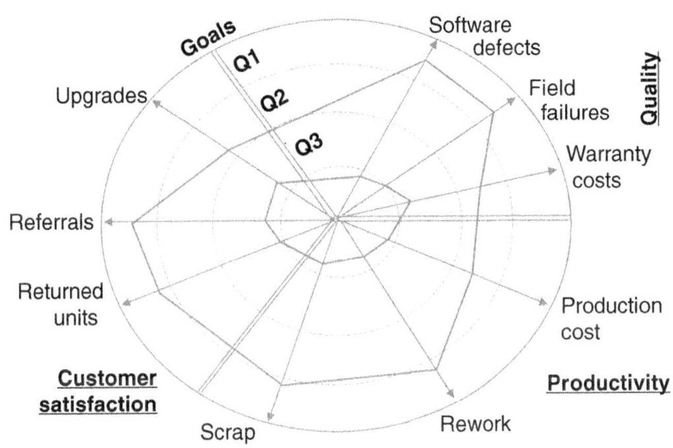

FIGURE 3.2 Project attribute tracking chart.

	Strengths	Weakness
High priority	Software defects rework	Warranty costs Field failures
	Referrals	Returned units
Low priority	Upgrades	Scrap
	Production costs	

TABLE 3.2 Benchmarking Matrix for Project Attribute Improvements

Most of the attributes in Table 3.1 can easily be benchmarked to help guide the company toward world-class quality, cost, and customer satisfaction. Global technology companies are competing fiercely and facing educated and demanding customers. The quality and cost metrics are not the distinguishing characteristics of companies and products; they are the entry fees to the new product success game. The need for continuously benchmarking products and processes is based on the following:

- Increased customer educational and product/technology knowledge
- Increased market segmentation
- Diverse and rapidly increasing technological advancements
- Customer expectations are continuously rising
- Customers are demanding:
  - Greater product variety and value
  - Comprehensive support
  - Rapid delivery
  - Fast and expert service

In a later section in this chapter, the structured tools to help in benchmarking customer need versus product specifications will be discussed.

3.1.2 Evolution of Customer Expectations

Defining customer needs and expectations is a moving target that requires constant updating, especially for technology-based products. The term "voice of the customer" or VoC has emerged to focus project managers on better defining customer expectations and, hence, ensure more successful outcomes for their projects. The process of identifying the VoC strategy has evolved into structured methodologies depending on whether the project is DTC or DTM. The customer is the driving force behind the product lifecycle versus technology innovation concepts discussed in Fig. 2.1. Marketing or program management also contributes to the product positioning using the concepts of the price performance curves in Fig. 2.2. The tasks of the PM in developing new product specifications to best meet customer expectations and maintain company profitability are:

- Position the new product correctly in the product lifecycle versus technology-innovations stages to properly size the development budget
- Decide with marketing on the new product position in the price performance curves to maintain company profitability
- Understand customer satisfaction levels and concepts
- Collect information on the customer product purchase decision

The first two tasks were discussed in Chap. 2. The last two tasks are best illustrated by the model developed by Noriaki Kano shown in Fig. 3.3. For the *X* axis in the model, Kano refers to a quality-achievement index, or the amount of physically fulfilled customer satisfaction. For the *Y* axis, Kano refers to the customer-satisfaction index or the amount of satisfied customers.

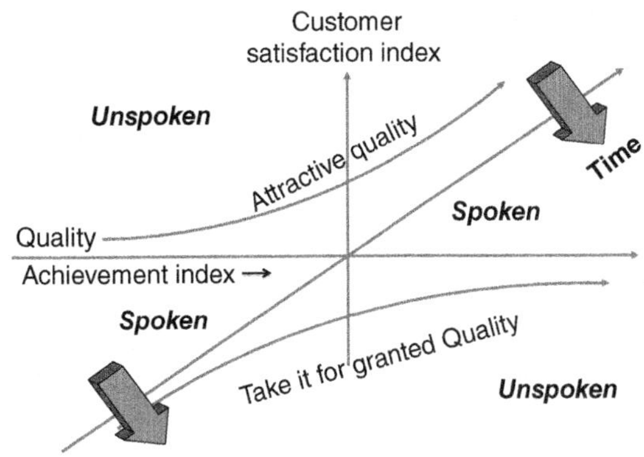

Figure 3.3 Customer satisfaction concepts.

3.1.2.1 The Kano Model Defined Three Levels of Customer Satisfaction Index

1. Expected (take it for granted) quality. These are product features that customers expect or take for granted. Customers do not specifically look for these features, nor do they ask for them, but they automatically assume the new product contains these features. If these features were omitted, it would result in negative customer dissatisfaction. Examples would be to check into a hotel and not find soap in the bathroom. This category is shown in the lower-right quadrant of the Kano model. These features cannot be satisfiers, but have the potential of being significant *dissatisfiers*.

   Many of these specifications are safety and comfort issues that can be called "unspoken" customer needs, such as a braking system in a car or a connection methodology for devices in a computer (USB or wireless).

2. Spoken quality. At this level, the customer will specifically ask for certain features or options. The customer also may or may not be satisfied with the new product performance. This level is the most challenging to the PM when deciding on the performance of a new product and requires more input from the customer through surveys and from competitive analysis of existing products. The customer-satisfaction level and purchase-decision criteria can best be modeled by the criteria-rating tool to be discussed in the next section. The spoken quality level is akin to customer *wants*, and the expected quality level is akin to customer *musts* in the decision matrix.

3. Attractive (exciting, surprising) quality. This level represents a significant breakthrough in customer satisfaction based on new product performance level or technology. Customers might not express their needs specifically, but might refer to the need in vague terms. This may be due to a lack of knowledge regarding the performance that can be achieved or simply because the customers did not think about it or expect it.

Customers generally accept this new attractive quality, and other customers hear about it and begin to demand it (spoken quality). Once the manufacturers realize this feature is important to the customers, they provide it in their new products. Customers now begin to expect it and stop asking for it (unspoken quality). New product design features or enhancements travel down from the attractive quality level, to the spoken quality level, and then further down to the expected (unspoken) quality level.

For example, the noise level inside a car could be improved significantly in a Toyota. A customer, who is expecting a certain level of road noise, would be pleasantly surprised by how quiet the Toyota ride is. The customer did not ask specially for a low-road-noise car, but would be pleased to ride in one. New car customers begin to ask for quieter-ride cars, and the auto manufacturers respond by making cars with quieter rides. Afterward, customers expect quiet-ride cars and stop asking for them. Car manufacturers start to offer new methods to lower noise in cars such as the noise-cancellation technology found in headphones. A visual example of this scenario is shown in Fig. 3.4, where the customer is conditioned to expect a slow evolutionary improvement of the noise specifications over time and successive product generations and becomes pleasantly surprised when the reduced noise level exceeds his or her expectations.

In the previous example, it can be seen how technology adoption, features, or specifications can be introduced in new products. The PM can then make the decision on how to implement new technology or features in the products, using all of the techniques discussed above.

3.1.2.2 Customer Product Purchase Decision

It is important for the PM to collect information about the customer purchase decision in order to understand the relative importance of the new product design features and performance specifications. The customer decision-making process can be broken down into many steps, with different attributes or criteria of the products making for different influences on the purchase decision. While more sophisticated tools such as quality function deployment (QFD) will be discussed next, the decision-making matrix using criteria rating is the most original and is the basis for many decision-making models that followed.

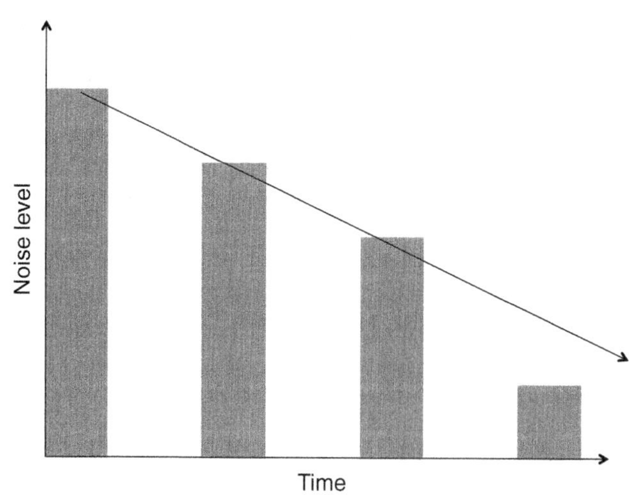

FIGURE 3.4 Noise level customer expectations.

Competing Suppliers					
Criteria	Rating	A	B	C	D
ISO 9002 Implementation	10	10	10	10	3
Adoption of Industry MFG Standards	10	10	10	8	8
Wide Use of SPC Charts	10	8	9	10	1
Use of Pareto Charts for Failure Analysis	10	10	10	10	5
Analysis Test Failures	10	8	7	10	5
Incoming Inspection Capability and Process	10	10	10	10	4
Failure Analysis Capability/Methodology	8	8	8	5	8
Obsolete Materials Disposition	8	7	6	8	4
Use of DFM Tools and Design Feedback	10	10	10	7	5
Continuous Improvement Process	10	10	10	10	5
Total	96	91	90	88	48

Table 3.3 Weighted Criteria Decision Matrix

Criteria-rating decision-making is a model that takes customer choice among different competing products and attempts to break down this decision into smaller independent decisions. The first step is to define the set of competing products that can minimally meet the customer needs (must have) in terms of price performance and quality. The competing products are rated on a set of hierarchical criteria that customers provide, through focus groups or from customer feedback. The customer input includes a list of criteria and the relative weight (rating) for each criterion. Each competing product is then rated against individual criteria, and a total weighed criterion is calculated for each competing product, as shown in Table 3.3.

In the example in Table 3.3, a decision matrix is shown for selecting a manufacturing supplier based on quality criteria. It is a part of a larger decision to select the supplier using many attributes of quality, cost, delivery, and supplier performance (wants), which will be shown in Chap. 6. The criteria were developed by the selection team, who also developed the relative rating for each criterion. The competing four suppliers were called A through D and were selected and preapproved based on *must have* financial and performance criteria.

It can be seen that the winner was supplier A. However, supplier A scored slightly higher than supplier B, pointing to a need for a second look at the criteria and their ratings or having the selection team look at other criteria not included in the decision matrix.

The steps for collecting information on the customer product purchase decision are as follows:

1. Collect information about customer needs through customer surveys, focus groups, feedback cards when customers buy current products, and opinions from major customers filtered by the marketing/sales force
2. Reduce the set of competing products to be included in the purchase decision through a threshold of *must have* levels of price, availability, performance, and quality
3. Analyze the customer needs to identify a set of criteria
4. Rate the customer criteria in terms of importance to the customer
5. For each criterion, rate each competing product individually against that criterion
6. Calculate a total criteria rating for each competing product

The PM, in conjunction with marketing, can examine the customer-purchase decision matrix as well as the Kano diagram to obtain a sense of the amount of technology adoption and features to include into the new product. In addition, they can use this information to set the proper product performance specifications that best meet current customer needs.

3.1.3 Concerns about Benchmarking

One of the dangers of benchmarking is making the wrong conclusion or using a different basis for comparisons of alternatives, leading to comparing apples and oranges. For example, management can compare the productivity of the engineers on the project team, or the productivity of each employee, by dividing the company revenue (in $) by the number of engineers or of all the employees ($/employee). The resulting number could be skewed by the fact that one company might have a much larger dependence on its supply chain and, therefore, the number of $/employee might be different. Benchmarking is susceptible to error if the base conditions are not accurately measured. Examples for project management could be as follows:

- Benchmarking new product development time might be different, depending on the amount of technology adoption in the product
- Design time per electrical engineering unit such as hours/Integrated Circuit (IC) and hours/PCB is dependent on the complexity and packaging density of the design
- Material and labor cost percentage of sales revenue in new products might be different depending on the amount of outsourced materials and processes
- Cost of quality of new products might be different depending on the definition and status of test times and whether the costs include repair and scrap costs
- Benchmarking is also susceptible to self-fulfilling prophecy; for example, if the quality of a productivity standard set for new products is based on benchmarking, then the engineers would alter their designs to match these goals:
  - A goal of software defects per line of code could be circumvented by writing too many lines of comments or segmenting single commands into many lines of code.
  - A goal of time/mechanical engineering assembly CAD drawing could be circumvented by the engineer having many smaller CAD drawings for individual parts that comprise the assembly.

3.2 Intellectual Property Concerns in New Technology Product Inception

The importance of intellectual property and patents has increased significantly in recent times. Historically, technology development in the technology industry was viewed as being so fast, it could quickly become obsolete. This would have rendered the efforts and cost to file patents and patent protection less important. In addition, technology companies were major innovators in developing a common supply chain. It made sense that the supply chain shared technology and innovation from a particular company's products to the benefit of all competitors in certain industry sectors. This could result in reducing costs and availability of multiple suppliers capable of competing for large-volume manufacturing of technology products. Once technology products reach the commodity stage as discussed in the Chap. 2, it is prohibitive for new competitors to enter the market, ensuring that only few companies can benefit from the supply-chain capabilities improvement derived from sharing innovations.

3.2.1 Intellectual Property Trends in High-Technology Companies

As *disruptive* or revolutionary products began to compete and gain market share from commodity products, technology companies realized they could not sit idle while disruptive technologies created a completely new market segment, eclipsing their commodity products. Some of these technology companies had enough resources to enter the market at any stage of development to regain their share of the market or enter new ones. In a defensive move, technology companies at the mature stage of commodity products began to consider intellectual property and patent protection as a competitive weapon to ward off competitors at the mature or commodity stage.

An example of this evolving patent viewpoint is the cell-phone industry, where RIM and then Apple *disrupted* the commodity cell-phone market by offering first a smartphone and then a smartphone with touch technology. Major technology companies such as Google and Microsoft are competing for software dominance of smartphone operating systems and are considering entering the design of phone hardware as well. Patent litigation has become a competitive weapon, especially in a down economy, as a defensive posture to protect products in the commodity stages from new companies entering the market. Consider:

- Apple bought Nortel Patents (6000) for $4.6 billion in 2011.
- Google bought Motorola Mobility patents (25,000) for $12.5 billion in 2011.
- Oracle and Microsoft are suing Google over Android cell-phone operating-system functionality.
- Apple and Samsung sued each other in Australia, Europe, and the United States, with mixed results, in the 2010s.

A PM should consider protecting the intellectual property of the new product by filing patents and taking steps to protect intellectual property when communicating and negotiating with suppliers and prospective customers. Possible patent infringement should be studied when developing a new product. When using an industry-wide supply chain, care must be taken to ensure that they do not inadvertently offer another company's proprietary technology to be used in the new product.

3.2.2 Patent Law and Issues of Filing a Patent

Patent law was changed in 2011 from first-to-invent to first-to-file. In a first-to-file system, the right to the granting of a patent for a given invention lies with the first person to file a patent application for protection of that invention, regardless of the date of actual invention. The obsolete first-to-invent process was comprised of two steps: (1) conception of the invention and (2) reduction to practice (i.e., showing a working embodiment) of the invention. When an inventor conceived an invention and carefully reduced the invention to practice (by filing a patent application or by practicing the invention, etc.), the inventor's date of invention was the date of conception. This change helps to reduce the legal implication of when the invention was originally conceived whenever patents are contested.

Patents involve the following:

1. Description of prior art and why a patent is unique. The prior art search is made to help ensure that the patent is unique. Patents could be filed in the United States as well as other countries. A cursory patent search can be done on any available free Internet site such as www.freepatentsonline.com or a Google patent search. The patent should have a preamble as to what the invention is trying to address and how it is different than prior art patents.
2. The patent might contain a description of a typical embodiment of the invention of the patent, with diagrams and descriptions of how the invention might be applied.
3. The last and most important part of the patent is the "what is being claimed" section. In this section, precise language is used to describe the invention in as much details as possible. The inventor has to think of all the possible combinations and conditions of use of the patent in different applications, products, processes, and through different methods. This section constitutes the prime intellectual property perimeter around the invention. Nuances in the use of keywords in the claims section can be critical in future litigations of the patent. The author has experience as an expert witness and notes that when definitions of engineering terms and structures such as pocket, notch, key, and channel are mentioned, they were important when patents were challenged in court.

Patents can be filed and managed with the patent office by a patent attorney or a patent agent representing the inventor. A communication stream begins once the patent is filed between the patent agent/attorney and the patent clerk in order to ensure the invention is unique and different from prior art. In some cases, patents are modified in order to clarify the claims or steer them away from infringing on parts of prior art patents.

3.2.3 Intellectual Property Infringement

Patents can be protected with legal means in patent or civil courts. If another product is infringing on the intellectual property of the company, or a company suspects another company of violating trade secrets, then the company has the right to take the violating company to court. In a patent litigation case, all communications and documents regarding the patent case have to be surrendered. All relevant personnel of the company can be deposed, and the two sides (plaintiff and respondent) can engage expert witnesses to write an opinion on the case. One famous case that led to significant impact on the defendants involved the intermittent windshield wipers patent where the inventor managed to sue all of the major companies, starting with small foreign manufacturers and then moving on to settle with the Big Three auto companies. Another important early technology suit was that of DEC suing Intel for allegedly infringing on its Alpha patents in designing the Pentium chips. As part of a settlement, much of DEC's chip design and fabrication business was sold to Intel in 1997. Later patent cases include the Apple versus Samsung cell-phone infringement case, which was settled for $1billion in 2012 in U.S. courts.

Most patent-infringement suits are settled out of court. Going all the way to a jury trial is risky, since the jury might get confused with the technical details of the patents. Once a suit is filed, a discovery process begins where the plaintiff and respondent parties collect all of the information relating to the patent and communications within the parties (pleadings). Expert witnesses are hired to bring their opinion into the case. The parties trade opinions and counter-opinions. After the discovery process, and before the trial, the parties can decide on whether or not to settle. If one party believes it has a strong case (or the other party has a weak case), then it can ask the judge for a summary judgment to dismiss the case. The judge can then agree and dismiss the case or allow it to proceed to trial. Patent suits usually go to trial if one of the parties believes it has a strong case or it simply wants to distract or punish the competition.

Infringement occurs if any of the defendant's accused product features violate any of the clauses in the patent claim. Infringement analysis is based on "claim-by-claim or element-by-element analysis" of the patent claims versus the violating product features. Many arguments revolve around the definition of what is "a person of ordinary skills of the art" commonly mentioned in patents, and the doctrine of equivalents (DOE). Under the DOE, the difference in whether the claim versus the accused product is "substantial" or not is based on the triple identity test, which is:

1. It performs substantially the same function
2. In substantially the same way
3. To yield substantially the same result

The defense argument is based mostly on two elements: (1) that the accused product does not infringe on the patent and (2) trying to invalidate the patent. Invalidation of patents involves the following arguments:

1. Invalidity analysis by respondent includes using prior art of patents, products, or publications (contested or admitted) that practiced the patent or claims before the patent file date
2. The patent is obvious to a person of ordinary skills of the art
3. Using a combination of prior art patents, a person of ordinary skills of the art would anticipate the patent inventions/claims

The defendant might ask the court to clarify the claims by asking the judge for "claim construction." The judge will have to rule on further explaining the claims as they relate to the original intent by the inventor. In cases involving the author, success or failure hinged on critical definitions of what is a flat surface and how much of a cover applies to a particular surface and what percentage of the area is affected. The claim construction process could be used to facilitate the arguments mentioned above.

3.2.4 Summary of Intellectual Property Issues for New Products

Most of the intellectual property issues for new products should be handled by professional law firms and their associates with formal legal training in patent law and litigation. The PM should perform *due diligence* to ensure minimum legal and intellectual property issues with their new product, thus:

- All new product reports, design meeting summaries, communications, decisions, specifications, concept sketches, drawings, and laboratory notebooks should be documented and classified in the project data base for future reference.
- All negotiations with suppliers and outside engineering resources should be preconditioned with the signing of nondisclosure agreement (NDA) forms to protect both parties from trade secrets and potential patent infringements.
- All project communications and specifications to suppliers and outside engineering resources should be marked "company confidential" and "do not reproduce without permission."
- Important project documents and new product specifications that might have to be shown to customers or suppliers before product release should be copyrighted and so indicated on the documents.
- Whenever a new design or process technique is adopted into the new product, a patent search should be made. If it is unique, then it should be patented by contacting the legal department or outside patent agents and attorneys. If it is already patented, then it should be discarded.
- The PM should be careful whenever a supplier offers a new unique part or assembly to be used in the product design or a new manufacturing technique to assemble or produce the new product. A thorough investigation must be undertaken to ensure that the supplier is not infringing on any patents with that unique part/assembly.

3.3 Voice of the Customer

Customer input into the design process for new products is typically referred to as VoC. Traditionally, PMs were handed specifications from marketing in DTM projects or were told to prepare a bid on a request for proposals (RFP) in DTC projects. In either case, the goal of the PM was to design the new product from the specification and then hand over the project documentation and parts list to manufacturing. This methodology works well when there is little technology enhancement in successive product generations and design procedures and manufacturing techniques are stable. The PM can accurately estimate project activities and costs and schedule them accordingly. This technique is typical in silo project management as discussed in Chap. 2.

In technology-based projects, small changes in specifications and performance, such as faster response or more accuracy in the new product could result in large variations in project time and cost. The level of technology adoption is also important to match the changes in performance and specifications. A high level of new technology adoption could lead to uncertainty in identifying the proper scope and time it would take to accomplish the project tasks, making project schedules highly uncertain. A more accurate methodology is needed to develop new products with the commensurate level of project investment versus the customer acceptance of the new products or the success rate in securing a bid from customer RFPs.

New technology product creation encourages the use of structured methodologies to accurately develop specifications that are closely coupled with customer needs. For DTM products, QFD is used to couple customer needs and the competitive position of the company to guide changes in new product performance specifications. For DTC projects, the emphasis of structured tools allows for a preplanned and controlled communications exchange between the customer and the bid proposal team to clarify customer intent, as will be shown in Chap. 4. An example would be to understand the customer's *unspoken* needs, or the customer perception on where the product is positioned in the product technology lifecycle.

3.3.1 VoC in Design to Market Products

The first phase of the VoC process in DTM involves identifying customers' needs and then gathering and organizing the data to define their needs and requirements. While the PM and marketing team might have a good idea of customer wishes, a fresh and timely look at the facts rather than opinions is important in setting the correct level of new product development priorities and specifications. A structured process of data analysis is superior to holding internal brainstorming sessions to evaluate customer needs. The process could be as follows:

1. Steps in Collecting and Analyzing VoC Data

   Customer data gathering usually starts with identifying an initial list of what the PM and the project/marketing teams believe are customer needs. This can be accomplished by a typical brainstorming session where the participants write down their thoughts on index cards or sticky notes. These ideas can be grouped into an affinity or tree diagram of major categories of theorized customer needs. The outcomes from this process can become the guidelines in actual customer interviews or discussions in focus groups as well as customer satisfaction surveys. Another source of VoC is the questionnaire of customer satisfaction that is packaged with the product registrations for warranty (either in returned mailed cards or online processes) or simply encourage customers to respond to a survey through the enticement of free products or prizes.

   In actual customer interviews, the tree diagram can serve as a guide for drawing out customer opinions via open-ended questions and discussions. The marketing team can conduct this via telephone interviews with customers representing different size organizations and product use. Some customers could be invited to face-to-face discussions in group settings (focus and users' group). It is important to use an outline of consistently phrased questions and shy away from precise technical details. Failure to adopt a consistent format and preplanning for phone interviews might result in repeat calls to clarify customer opinions.

   A face-to-face meeting with representatives of customer companies is also an important opportunity for the new product team and project managers to understand how the current products are used in customer installation. They include the reaction of the users to the product's performance and to help obtain a better understanding of the purchasing decision at all levels and functional groups of the customer company.

Once the customer input is collected and classified, a reality check should be made as to the variance between the company perception of its customers' needs and the actual customer expectations. In most cases, customer input does not correspond one-on-one to the initial internal list drawn by the team. The data analysis should record the ranking or importance of each customer need.

Customer statements or quotes can be collected, then expressed in a relationship matrix with the internal list of customer needs compiled by the development/marketing team. A system of weighted criteria can be assigned to each customer statement as to the level of their relationship to the internal list. A simple low, medium, or high relationship and frequency of mention of particular needs could help in understanding the importance level of that need. Statements or quotes from influential customers who have more technical knowledge, experience, or influence in the marketplace can be scored higher in the weighing factors. Figure 3.5 shows what a portable-device VoC matrix could look like, where multiple customer quotes from the input process can be effectively translated into customer needs, with a relative weight applied to each need.

Individuals and sub-teams of the marketing/development team can complete separate sections of the matrix. The matrix can be outlined in a bar graph showing each customer need and its relative weight.

2. Results of the VoC Analysis Process

In many situations, the VoC relation-to-needs matrix analysis can reveal whether the preliminary assumptions in the marketing/development team about customer needs were correct. If a particular need was deemed of low importance to the customer, then the design focus or materials selection can be reduced. The specifications of the design relating to that need can also be adjusted.

The matrix can also reveal that customers might de-emphasize certain essential functional needs for the product. The tendency of customers to communicate their needs incompletely is explained by the Kano diagram in Fig. 3.3. Care must be given in that situation, as some of the important functional requirements could be unspoken. This underscores the importance of visits to the customer to observe and record the customer's use of the product in his or her own environment.

After the conclusion of the VoC analysis process, a set of customer needs can be agreed upon by the team to push forward into the next step of matching these needs with the proposed technical requirements of new products. By examining the weighted customer needs in Fig. 3.5, some initial internal customer needs could be eliminated, while others could be retained, even if they had a low score.

In summary, gathering and analyzing the VoC in DTM products entails the following:

1. Gathering Customer Data for VoC

- Use VoC early in the product definition before the first design concepts are developed
- Involve all team members from marketing and development into gathering the customer input and data

Relationship matrix ◎ Strong = 5 ○ Med = 3 △ Weak = 1 Customer needs quotes	Internal list of customer assumptions # of mentions Quote weight	Instruction manual	LED panel size	Battery size	Wireless connection	**
**	*					
Easy to install	2	◎				
Remote operation	1			△	◎	
Long operation	5			◎		
Large display	3		◎	○		
**	*					
Weighted customer needs		**	**	****	*	

Figure 3.5 Relation to needs matrix.

- Brainstorm an initial set of customer needs and expectations
- Organize the initial set of proposed customer needs into an affinity/tree diagram
- Use tree diagrams to develop a coherent set of customer questions
- Phrase questions consistently and use the same topics when conducting customer surveys and interviews
- Collect input from product users' groups, customer focus groups, warranty registrations process, and direct customer telephone interviews
- Encourage and schedule face-to-face meetings with customers to collect more data and interactions
- Understand the purchase decision-making process at customer sites at all management functions

2. Analyzing Customer Data for VoC
   - Collect relevant customer quotes and distribute them in similar groupings
   - Rank quotes based on frequency of occurrence or source from technical/high-influence customers
   - Match customer quotes to an internally generated list through a VoC relation to needs matrix (Fig. 3.5)
   - Decide on a relationship-strength level (low, medium, and high) between quotes and internal list of needs
   - Calculate a weighted customer-need index value for each need to input into the next VoC analysis step
   - Consider and include unspoken functional requirements from the Kano diagram analysis

3. Defining Product Technical Requirements from VoC

The step for analyzing the customer-needs data and prioritizing or ranking them is a qualitative analysis. This is usually referred to as the *what* area of the customer-needs versus the technical-requirements matrix. These needs can be matched with the technical-performance requirements of the new product, also known as the *how* area of the matrix. The technical requirements are quantitative, design-independent product performance levels that result in customer satisfaction by best matching the qualitative customers' needs.

Technical requirements can be identified using internal product specifications, awareness of the level of technology adoption and technical product knowledge as well as team brainstorming sessions. It is important to limit the number of customer needs and product requirements in order to facilitate data collection and discussions and complete the analysis in a timely manner. For example, quality, reliability, and customer-installation issues can be treated as a single item. The number of items in the product technical requirements should be limited to only those for which the team needed guidance in developing new specifications and not the total knowledge base of the product.

A weighted-criteria analysis can then be made to produce a set of weighted values for each of the product's technical requirements, using a three-level relationship (low, medium, and high). These levels can in turn drive target values for each technical requirement. These target values can be the current product performance levels and how they should be altered for the new product, based on the analysis. An up arrow could be added when the weighed value is high to improve the specification to match high customer expectations. A down or horizontal arrow can be added to show that the technical requirement scored low based on customer needs. Subsequently, the design methodology or specifications do not have to be altered or emphasized in the new product. Alternately, the team could set a world-class target value if it is achievable without having a significant impact on the project scope and schedule time. An example of a VoC customer needs versus technical requirements matrix for portable electronic devices can be seen in Fig. 3.6.

Relationship matrix ◎ Strong = 5 ○ Med = 3 △ Weak = 1	Target values (How)	100 pages ↑	3" × 5" ↑	20 Wh ↑	IEEE 802.11 →	
	Technical requirements	Instruction manual	LED panel size	Battery size	Wireless connection	**
Customer needs (What)	Importance					
**	*					
Easy to install	2	◎				
Remote operation	1			△	◎	
Long operation	5			◎		
Large display	3		◎	○		
**	*					
Weighted technical requirements		**	**	****	*	

FIGURE 3.6 Customer needs versus technical requirements matrix.

An important addition to the customer-needs versus technical-requirements analysis is to complete another matrix showing potential interactions among technical requirements, which could lead to design compromises and technical trade-offs. In Fig. 3.6 example, the customer needs for a larger display will interact with the battery-power specifications, hence requiring that the battery-power-rating increase above the current 20 Wh.

4. VoC and Competitive Analysis

An additional refinement to VoC technical requirements is to conduct a competitive analysis of the current product portfolio of the company as well as other competitive products in the market. This would determine customer satisfaction with current company products versus its competition. The data for this analysis could be collected at the same time as other VoC data and tabulated for each customer's need. If the satisfaction of the current company's products is below that of its competitors, then an additional multiplier weight could be added to the importance column, forcing the technical requirements that match that customer need to be further elevated. This multiplier could be in the range of ×1.2 to ×2 increase in importance, depending on team discussions.

Another advantage to competitive analysis is to evaluate the design concepts of the new product to the competitors' products using the criteria-rating methodology. Figure 3.7 is an example of competitive design-criteria analysis for a new connector that will be soldered on a PCB. Important criteria include the number of connections, opportunity of manufacturing the connector overseas, whether special tools are required to install the connector to the PCB, number of parts in the connector, and stability of the connector during the soldering process.

The design-criteria analysis begins with the current product design and lists the important criteria for that design. Then it compares each design criteria with that of the new product and the competition. The comparison results are listed in terms of the design of the new product or the competition exceeding the original product design criteria (+), satisfying the design criteria (S), and not fully meeting the design criteria (–). The numbers of (+/S/–) rating for the new product and three different competitors are added, and the best design is ranked based on the maximum (+) and minimum (–). In the case of Fig. 3.7,

Original design (Datum)	New design criteria	Competitive design rating New product	Design #2	#3	#4
4	# Connections	+	+	+	+
$4.50	Cost	S	+	S	S
No	Overseas built	S	–	+	+
No	Special tools	S	–	S	S
3	# Parts	+	+	+	+
Yes	Parts availability	–	+	–	+
Yes	Stability	S	S	S	–
	Σ +	2	4	3	4
	Σ –	1	2	1	1
	Σ S	4	1	3	2

Figure 3.7 Competitive criteria analysis.

competitive design #4 scored the best design features that meet or exceed the criteria. The design team can attempt to redesign the new product based on the competitive analysis. Care must be exercised to ensure that the competitor's intellectual property is not violated.

3.3.2 Quality Functions Deployment

QFD is a structured approach to VoC, with an approach that is similar to the VoC outlined above. QFD is a defined process that combines the matrices outlined above in a singular construct called the "house of quality." The "house" term is named for the triangular structure above the quality-characteristic matrix used for quantifying the interactions among the technical requirements. In addition, the competitive analysis is integrated into the house of quality and can have a direct impact on the matrix results.

QFD can be described in four phases:

1. The Organization Phase. During this phase, a product or service is identified, and the PM and development team is selected. The extent of the QFD study depends on the product or service complexity and the type of improvements sought. The PM and team involved in the study need to identify the goals of the study, the amount of funding required, and the expected schedule for the project. When a project is selected for QFD analysis, the PM has to clarify the reasons for the project, what breakthroughs are needed, and where the breakthroughs are likely to be.
2. The Description Phase. The project team describes the product or service selected for QFD analysis in terms of actual customer words, function, cost, reliability, and characteristics.

   The description phase begins with an investigation of customer expectations. It is important to understand *what* are the customer needs, musts, and wants and to establish measures to ensure fulfillment. It is also important to understand *who* the customer is: internal or external customers, end users, retailers or wholesalers, purchasing agents or suppliers. This is because different customers have different wants and needs that are particular to their function in a business sector.

   Customer needs should be ranked in order to guide the development team into breakthrough opportunities by changing the quality characteristics or design/technical requirements. Customer needs should be:

   - Traceable to customer quotes
   - Related to use, not to price, since price refers to existing products
   - Descriptive, such as a verb followed by a qualifier: "lasts a long time" or "easy to use," or a noun followed by an adjective: "small size" or "long shelf life"
   - Non-numerical
   - Positive; avoid using "screen not bright"

   The quality characteristics or the technical requirements of the design should be measurable, such as technical performance in temperature, size, weight, response time, and accuracy. There should be at least one technical requirement for each customer need. However, each technical requirement may have multiple relationships in addition to the prime customer need. In summary, the technical requirements should be:

- Measurable
- Have at least one requirement for each customer need
- A single requirement can be related to multiple customer needs
- If there is interaction between the requirements, it should be shown in the *hat* on top of the house of quality
- The relationship between a requirement and a customer need can be identified as zero (no relationship, score = 0); low (triangle symbol, score = 1); medium (circle, score = 3); or high (two concentric circles, score = 5)
- The output of the QFD matrix is a rating value for each technical requirement showing the relative importance of that requirement to the customer; the specifications should be moved up or down by a relative value commensurate with the rating of each requirement
- An arrow should also be added to the QFD matrix to indicate the direction of increased goodness in the specifications

3. The Decision Phase. In this phase, the team selects areas for improvements through investigation of new technologies, design concepts, improved quality and reliability, cost reduction, and process improvements. These improvements are directly coupled with satisfying customer needs through the use of a structured methodology such as QFD.
4. The Implementation Phase. The project team describes the changes in technical specification and target values for specifications of new products and services.

3.3.2.1 QFD Process Improvement Example

QFD is an excellent methodology for developing the proper technical specifications that best suit the market or the customer. Though it is a formalized tool, it contains all of the elements necessary for successful identification of the VoC. It is a process of collating and ranking the customer needs and matching them to the technical specifications and conducting a competitive analysis for the new product. It can be successfully used for either process/design improvements or new product development.

A process-improvement example is a QFD matrix for an electronic assembly (shown in Fig. 3.8). In this example, the process team decided to conduct a DoE to improve the quality of soldering for PCBs, but was not sure where to begin. It was decided to conduct a QFD analysis of the PCB assembly process in order to determine which step would be the best candidate for process improvement.

In Fig. 3.8, it can be seen that the QFD analysis is a typical house of quality construction, with the interactions to be indentified on the top of the QFD matrix, and the competitive analysis to be conducted on the right side of the matrix. An additional column of relative Wt (weight) is shown to account for counter-balancing the customer perception that the process to be improved does not meet a particular customer need versus its competitors. In that case, the relative importance weight of that need can be increased.

Solder process matrix

◎ Strong relationship = 5
○ Med relationship = 3
△ Weak relationship = 1

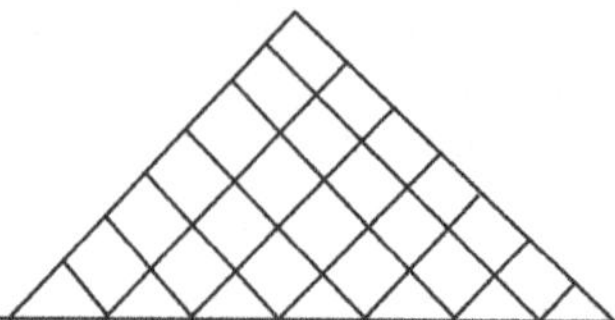

Solder process improvement

Customer needs: Primary	Secondary	Screening	Pick/place	Loading sys.	Reflow	Vision Sys	Supplier		Importance	Relative Wt	Competitive analysis (Worse – Better)
Quality	No solder shorts	◎			◎				9		
	No opens	◎			◎				10		
	No loading errors		◎	○		◎			8		
	Paste height	◎					△		7		
Delivery	Low downtime		◎				△		6		
	Consistent output	◎	△			○			5		
Rating		155	71	24	95	55	13				
Target values											
Technical analysis											

(Header corner: Technical requirements design / Customer needs)

Figure 3.8 Solder process improvement using QFD.

In this example of process improvement, the relative weight and interaction (hat) analysis were not conducted since this was an internal process, and the team felt that the interactions were minimal. A comparison to world-class quality could have possibly altered the outcome of this analysis, but the team felt that it would take too much time.

The team was comprised of the PCB assembly management, manufacturing process engineers, and test engineers, who are the customers of the assembly process. The team selected the customer needs (quality and delivery), organized them into primary and secondary attributes, and ranked these needs in terms of importance to the customers. Then the team identified the technical requirements, simply as steps in the soldering process. A relationship matrix was developed between the needs and the technical requirements. The ratings for each requirement were calculated by multiplying each importance by the relationship in each requirements column. The screening step had the highest ratings, followed by the reflow step.

The process QFD analysis was a great team effort, with the participants highly engaged and motivated. QFD offered the team a structured method for achieving a technical decision. The team felt that the conclusion was sound and created good consensus. The result of the QFD analysis was to select the screening step in the soldering process for further improvement using the DoE techniques.

3.3.2.2 QFD New Product Specifications Examples

QFD can also be a powerful tool to set the direction for optimizing the specifications of new products to best match customer needs. A proposed power-generating backpack

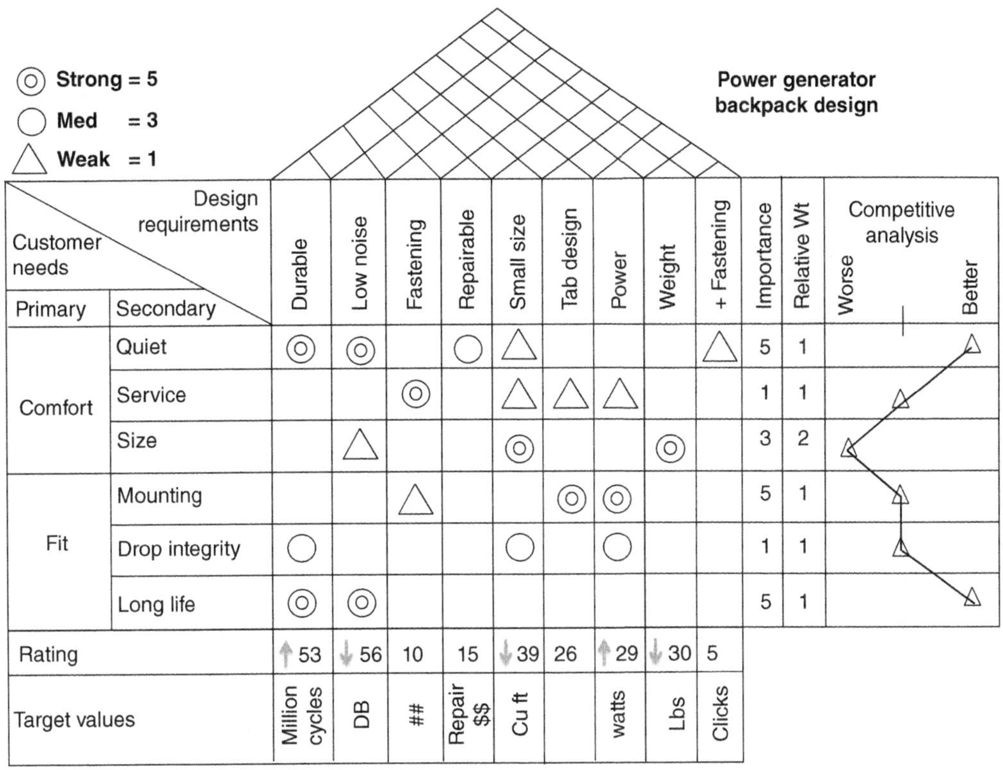

FIGURE 3.9 Power-generating backpack design using QFD.

design using QFD analysis is shown in Fig. 3.9. The customer needs are shown, grouped in two primary categories—comfort and fit—and with three secondary categories for each primary need. The development team then identified the design (technical) requirements, which included elements of durability, noise, reparability, weight, power, and fastening. The fastening was considered to be important, and three subcategories were designated for analysis: tab design, fastening method, and the notification to the user that the fastening system was engaged (+ fastening). At the bottom of the matrix, the ratings were calculated for each design requirement, with a target value indentified for each. In addition, the team added arrows showing the direction of changing the specifications in the direction of more positively meeting customer needs.

In Fig. 3.9, the competitive analysis is shown to the right of the matrix. One of the important customer needs is backpack size, and the VoC determined that the company's backpacks sizes are perceived to be of lower satisfaction than those of the competition. The result of the competitive analysis is to have a multiplier of two in the relative weight, which amplified the importance of that customer need. This is to focus the PM on the negative perception of customers to the current product size and increasing its importance in the design of the new product.

The top design requirements were durability, noise, size, and weight. The PM and the team can decide on the relative movement of each specification, considering its impact on the project cost and schedule. Other design requirements were found not to have scored high in the target value, such as the fastening-system components and repair costs, and thus the PM can decide on using the same current fastening system and spare parts replacement in the new product design.

3.3.2.3 Beyond the House of Quality

After completing the QFD steps of identifying the product technical specifications, the QFD process can proceed to identify the individual part material selection and dimensions and rank these part requirements according to customer needs. This could point the design team toward selecting the proper material and specification limits for each part and identifying opportunities for greater impact on customer needs, whether by conducting a DoE or finding alternatives to the design material and tolerance.

This process can continue with defining general process requirements and the production control structure needed to maintain the process requirements, as related to customer needs, as shown in Fig. 3.10. An example could be that automobile customers ranked rusting on car doors as a high priority in customer surveys. Using this cascaded method, the painting process for car doors could be targeted for improvements. The primer-or paint-process selection could be optimized through a DoE. In addition, the painting line-operation could be investigated and quality monitoring improved directly based on customer-needs input.

3.3.3 VoC Structured Methods in Design to Customer Projects

It would seem that the VoC structured methods discussed above, such as QFD, would not work in the DTC products. Since the customer normally proposes a new product or a project based on exact specifications, the VoC methods discussed in this chapter

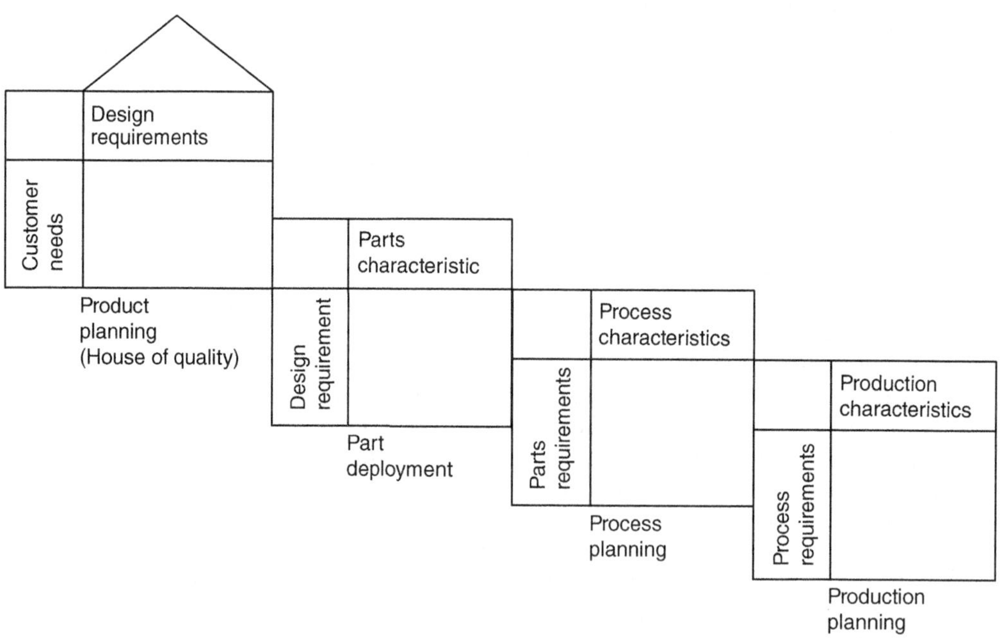

Figure 3.10 QFD cascade of matrices.

would be redundant. However, VoC is an excellent methodology to clarify customer intent. The process is best used in the proposal development, when the PM and the customer representatives meet to clarify customer intent. The purpose of the structured approach is to increase the success rate of bids to the customers by the following:

1. Collect Customer Satisfaction Information. By having a collective dialogue with the customer prior to bid submission, the team can collect important information to clarify and analyze customer expectations. The team should meet the customer representatives on a one-to-one basis. The major issues to address in this stage are:
   - Increase understanding of customer requirements (contractual and technical)
   - Understand the customer position on the technology development and maturity of the proposal (Fig. 2.1) in order to bid for the correct level of technology adoption and development costs
   - Be aware of any *unspoken* customer needs (Kano diagram)
   - Attempt to move back from the specification to the true customer intent
   - Discover if there are any discrepancies between the proposal and the customer expectations in technology selection, part/system specifications, customer supplied information or products, and the proper content and level of output (deliverables) measures
   - A proper document review of all statements of work (SOW), technical specifications, project schedules, and costs
2. Clarify Customer Intent Through Controlled Communications. In this phase, the VoC process for design to customer (DTC) is focused on an actual meeting or communication with the customer to clarify intent and to adjust the proposal for maximum customer satisfaction, thus:
   - Establish a set of discussion items at several levels and prepare an agenda for topic discussions: general (nonspecific) questions, followed by follow-up (more detailed questions) on each topic; in addition, prepare a set of investigative questions to probe customer positions
   - Each customer declaration should be verified by relaying back the statement and asking the customer if he or she concurs with the statement repeat
   - This clarifying phase should be short enough to capture the essence of customer needs and any qualifier that might arise from the communications
3. Analyzing the Customer Needs. In this phase, the results of the communications are analyzed and grouped together into an affinity diagram. The grouping of the customer statements can then be used as an input to a structured process for VoC, such as QFD to obtain the optimum design requirements that meet customer needs. A case study of the VoC for DTC is given in the next chapter.

3.4 Conclusions

This chapter outlined the essential steps and processes that are needed to be completed before the start of project approval, whether by the customer accepting the project proposal or the company management giving the go-ahead for the project to proceed from concept to development. As part of the project investigation phase or proposal development, this chapter outlines the methodology to propose a viable project specification based on many tools discussed above. They include benchmarking of competitive products, processes, and technology and investigating the intellectual property issues as well as following a structured approach to scope specifications correctly through using the tools of VoC.

An important part of the project-inception phase is the business plan, which is the outline of the ROI for developing the project. A business plan is needed for potential investors, company management, or the customer to make sure that their investment in the new product development will generate reasonable returns. Elements of the business plan involve marketing analysis, benefit cost ratios, and ROI analysis. These concepts will be further explored in Chap. 5.

References and Bibliography

Bettig, R., *Copyrighting Culture: The Political Economy of Intellectual Property*, 9–32, Westview Press, Boulder, CO, 1996.

Bogan, C., and M. English, *Benchmarking for Best Practices: Winning Through Innovative Adaptation*, McGraw-Hill, New York, 1994.

Boxwell, R., *Benchmarking for Competitive Advantage*, 225, McGraw-Hill, New York, 1994.

Brenkert, G., (Editor), and T. Beauchamp, *The Oxford Handbook of Business Ethics*, Oxford University Press, Oxford, 2012.

Burchill, G., and C. Hepner, *Voices into Choices*, Joiner Publications, Madison, WI, 1997.

Dwyer, D., *The Judicial Assessment of Expert Evidence*, Cambridge University Press, Cambridge, 2008.

Ficalora, J., and L. Cohen, *Quality Function Deployment & Six Sigma*, 2nd ed., Prentice Hall, Englewood, NJ, 2009.

Griffin, A., and J. Hauser, "The Voice of the Customer," *Marketing Science*,12(1):1–27, 1993.

Hahn, R., *Intellectual Property Rights in Frontier Industries: Software and Biotechnology*, AEI Press, Washington, DC. 2005.

Hauser, J., and D. Clausing, "House of Quality," *Harvard Business Review*, May 1988.

Kaiser Associates, *Beating the Competition: a Practical Guide To Benchmarking*, 176 Kaiser Associates, Washington, DC, 1988.

Lindberg, V., *Intellectual Property and Open Source: A Practical Guide to Protecting Code*, O'Reilly Books, Sebastopol, CA, 2008.

Maskus, K., "Intellectual Property Rights and Economic Development," *Case Western Reserve Journal of International Law*, 32:471, 2000.

Mazzone, J., "Copy fraud," *New York University Law Review*, 81:1027, 2006.

Meurer, M., and C. Nard, *Invention, Refinement and Patent Claim Scope: A New Perspective on the Doctrine of Equivalents*, Boston University School of Law, Boston, 2004.

Rigamonti, P., "Conflicting Theories of Equivalence: 35 U.S.C. § 112, para. 6 in the Federal Circuit and the Supreme Court,"40 *IDEA: The Journal of Law and Technology*, 163, 2000.

Pugh, S., and I. Morley, *Towards a Theory of Total Design*, University of Strathclyde, Design Division, 1988.

Smith, D., *Being an Effective Expert Witness*, Thames Publishing, Thame, England, 1993.

Ulwick, A., "Turn Customer Input into Innovation," *Harvard Business Review*, 1:80, 2002.

Discussion Topics

1. Benchmarking. Give a definition and examples of a world-class company's performance in service, product features, or technology that other companies would aspire to, such as Motorola Six Sigma.
2. Show examples of a product feature offering through the Kano diagram methodology. Can you trace chronologically when the technology came in and through the luxury models and then into all economy cars? For example, features in the auto industry such as:
   a. Anti-lock braking (ABS)
   b. Rear backup cameras
   c. Electronic stability control (ESC)
3. Give examples of patent wars in the electronics/communications industry. Why are they happening more frequently today?

Problem

3.1. Develop a QFD matrix for a household or personal item or product you use. Think of the possible customer and design requirements for this item. Complete the following:

a. The top level for the house of quality matrix for the item/product, including the interactions (the *hat* on top of the house) and the competitive analysis (please name the competing products) in the matrix.
b. Show with arrows (up or down) what direction the new product specifications should take (tighter or looser).
c. Develop the competitive analysis of similar products (please name the products). Shows how the competing products do not meet, meet, or exceed the design criteria (+/S/−).

Participants are encouraged to work in teams to have the full experience of communicating and deciding on the content of the QFD matrix. Some ideas for household items that you can participate in designing are given below. Please complete a QFD matrix for one product only from the list below or choose your own product to analyze:

a. A wallet or purse; some of the features you might consider are the size, the stitching, the material, number of credit cards holders, coin pocket, picture ID, etc.
b. A personal multitask printer (fax, copy, print, scan, speed, black and white or color, number of cartridges, control panel touch screen, print photos, etc.)
c. A high school calculator (functions, solar/batteries, clock, display size, aspect ratios, functions, touch display, etc.)

You can use a template designed by Christopher Battles, which is easy to understand and decipher. www.schrodingersghost.com or watch a video www.youtube.com/watch?v=u9bvzE5Qhjk and a tutorial www.webducate.net/qfd/qfd.html.

CHAPTER 4

Voice of the Customer Case Study

The concept and practice of voice of the customer (VoC) is represented in a wide variety of techniques and processes. VoC is a key process within requirements management. VoC is a principal process and represents a key primary output with the customer in the Project Management Institute (PMI) and the *Project Management Book of Knowledge* (*PMBOK*). As developed and adapted by legacy Goodrich Corporation (now part of United Technologies Aerospace Systems), it involves key steps that include actively listening to the customer and, if possible, capturing the customer's words verbatim, without embellishment or filtering. Captured words are then affinitized and placed into various categories or themes. Each of the themes reflect additional customer needs that are different from what has already been found in the initial documentation such as draft requirements, statement of work, and other documents. The VoC can help determine conflicts or gaps between the stated (written) requirements and the unstated requirements. These new requirements captured during the VoC can be summarized into new customer intents that then can be addressed, whether in an impact assessment-style matrix or directly into a project plan.

In 1999, Goodrich Corporation developed and continued to evolve its own methodology and processes to capture the VoC. While the VoC can be used in a stand-alone manner within the Goodrich methodology, it works best linked with other lean product development tools and processes that Goodrich adapted, which are described in the Fig. 4.4. In July 2012, United Technologies Corporation (UTC) acquired Goodrich Corporation. UTC is a diversified company providing high-technology products and services to the building and aerospace industries.

In mid-2012, two industry leaders, Hamilton Sundstrand and Goodrich Corporation, combined to form UTC Aerospace Systems (UTAS). UTAS is comprised of two segments: Aircraft systems (AS) and Power, Controls, and Sensing Systems (PCSS). Aircraft Systems includes Actuation Systems, Aerostructures, Air Management Systems, Interiors, Landing Gear, Propeller Systems, and Wheels and Brakes. Power, Controls, and Sensing Systems (PCSS) include Electric Systems, Engine Components, Engine and Control Systems, Kidde Fire Systems, Intelligence, Surveillance & Reconnaissance (ISR) Systems, Sensors & Integrated Systems, and Space Systems. This combination of two industry leaders has

---

This chapter was written by David A. Nolte.

created an organization with key positions on a wide range of aircraft flying today and substantial content on various unmanned airborne vehicles, satellites, and ground and naval vehicles.

The case study presented here will focus on the use of the voice of the customer (VoC) tool process originally developed by legacy Goodrich Corporation, which is currently used at ISR Systems, a business unit within PCSS. While there are many forms of VoC developed by a wide range of industries; for this case study, VoC will refer to the methods and processes developed by Goodrich and currently practiced at United Technologies Aerospace Systems, ISR Systems.

4.1 Voice of the Customer Methods and Techniques

Our perceptions of what customers want can be different than what our customers actually require. At times in their written documentation, such as contracts and technical specifications, customers can have preconceived notions of solutions that make it difficult to understand their desires, even for themselves. In order to understand the greatest array of solutions to an opportunity, it is vital that one captures and understands actual customer intent. Knowing this intent, one may use that knowledge in planning any countermeasures that may address those intents.

Why do we want to understand the voice of the customer? Because it will lead to more solution possibilities and, hence, a greater probability of success in truly meeting the customer's real needs, not just what was written. It will also give a company an additional set of customer intents, permitting better understanding and an opportunity to properly respond to all of the intent behind a statement of work (SOW) or customer specification.

Voice of the customer is an objective, structured, deliberative, and disciplined active-listening process to glean information from the customer, clarify it, and synthesize that raw data into discreet themes against which one can later plan either to provide for or to develop countermeasures used either to take advantage of an opportunity or to reduce a project risk. It also gives us the opportunity to:

- Review our behaviors in past programs with our customers to identify where we might not have fully understood their intended customer requirements
- Determine if there are any clear trade-off factors between key intents that a customer is prepared to make, such as cost-to-weight penalties or rewards
- Hear a customer's true intents as a single, unambiguous voice, despite being made up of a collection of groups, each with potentially different perspectives and objectives

Examples of where voice of the customer process can be used are:

- Pre-proposal and proposal efforts
- Business-acquisition activities
- Development of key suppliers
- Internal customers such as one business unit working with another

- Start of the product and process definition process
- Major changes in company activities
- Developing a product, service, or need to understand intent

It is a myth that all projects rely on stable, nonchanging requirements from the customer. More traditional life cycles did focus on getting requirements right, as early as possible, with the assumption that design and development would be successful. This was nice in theory, but impractical. Often customers or end users do not know what they need until they actually see a sample or prototype. This is especially true for software or other high-technology projects. Requirements naturally evolve and change, and the project manager must plan for change. Using the VoC process at several project phases will help develop more accurate responses to those evolving requirements.

Note that requirements can be identified in many ways:

- Technical or physical, such as size, weight, color, and quantity
- Functional or performance, often used in acceptance criteria by the customer

Prior even to a voice of the customer, the project may consider the use of several methods to gather the customer's written requirements. The gathering of the requirements must be captured in a draft and decomposed to a level such that the integrated project team (IPT) can analyze in some kind of compliance matrix. *PMBOK* calls this a "requirements traceability matrix." If the IPT completes the matrix early on (even during the proposal phase), it can help clarify ambiguous or conflicting requirements and fill the gaps in understanding to have a stronger set of customer requirements early on.

Methods to gather customer requirements include the use of focus groups, market feedback analysis [a UTC tool, which is part of the Achieving Competitive Excellence (ACE) toolset], or interviews in teams of two or more to probe for understanding of requirements as part of the pre-proposal research. The project team might speak with end users to get better knowledge of their needs. For example, ISR has managed several working group sessions with end users to better understand how the systems are used in the field environment, which leads to development of new functions, capabilities, and applications.

In addition to voice of the customer there are other methods such as nominal group technique (NGT), joint application development (JAD), and rapid prototyping that can be used to firmly help establish requirements.

- Nominal group technique
  - Gather requirements: Each team member develops a list of requirements without discussion. Requirements are listed on a flip chart or board; each requirement is then discussed among the group and clarified; assumptions are stated.
  - Prioritize requirements: Each team member then ranks the requirements, putting the most important first. Team ranking becomes the consensus.
- JAD: A process that particularly helps information-systems staff work effectively with users to speed up the requirements-gathering process. It avoids the pitfalls

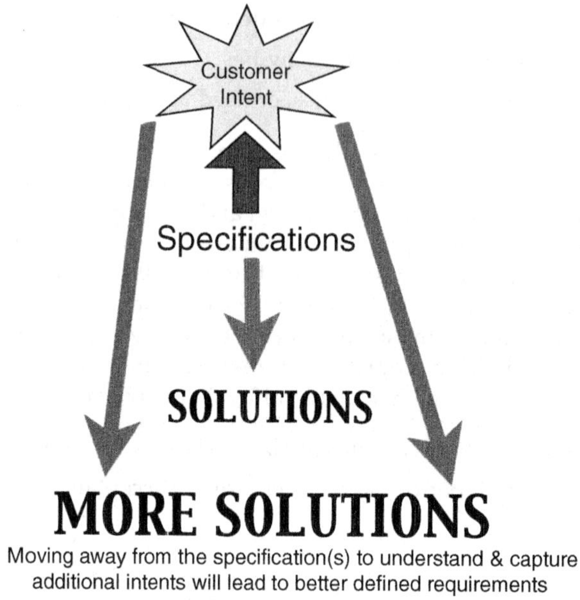

FIGURE 4.1 Moving away from the specification.

of the traditional project life cycle that results in building to specifications, little or no customer involvement during development, and customers saying they want something different at the end.

- Rapid Prototyping: A methodology to assist in customer-requirements development. It includes, for example, planning for a known number of iterations of prototypes are made and continual involvement by the customer. This method reduces the risk of a non-working solution used to respond to abstract requirements.

VoC as used at ISR is to help the integrated project team (IPT), which is comprised of the various technical leads, program management office (program manager, business manager, contracts), operations, supply chain, quality, and post-delivery support disciplines to move away just from the provided (and often incomplete) specification(s), as shown in Fig. 4.1.

ISR Systems use a series of lean product development tools (LPD) that are applied by the IPT as early as possible when addressing new proposals or managing projects. These tools include the use of document review as part of the voice of the customer, an impact matrix, a deliverables map, a least-waste-way derived project plan, and a risk register. These tools are used throughout the project life cycle.

4.2 Voice of the Customer as Part of the Lean Product Development Tools and Processes

At ISR Systems, proposals utilize LPD tools and processes (Fig. 4.2). In order to do a complete VoC, the IPT must start with a document review. The document review usually takes one to two days, depending on the size of the project, but in some cases can

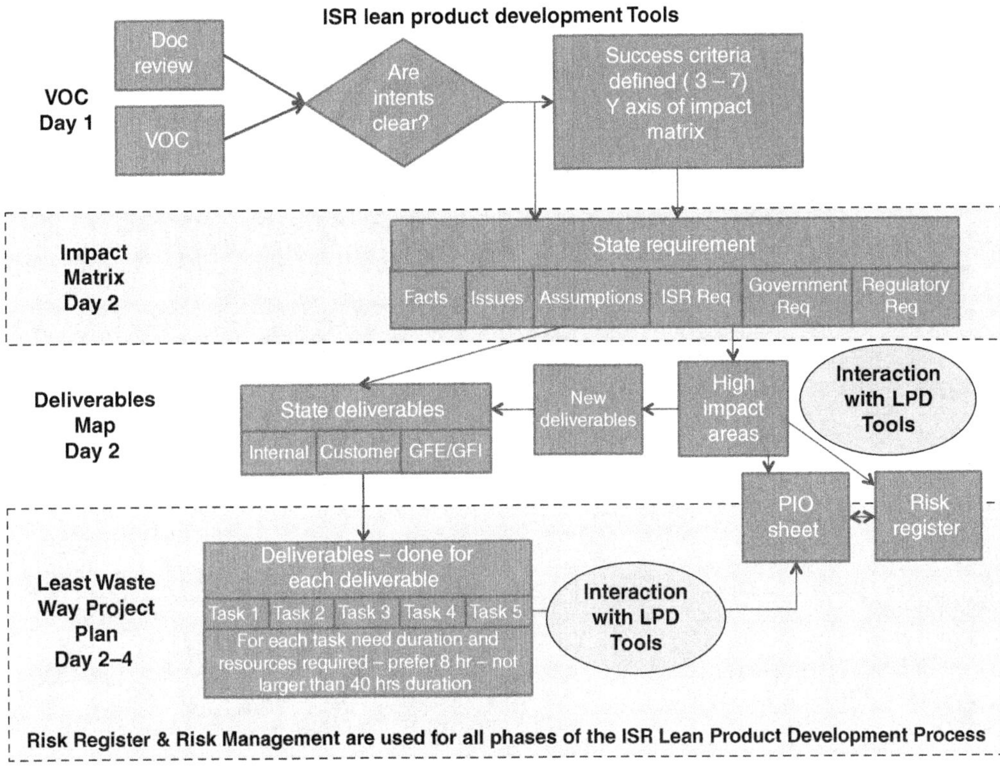

FIGURE 4.2 Lean product development tools used at ISR Systems.

take longer. It is done by the IPT and includes a thorough review of the specifications, request for proposal, statement of work, contract, project charter, or other available information. During this review, the IPT tries to understand what is known and what is unknown. Too often, initial specifications or statement of work documents include far too many "To Be Determined" (TBD) statements. It is important to eliminate these TBDs, as they pose uncertainty and risk. During the document review, the IPT creates questions to ask the customer during the VoC in order to answer the unknowns and uncertainties. These are in the form of probing and clarifying type questions.

Another outcome of the document review should include the initial capture of deliverables, both internal and external. As the IPT discusses the reviewed material, there is also an opportunity to capture risks into the risk register. The risk register, as used at ISR Systems for each project, is an Excel-based worksheet that documents and tracks project risks. The risk register is more than a simple capture format. It is used extensively throughout the project life cycle and is an important element of project monitoring and control. Last, as the team reviews the available material, contractual deliverables and specific contractual requirements can be identified and tracked.

Once the document review is complete, the IPT can prepare and host the voice of the customer. The VoC is a means to work with the customer (internal or external customer) to eliminate any specification, requirement, schedule, or contractual misunderstandings

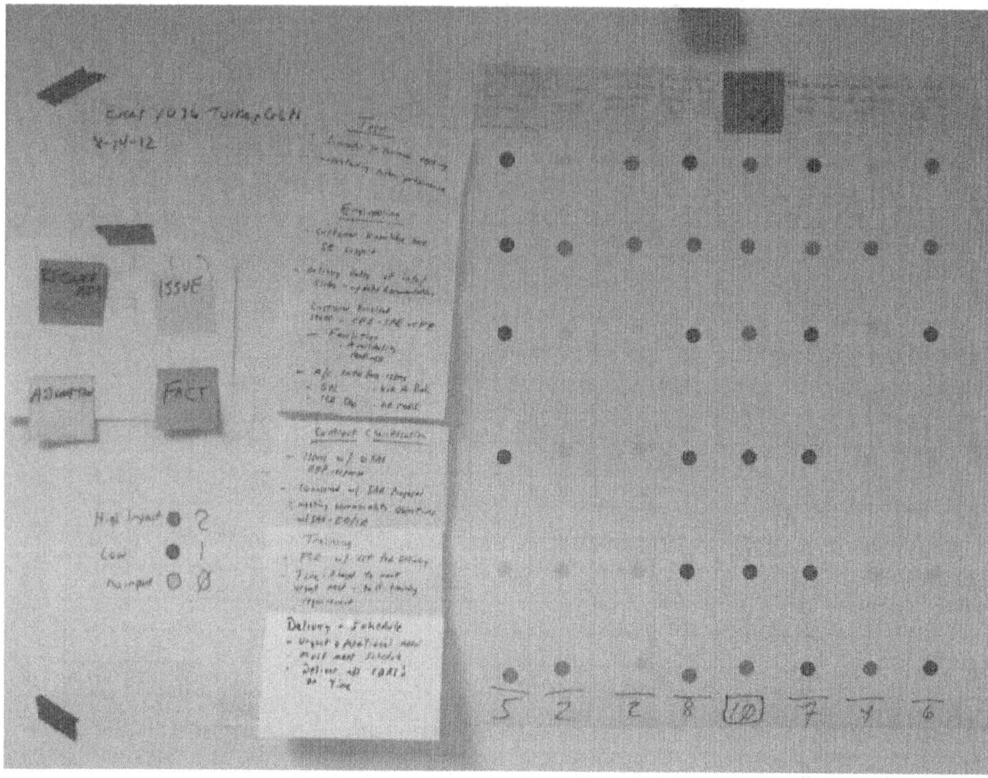

FIGURE 4.3 Impact matrix.

or uncertainties. The VoC, including the document review, feeds the impact matrix, which is used to assess the impacts of customer intents to the project or product requirements such as schedule, cost, and risks. Inputs captured during the VoC are categorized into facts, assumptions, issues, and requirements (FAIR) items and placed on the impact matrix for analysis (shown in Table 4.1). The impact matrix gives the IPT its first overall analysis of customer intents and the impacts on the proposal or project. The impact matrix can also provide inputs to the deliverables map, the least-waste-way plan (LWWP), and the risk register.

The deliverables map is linked to the work breakdown schedule (WBS), and it is similar to a project-schedule network diagram as described in *PMBOK*. It is a scheduling tool that helps to establish all of the high-level deliverables both internal and external for the project. Deliverables are shown against a milestone-based timeline. The timeline is deliberately not detailed as in a project-schedule timeline, merely providing a general milestone-based timeline to logically flow all of the deliverables. This frees the IPT from struggling initially over the notion of time and to concentrate on the identification of deliverables that meet the customer's requirements. Behind every deliverable are tasks in the form of labor and materials. Missing deliverables means missing labor and material, which could affect the accuracy of the project plan.

Fact	Assumption	Issue	Requirement
Statement of certainty that can be proven	Something taken to be true without proof	An important question that is in dispute as believed by the originator and must be settled	A specific want or need
"Carbon composite is lighter in weight than steel."	"Sure, we can engineer a lighter solution using carbon composite rather than steel."	"The customer wants us to use a carbon composite for the sensor frame, but we have limited experience working with this material."	"You must meet the scheduled delivery date of August 25, 2015." [customer requirement]
			"Meet MIL-SPEC 2117A." [regulatory requirement]
			"Compliance to target cost." [internal ISR requirement]

TABLE 4.1 FAIR Definitions

The deliverables map shown in Fig. 4.4 feeds the LWWP. At ISR Systems, plans are initially done using a paper Gantt chart-style format then moved electronically into Microsoft Project© or other project-management software. The plan reflects quantitative data and contingency plans to address risks. When executing the project, the IPT executes tasks, monitors progress, and addresses risks on a continual basis.

The use of these tools is not a one-time process. A VoC may be used at the proposal stage and later at the award stage or at other times if there are large elements of uncertainty. During the execution, monitoring and control phases of the project, the IPT may update the impact matrix with the following: a change in scope activity, adding new deliverables to the deliverables map, updating the least-waste-way project plan, and addressing risks in the risk register.

4.3 Preparing for the Voice of the Customer

A voice of the customer session requires preplanning and coordination with the customer(s). Projects at ISR often use a three-week checklist to help prepare for the application of LPD tools to the proposal process. This includes discussing with the customer what a VoC is, its benefits, and its importance to the project. VoC standard work is also sent to the customers for their understanding and time requirements.

For some proposals and contracts, it may not be feasible to do a VoC with the customer during the proposal stage. One technique that can be used is to create a *surrogate* customer using business development, contracts, and project-management members who have direct contact with the customer. They role-play as the customer during the VoC event, which would be done at the proposal stage. It is strongly recommended that the project manager and IPT plan a second VoC with the actual customer after an award. Figure 4.5 illustrates the process for executing a voice of the customer.

Figure 4.4 Deliverables map-capture project internal and external deliverables in a network-diagram-style chart.

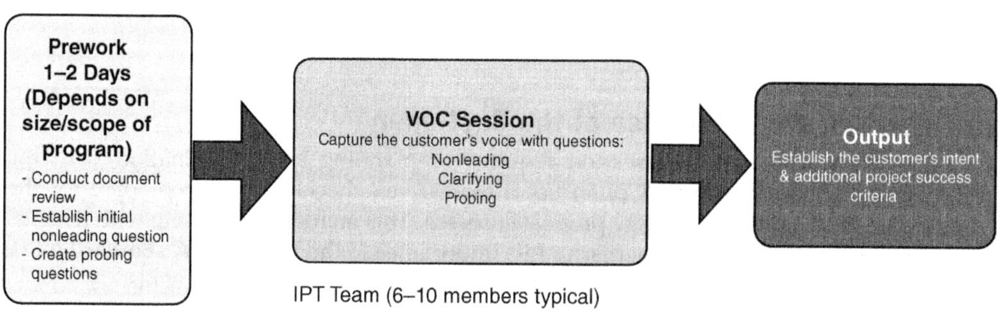

IPT Team (6–10 members typical)

1–2 days of Prework/Document Review

Typical 1-day VOC Event

Handoff: Used with Impact Matrix Tool

Figure 4.5 The VoC process.

4.4 Initiating the VoC; Summary of the Key Steps

Once a customer agrees upon a date for the actual VoC, the IPT needs to meet prior to and prepare for the event. It is recommended that the IPT schedule the one-to-two day document review approximately one week prior to the event. That gives the team enough time to understand the documents under review, prepare probing questions (that can be shared in advance with the customer), and develop the VoC event agenda. The key steps in executing the voice of the customer after VoC preparation and approval with the customer are:

1. The integrated project team reviews specifications and requirements, including any existing paperwork, emails, or other notes that relate to the product under consideration. A draft compliance matrix should be developed with questions/issues highlighted for follow-up. This compliance matrix can also include a column for deliverables and a column for risks that are uncovered during the document review.

   Prior to conducting the VoC, the process is shared with the customers. It is important that the customers understand that, although an observation from them may be captured, grouped, and analyzed, it does not immediately become a requirement or an action until it is agreed upon by both parties. This nuance is important, as the VoC event is not meant to be a contractual, binding process. Rather it is an effort to capture and understand all of the customer's wants and needs.
2. The project team conducts a VoC meeting. In general, the facilitator will review the VoC process, ground rules, and the expected outputs and results of the meeting. An update on the current state of the project and review of product/process might take one or two hours of the session. It can also be used in place of starting the session with a *non-leading* question as described in step 3.
3. The process begins with a key non-leading question. This first question is broader in scope than the probing questions. The customer can immediately see how the VoC process will work. Sometimes the customer responses to this first, broader question may provide additional information not expected and can even provide the IPT with additional avenues of questions during the VoC event. Next, the IPT would continue the VoC with the clarifying and probing questions, all captured as close to verbatim as possible.
4. Review the output responses as a group. This is often performed by the ISR Systems team only, but it can be useful to include the customer if time allows.
5. Select a grouping of customer responses that are aligned together and describe it in a few words or a short phrase. The IPT team should then clarify and combine the phrases into a summary description of the grouping.
6. Review the summary and content of the grouping.
7. Prioritize these additional customer intents/project success criteria. Usually the top three to five criteria would be added to the impact matrix.

4.5 Skill Sets Required for the Host IPT Team

The definitions for the customer include the group or people whose intents you are trying to satisfy. They could be internal or external. The host IPT team is providing the solution or service to the customer. These people are responsible for project execution and are comprised of staff whose disciplines support the project throughout the project life cycle. At ISR, these disciplines include project management, business management, contracts, supply chain, quality, and the engineering disciplines (mechanical, electrical, software, systems, imaging etc.) The IPT team should be aware of these other skills that are helpful in completing a successful VoC:

- The host IPT team needs to be open-minded.
- They have the ability to focus on the *big picture.*
- Some team members preferably are graduates of project-management skills and soft skills courses.
- Team members represent the various project disciplines, e.g., mechanical, electrical, software, systems engineering, and support groups such as finance, supply-chain management, and quality.
- They are familiar with the voice of the customer ISR Systems training module.
- They are aware that when developing and asking non-leading and clarifying questions that they prevent the introduction of potential ISR Systems (UTC) biases to our customer.

4.6 Supplies Needed for the VoC Activity

- 8.5-by-11 inch pads of paper (typically used in a landscape orientation)
- Blue, black, and red markers
- Blue masking tape (protects the walls)
- Flip chart
- Pushpins
- Additional breakout room if needed by customer or host team
- Refreshments (be a good host)

4.7 Steps in Understanding VoC

There are two steps to understanding VoC to achieve better results: (1) do your pre-work and (2) the document review process.

These steps should be completed before the actual VoC session. For projects that are driven by programs and are primarily for external customers, the documentation will normally be quite formal. Some of the documents to review might be:

1. Statements of Work (SOW). SOW is associated with the contract for the work. The statement of work is normally quite detailed and lists each party's responsibilities for the project. Information gathered during the VoC process may be in conflict with the SOW in many ways. For example, new work may be added to the project that was not in the original SOW, or work originally the responsibility of the customer might now be expected of ISR Systems. Such conflicts must be identified, answered and mitigated by the IPT. Some issues may become unique opportunities that would be of benefit to the customer and ISR.

2. Deliverables Map. In some cases, for a previous project and contract award that utilized LPD tools, a deliverables map for that project may have been captured. Teams can use project deliverable maps as starting points and, as the document review unfolds, check off deliverables or annotate new ones. This will accelerate the creation of a revised deliverables map for this project. The team should exercise caution if doing this. Merely *cut and pasting* the previous deliverables map into the proposal process can result in the team overlooking deliverables.
3. Risk Register. A previous project and contract award that has utilized the LPD tools may have an existing project risk register. This risk register may provide a starting point for risk capture. During the document review, the IPT can add additional risks as it reviews the documentation. Team members should review similar program-risk registers to familiarize themselves with past risks, the countermeasures proposed, and what risks that actually were realized during the life cycle of the project. By doing this prior to the document review, team members' awareness of possible risks are increased, resulting in better scrutiny of the documents under review.
4. Contract. In the case of the contract in general, there may be other sections that are not in alignment with the original agreed-upon terms. For example, the terms and conditions of the contract might indicate a delivery penalty that is now more likely as a result of a new requirement resulting from the VoC session. In this case, the customer needs to reexamine which side of the contradiction is most important and adjust accordingly. The contracts IPT member should note differences and work within the program management office (PMO) to resolve with the customer.
5. Technical Specifications. In many cases, the product will have a product description and associated specification referenced and/or provided by the customer. It may be that the VoC session has surfaced an error in the specification or a contradiction or even an oversight. All of the issues must be documented and worked out with the customer before proceeding into the next program phase. Failure to resolve early issues on specifications and requirements will most likely result in additional conflict and mistakes further into the project phase. This means costlier schedule and cost impacts to the project.
6. Schedules. Last, the scope of the work may have grown as a result of the VoC process that delivery dates, milestones, etc. are immediately in jeopardy or unrealistic. These disconnects need to be surfaced and worked out. Often there may be additional undocumented dialogs between customer and host team engineering, for example, that may suggest schedule date changes that the PMO might not be aware of. This often results in miscommunication with the customer and changes in customer expectations.
7. Memos. Any project memos or guiding documents created to provide the basic guidance for the project should be checked for consistency against the VoC results. Early expectations from the customer may change.
8. Emails. This is an especially challenging area because emails are frequent sources of scope creep. All emails with new scope should be reconciled with existing scope, and all VoC data should be reconciled as well. IPT members should also be extremely vigilant about reliance on email communication. There is risk that certain

intellectual property will have an inadvertent transfer or that an International Traffic in Arms Regulations (ITAR) violation could occur.

9. Written Requests. Same information as the two items above.

   The key issue with internal projects is to maintain some sense of formality to ensure agreement between both parties in the project as well as to limit scope creep. Changes in scope, schedule re-baselining, or other major impacts to the project should be well-documented from the contracts area.

Why do we need to do this pre-work? Our experience with our customers has shown that often documented requirements appear to be compiled by people with specific *functional requirements and responsibilities* but inadequate coordination between themselves and their project management structure. Customer wants and needs may be identified in documentation but may be incomplete or easily misinterpreted. The voice of the customer process aids in identifying other needs and intents and helps to align those wants and needs.

Deconflicting the initial requirements is important. There can be many types of requirement conflicts that the IPT must deal with:

- Performance characteristics given by Customer Technical Folks A conflict with other requirements given by Customer Technical Folks B
- Quality and ITAR requirements may conflict with certain other requirements
- Worse yet, the customers just leave it all to you; don't bother them with all of that detail until you are ready to deliver, and then it's too late.

By using the document-review steps along with the voice of the customer step, there is reconciliation and a better understanding of project requirements. Any new requirements captured during the VoC can be checked for conflicts with the existing documented requirements. If a conflict is found, it can be de-conflicted with the customer now, rather than later. The VoC, together with a deliverables map, will confirm what is to be delivered to your customer before you create the plan to deliver.

The deliverables map will represent the complete range of internal and external deliverables that are required to meet the proposal or contract. For every deliverable, there will be tasks (labor) and possibly material behind it. It is critical, then, that after the VoC any new deliverables are captured and added to the deliverables map. Missing a deliverable means missing labor and spending project dollars on labor that has not been captured in the contract.

A second most powerful aspect of this pre-work process is that it is done with the integrated project team, a collaborative process that involves all of the team members, which provides a diversity of experience, functions, and skills to the proposal and project. This will help clarify the initially documented requirements and strengthen the proposal response and project execution.

1. Select Proper Representatives from the Customer

   Ask the customer to ensure that one decision-maker from each of the key areas of the customer project is available in order to get the best representation of that customer's *one voice*. Typically these are the *buck stops here* key stakeholders. You should also discuss with your customer additional customer staff whose input, if not heard by the host team, is likely to cause significant risks and redo loops later. Ideally, the customer team matches one-on-one the host IPT team in terms of disciplines represented.

2. Select Proper Representatives for the Host Team

   The IPT should provide sufficient, equivalent members for the host team that match the customer panel one-for-one. Avoid outnumbering the customer group. Again, the host team members are those IPT members who are actually responsible for the work, not a *proposal team* that is not responsible for execution of the project.

3. Select a Facilitator

   It is important to have a facilitator who is skilled in communication and has no stake in the proceedings. This person will be important in ensuring that host and customer teams follow the process, policing the non-leading questions, preventing sidebar conversations, and keeping the question-response-capture process on the subject. The facilitator must be comfortable with doing things such as interrupting senior management or subject matter experts if necessary to keep the process on track. The facilitator also can ensure the comfort level of the customer's panel, polices for dialogue, and address in real time any in-process issues.

4. Prepare the Opening Question

   After training, the host team should brainstorm only the first, non-leading question to the customer panel. It should be balanced between being not too broad and not too specific and should, if well designed, provoke the customer panel to respond initially and continuously throughout the time set aside for the event, across a broad range, and in detail. Consider the non-leading question an *icebreaker* to get the customer talking and at the same time offer a *safe* question the customer can respond to. As the host team captures the response, the customer can see and understand the *mechanics* of the process as well. In addition, the IPT should have its list of probing questions. These questions should be grouped under general themes and prioritized in a sequence of importance to the team. Sometimes, one probing question response by the customer may answer a different probing question on the host team's list.

   The IPT should have, during the document review, created its initial list of probing questions that are linked to what the team does not understand. These questions represent a more specific attempt to capture understanding of TBDs, vague requirements, mistakes in specifications, assumptions, and other potential issues. Often, the IPT may have a list of 20 to 50 or more of these questions. Grouping questions into themes and prioritizing them ahead of the VoC will provide the IPT with a strategy to get a complete understanding from the customer. The IPT should link this directly with the compliance matrix for source requirements. The expectation is that the results of the VoC will be reflected in some level of changes to written requirements.

   Why not just send the customer the list of questions and wait for a response? That may appear to be a more efficient means of getting answers, but

   - You will not know who is really responding
   - You will have the opportunity for every member of the customer's team to respond to the question
   - There will be no body language for you to observe

- You will not have an opportunity for an immediate, clarifying question or additional questions if required
- It disconnects both the IPT and customer from each other, which may result in a lack of confidence and trust-building between both groups

5. Arrange the Room

   Customer and ISR Systems participants shall face each other so that the customer is able to see the comment pages posted on the walls. The customer needs to feel comfortable and nonthreatened. It is an excellent idea to have an adjacent room ready for either the customer or host panels to use during a *time-out* when required. For example, the host team may see a need to regroup for a few minutes to either reprioritize its questions or add new questions. The customer panel may need private time to formulate responses or to further discuss a response. Ensure the room also has a phone that the customer may use. A room arrangement is shown in Fig. 4.6.

6. Establish and Execute the Agenda

   Under ideal circumstances, you would have your customer speak openly to you until all the subjects have been exhausted, and there is nothing left to say. However, because the customer panel is likely to be made up of important or busy decision-makers from across its company profile, we need to be respectful and maximize the benefits of the time spent as well as the customer's comfort.

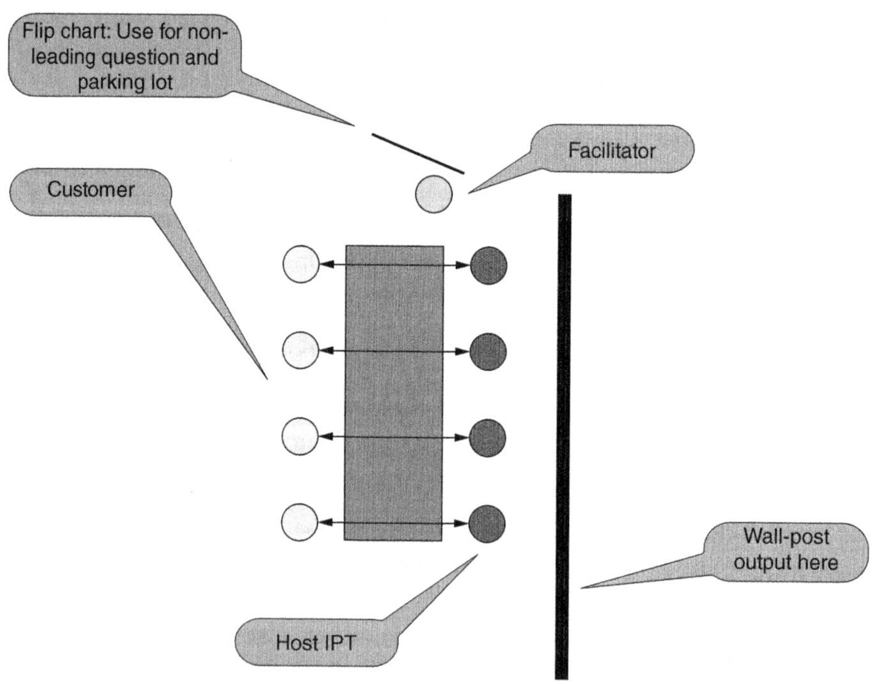

FIGURE 4.6 Room arrangement for VoC.

Therefore, setting an agenda with a time goal in mind is useful. The facilitator can monitor the pace and work with the customer and host team to ensure the proper amount of time for the VoC is available.

An agenda could be as follows:

Coffee while the panels gather:	8:30 to 9:00
VoC kick-off; welcome and introductions:	9:00 to 9:15
Brief overview of agenda and VoC process:	9:15 to 9:30
Start off the VoC with the opening question:	9:30 to 10:30
If needed, step-back review and process check:	10:30 to 11:00
Continue VoC, with break(s):	11:00 to 12:00
Lunch:	12:00 to 13:30
Continue VoC, with break(s). Probing questions:	13:30 to 15:45
Closure of the VoC:	15:45 to 16:00

7. Discuss Expectations

   As an opener, this step is important. It sets the level of expectations from both groups.

   - Bring your customer into the room
   - Have everyone introduce themselves, including their names, functions, and roles on the team; the customer program manager may want to make brief opening remarks

8. Let the Customer Explain *Their* Expectations Regarding This Event

   - Give an overview of the process
   - Make sure that the customer is aware of all of the subsequent steps

9. Explain Why We Are Doing This Voice of the Customer

   The overview might sound something like this:

   Host Project Manager: "We want to listen to you as our customer without introducing our biases. We will collect and evaluate the information generated and take appropriate countermeasures."

   "We want to listen to you."

   "We will only be asking non-leading, probing, and clarifying questions."

   "We might generate additional questions for you as the process unfolds."

   "We want to understand your view of the requirements at all levels of the project of the opportunity that brings us together today."

   "We want to understand how you view our past and current performance."

   "We will evaluate the information and take appropriate countermeasures."

   "We will advise you of our countermeasures."

   "We will not, however, answer any questions today."

10. Explain What You Want from the Customer

    An example of this introduction might go like this:

    Host Project Manager: "Please tell us about this opportunity and about our company/service."

    "Tell us anything you wish—the good, the bad, and the ugly."

"Be frank, candid, open. There is nothing you can say that will embarrass us."

"Please have your comments cover our entire relationship."

"We will pursue a particular topic until we have a clear understanding of what *you* mean and then we'll move on to the next topic."

Stress these points *without* using examples such as quality, delivery, or performance to avoid *leading the customer*.

11. Tell The Customer What You Will Be Doing and What You Won't Be Doing

    We will

    - Be proactive listeners
    - Ask only non-leading and clarifying questions
    - Ask if a topic has a trend—better, worse, new, old, etc.
    - Write everything down (on 8.5-by-11 paper, landscape)
    - Ask how we compare to the competition on each topic (relative comparisons are okay and competitors need not be named)
    - Try to make you as comfortable as possible with the process and the proceedings
    - Encourage the customer team to caucus, privately, should there be any disagreement as to providing a *single voice*

    We will not

    - Be doing any problem solving
    - Be answering questions as part of this process; this can be useful to quickly resolve a conflict or expose an unclear requirement
    - Be entering into any dialog with the customer

12. The VoC Capture Steps

    The host team should explain the mechanics of how the voice of the customer will be captured. It is a good idea to have the facilitator quickly review the *mechanics* of the process. Have copies of the standard work for VoC there. Try to capture more than one customer comment at a time. By doing so, the facilitator should ensure that each of the customer representatives have an opportunity to give their response to the question, one at a time.

    Host team members will be selected as scribes. There are several approaches for this. For one approach, whatever the question is asked, the host person equivalent to the customer person responding acts as principal scribe. If there are multiple customer representatives, the team may select/match a scribe to write comments given by a specific customer representative and a different scribe who will write comments from different representatives. This method will help capture the comments more efficiently, as many times the customers will be talking rapidly and simultaneously. Other host team members should actively listen and capture as well. This is a *backup* listening mode and demonstrates the body language that the entire team is listening to its customer. Another approach is to have one to three of the host team serve as scribes and ensure that all of the questions are asked during the day. This does put more pressure on the scribes to ensure efficient capture, however. During the capture process, the customer will tell the host if his or her responses are accurate and correct. The capture process is shown in Fig. 4.7.

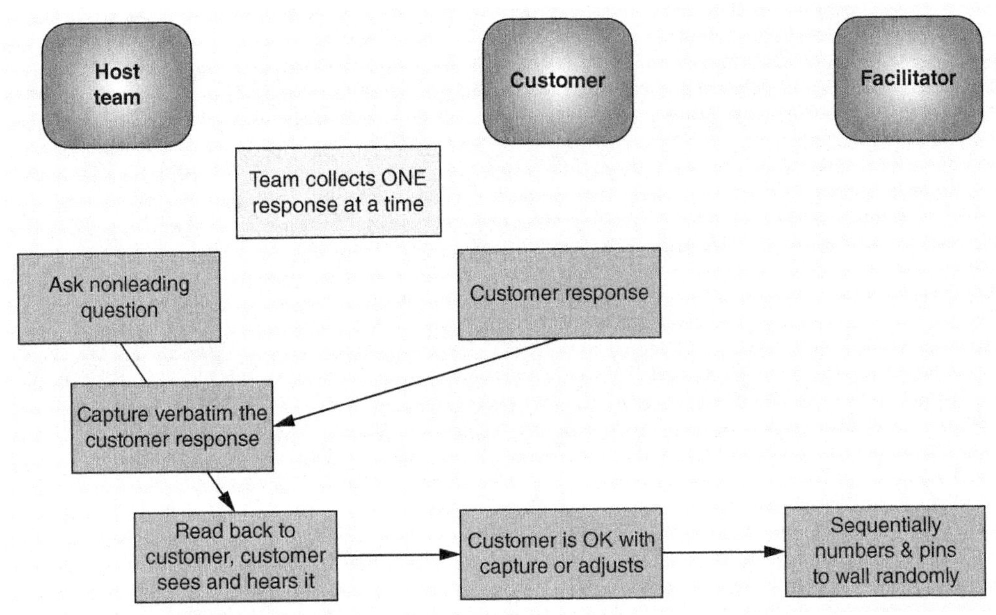

FIGURE 4.7 VoC capture process.

The capture mechanics are simple. Capture comments in a horizontal layout on 8.5-by-11 sheets of white paper. Use a permanent fine-point marker. It's important that whatever host member is doing the capture that his or her writing is legible and large enough to be read from where the customers are seated. Use color-coding for capturing comments. The host team can decide how to do this via using sticky dots, colored pens, or colored paper. An example is shown in Fig. 4.8. If color pens are used, then the capture can be identified as follows:

- Verbatim comments should be recorded in BLACK ink or other color schema
- Clarifying comments should be recorded in BLUE ink or other color schema
- Probing comments should be recorded in RED ink or other color schema
- Putting the initials of the customer who is responding is also helpful for the later grouping process

This color-coding will aid the host team later, during the grouping process, shown in Fig. 4.8.

As the responses are happening, these comment pages shall be posted randomly on the wall (facing the customer panel) or on butcher paper.

13. Initiate the Customer Responses

The facilitator begins the capture by sharing the starting non-leading question with the customer. This opening question will have been prepared ahead of the VoC event as part of the pre-work. Start with asking the customer PM to respond

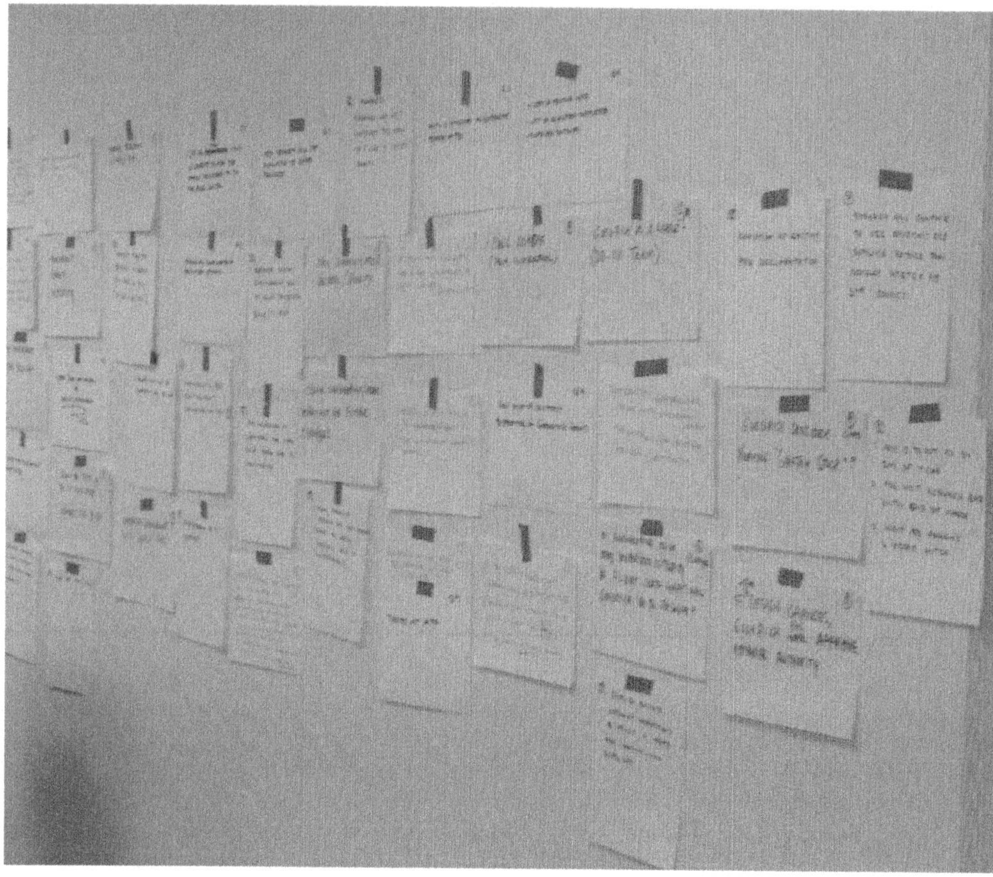

Figure 4.8 VoC capture in progress.

first, followed by the other team members. The host PM listens and captures the customer's exact words. Other IPT members also listen and capture to ensure the statement is captured fully and correctly. Capture one response at a time and record the comments as close as verbatim as possible, using whatever color-coding the team has decided upon prior to the event. For this example, the comment is captured using black pen on 8.5-by-11 paper in landscape format.

The host member doing the capture should not paraphrase or summarize what she or he is hearing. Ask the person to state it again if any details were missed. If the capture is summarized or paraphrased, there is an immediate risk of interpretation and filtering by the listener, which and will not reflect accurately what the customer has stated.

Continually show the customer the sheet holding the written response and ask the question: "Was that what you said?" This step not only checks to see that you have listened accurately, but that if one or more customer representatives are present that they are in consensus. If the customer is not in consensus, one approach is to have another sheet written by the customer with the different perspective or the customers can caucus amongst themselves to come to an agreement and then provide an updated response.

Posting customer comments is shown in Fig. 4.9. If the capture is OK'd by the customer, put it on the wall, progressively numbering each sheet as it is collected, confirmed, and posted. Do not assume that you know specific definitions of words. Ask clarifying questions such as, "What do you mean by ____" or "Tell me more about ____." Again record exactly what they say, this time using a blue pen and post it on the wall.

Here is a sample dialogue using non-leading questions to draw out the customer's intent:

Host: "Please tell me about Acme, Inc."

Customer: "Sometimes your performance is terrible."

Host: "What do you mean by performance?"

Customer: "Delivery." (BLUE ink)

Host: "What do you mean by delivery?"

Customer: "Well, it comes on time, but it goes to the wrong plant." (BLUE ink)

Host: "What do you mean by 'terrible?'"

Customer: "Sometimes when your part arrives at the wrong plant, it causes us a lot of headaches tracking it down and moving it to the correct facility." (BLUE ink)

Host: "Sometimes? What do you mean?"

Customer: "Sometimes you do it right, sometimes wrong. Your shipping paperwork is not always provided, so we spend extra time having to call you and obtain." (BLUE ink)

Figure 4.9 Posting customer comments.

Host: "How long has this been going on?"

Customer: "Forever, well at least over the past year. We've noticed a lot more of these since you expanded your facilities two years ago." (RED ink)

Host: "Do any of your suppliers do it right, all the time?"

Customer: "Most of them." (RED ink)

Host: "Anything else on this topic?" (responses, if any, in RED ink)

Silence may follow for a few moments . . . (be patient for up to two full minutes if necessary)

Host: "OK, then, what other topics do we need to address?"

14. Clarify and Probe the Customer

While a clear comment is written in BLACK, if a customer comment is unclear to the team on its surface, and the team needs to clarify the initial response, and the scribe shall write the clarifying response in BLUE. If the team needs to further prod the customer or sees a trend developing, the scribe shall write that response in RED.

When this question has been covered by each of the customer's team members, move on. A customer team member may decline to respond or simply state, "Already captured." Note and move on. Note the *relative importance* of the subjects; throughout the discussion, each individual on the host team should note the following:

- The length of time spent discussing the issue
- Who raised the issue?
- Was there notable passion or emotion from the customer?

At least one IPT member should be tasked with tracking the above. Don't be shy! If a particular topic has not been brought up during verbatim and clarifying responses, you are obligated to ask about it. For example: "You didn't mention quality." Remember, this is the team's chance to ask such questions directly with the customer in a more focused manner.

Responses to this probing should be written in RED.

Say you have been through a long discussion, but at no time did the subject of *quality* arise. You feel it needs to be addressed.

Host: "So you did not mention quality."

Customer: "No need to mention quality."

Host: "Why not?"

Customer A: "Your quality doesn't matter."

Customer B: "Quality is not a problem"

Host: "What do you mean?"

Customer A: "The quality is always found to be exactly to part print by our 100 percent receiver-inspection quality department."

Customer B: "Quality always works."

Continue the responses until non-leading (probing) questions have been asked and no additional questions are generated for the host team to ask.

15. What the Host Panel Should Do

The host team needs to stay focused during this process. Be proactive listeners. Listen and write everything down. The team should be engaged and focused. Avoid trying to *problem solve* or answer questions through dialog or asking questions that take the responses off track. Definitely do not challenge the customer's intent. Self-police for dialog amongst yourselves; don't rely just on the facilitator to manage the process. It's important to permit the process to work. It may seem slow at first, but the VoC is a deliberative process. Host members should avoid asking a further question while the previous response to a question is being captured: just write it down for later. Typically, the customer tunes into the process and will provide responses that are easier to capture.

Ensure that the customer can read what is written down and displayed. All of the responses will be put up on the capture area. It is important that all of the participants can read what is captured so there is no misunderstanding of the statement captured. Ensure that only one response per page is captured. If you start capturing multiple responses on the single sheet of paper, you cannot efficiently group and affinitize the response later.

If the customer is giving the team long run-on answers, the facilitator can gently interrupt, "We need to capture one element on one sheet at a time . . . you gave us several. Can you please repeat them?" Last, as a host group, you may encounter periods of silence from the customer. Let it happen; sometimes the silence will prompt an additional response from the customer. The facilitator can decide when to move on.

What should the team do if the customer responds to your question with something completely unrelated? The team should capture it, as it might be an important concern or issue for the customer. It may be that the customer is sharing his or her experience with you. It might just be that the customer is happy that someone is listening. Often the host team can follow up such responses with a clarifying question or a later question.

16. What the Customer Panel Should Do

    Customers should tell the facilitator if they are not comfortable with the environment and that they can hear and read the capture of their responses easily. It is important that the customers understand that they need to be frank, candid, and open. Share the good, the bad, and the ugly. Nothing said will embarrass the host panel. The customer should understand that the process will cover the entire relationship. If the customer raises a theme that has not been covered by the initial list of probing questions, the facilitator can call a *time-out* and allow the host team to meet and generate questions.

    A particular topic (or theme) of questioning will continue until the host panel has a clear understanding of what you, as the customer, mean before moving on to the next topic. The facilitator is there if you, as the customer, need a moment with your team to engage in a dialogue.

17. What the Facilitator Should Do

    Sometimes, especially at the start of the VoC, the customer may give a run-on answer. Early on, the facilitator should gently remind the customer that the host team needs to capture everything said by the customer. Some example *control measures* by the facilitator might include, "This is important . . . let the host team capture that first point," "Help us fit that onto one piece of paper," or "Sounds like a few ideas. Can we get each one on a separate piece of paper?"

18. When to Excuse the Customer

    Once all of the verbatim, clarifying, and probing responses have been collected; or there have been sufficient, silent pauses; or the TAKT time/agenda for the VoC event has been reached/completed, the facilitator can ask both groups if they are ready for closure. The facilitator and host PM can thank the customer representatives for their participation and excuse them from the meeting room. The customer should understand that the results of the groupings and potential generation of additional customer intents and project success criteria will be shared with them, if not that day, as soon as possible after the VoC.

4.8 Start of Affinitization When the IPT Team Does the Groupings

This step starts the distillation of customer statements into more meaningful groupings that then will lead to capture and/or affirmation of customer intents. The team should remember and reflect on the customer's emotions during the VoC process.

- Intensity: What probing questions did the customer give the longest answers? Were there additional clarifying questions? Did one *theme* of questions create additional questions?
- Passions: How was the body language of the customer? Was the customer more passionate about some themes versus others?
- Focus: Did the customer *drift* during any of the questions or themes? Were responses brief or limited?
- How long did the team spend in any particular area?

These are all important.

The IPT, as a group, starts to silently group the comment sheets into themes. This is not always easy as team members may start dialoging over the responses. The intent to keep silent is to avoid having some team members influence others. The pairings should be done by the team as a whole. The facilitator can help limit the conversations during this step. The host team members will need to approach the wall with all the comments pinned to it as shown in Fig. 4.10.

The team needs to start arranging the papers into groups. Some comments will clearly belong together; some comments will be difficult to match with others. Don't worry about any outliers at this stage; continue to pair and group responses. The team needs to go on with this until it nearly stops. At first there will be a lot of movement, and things will go back and forth a few times.

Individual team members can start grouping immediately or wait until other team members have done some grouping. All team members have the right to move comments to different groups or create new groups during the silent grouping. Eventually, groups will begin to appear, shown in Fig. 4.11.

The opportunity to discuss the groupings will occur in the next stage.

Don't overthink *outliers*. Outliers might be either countermeasures or those comments that don't appear to fit into any particular group. The next step will take care of them.

FIGURE 4.10 Grouping the VoC capture.

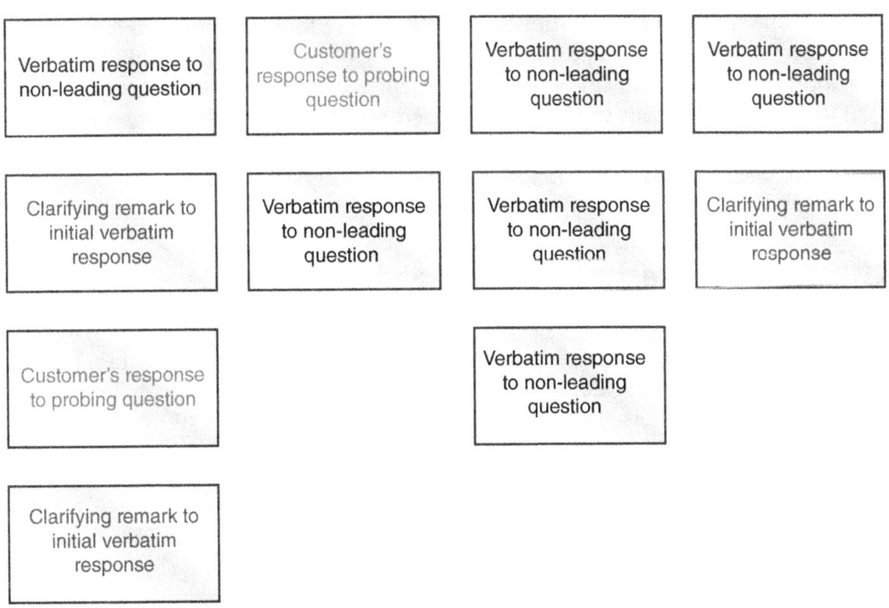

FIGURE 4.11 Example Groupings of the VoC.

4.9 Label the Groupings

The team should look at the collection of groupings and put a label above each.

4.10 Analyze the Groupings

Now the analysis of the customer statements can begin. The IPT should review each of the labeled groupings to identify solutions, countermeasures, and opinions. Have someone read each paper out loud. Agree on the intent of each comment—some are solutions, some are opinions, some are philosophy, some are just rants. Remember, each of these statements is verbatim from the customer. Discuss contradictions you see. Discuss trends that emerge. This step creates a common awareness in the host team.

Things the IPT should ask themselves and look for:

- Is there a common theme or are there several themes?
- Is there a clear subject here? Write label headers on another sheet of paper and hang them above the group. The labels should include, WANTS, NEEDS, INTENTS, and REQUIREMENTS
- Are there illegitimate needs? Customer wants that we clearly cannot support?
- Where is the waste? Where is the opportunity?
- Review all the ones under the heading; move if required.

Now, quickly review each 8.5-by-11 and note on it if it is a FACT, ISSUE, ASSUMPTION, or REQUIREMENT. The team can attach colored sticky dots to help FAIR the statements. Table 4.1 summarizes the IPT understanding of fair items.

This will be useful in supporting other LPD tools and visuals such as the impact matrix, deliverables map, and least-waste-way program plan—all of which are lean product development/project management tools and processes used at ISR Systems. The four groups with the headings added and statements FAIR: green (circle) dot is a FACT; yellow (square) dot is an ASSUMPTION; red (triangle) dot is an ISSUE; and orange (rectangle) dot is a REQUIREMENT (shown in Fig. 4.12).

4.11 Capturing Customer Intents and Additional Project Success Criteria

Take a step back and decide what the wall (and each of the groupings) is telling you. Make observations (bullet statements) and write them down on a flip chart in a team brainstorming exercise (shown in Fig. 4.13).

These items now stand as the *success criteria*. Some examples of *good* customer-intent success-criteria statements might be:

> "Recover a positive margin from the current situation of a 73 percent loss on the ABC-001 camera."
>
> "Demonstrate new innovative solutions, including use of shortwave infrared technologies."

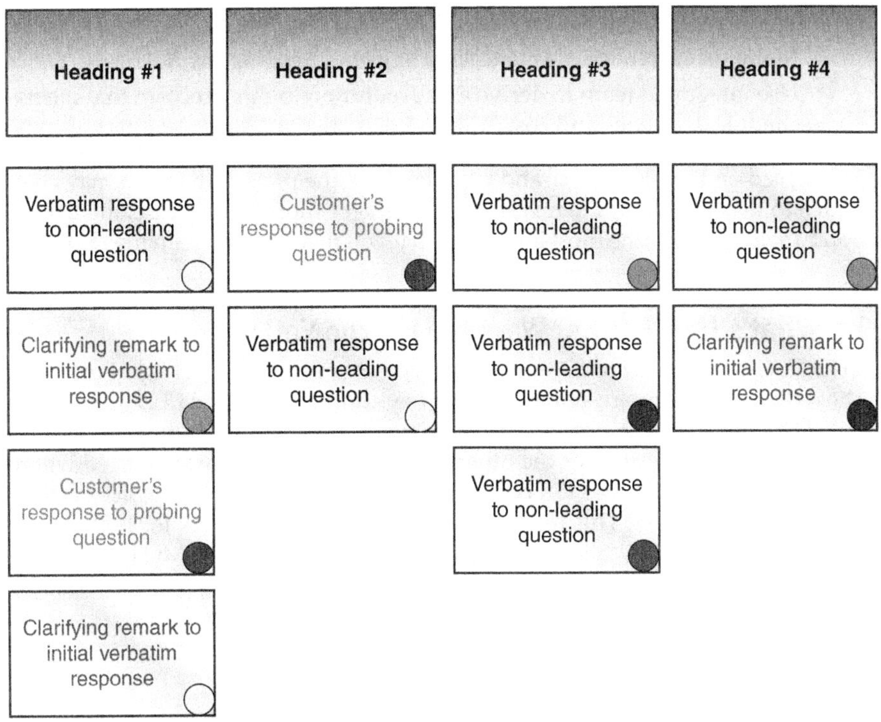

FIGURE 4.12 Grouping of VoC using FAIR definitions.

FIGURE 4.13 Defining additional project success criteria by affinity.

"Identify best value for money solutions that meet the Concept of Operations."

"Maintain current performance as a supplier to company X."

"Use integrated team to derive a total solution to this problem that meets customer schedule deadlines and cost."

Based on the size, scope, and complexity of the project, the team may come up with three to five additional intents/success criteria or more. In this case, the IPT will need to prioritize which are the most important that will then be used with the impact matrix tool.

4.12 What's Next? Other Ways to Use the VoC

The VoC can be used to capture other types of customer feedback. For example, if the host (ISR) team is interested in aligning key suppliers into the LPD process, an excellent way to start is to host a dual VoC. In this case, half the day is devoted to the ISR (host) and customer role, and for the other half of the day, roles are reversed where the supplier is acting as host and ISR is the customer. This is known as a voice of the customer/voice of the supplier. This means additional preparation work for both groups, but the results can be worthwhile. VoC can also be used internally. PMO can serve as the customer and engineering can serve as the host.

4.13 Lessons Learned from Use of the VoC

The VoC can be performed at various flight levels of detail and at various phases of the project. One example is having information at a higher flight level for the initial VoC with future VoCs planned for lower-level details. At a proposal stage, often requirements are defined at a higher level or are initially *to be determined*. A VoC at the proposal stage may draw out other customer intents, but a second VoC may be necessary to ensure complete understanding of customer intents after final negotiations.

It is important that the host team *sells* the customer on the process and the time commitment. This can be done through pre-VoC planning. Share the standard work and TAKT time requirements with the customer ahead of the event. Answer customer questions: use standard work so that the customer understands your commitment to time and efficiency of the process.

If your host team is new to the VoC process, a *dry-run* approach can be used: ask your business-development people to role-play as the customer. It is important that the host team understand it is not at a VoC to have a dialogue but to actively listen. Train the host team in advance with emphasis on *not* challenging the customer intent and allowing the customer time to *talk*.

The document-review step is extremely important. The IPT should carefully review the proposal, statement of work, specifications, and other documents. From the review, the team should be able to:

- Identify understanding gaps: the TBDs
- Identify assumptions and risks
- Identify initial requirements—those that are completely understood, and those that might be less specific and need clarification
- Capture internal and external deliverables

- Develop the non-leading initial question and probing questions to close the understanding gaps

Ensure that prior to the event, the customer is educated about the event process, steps, and time required. This can be done through pre-event discussions and by sending VoC standard work to the customer. Business-development staff should be an integral part of the host panel. In many cases, they have already either been working the proposal opportunity and have substantial contact with the customer prior to the event.

Sending the customer your list of probing questions and themes ahead of time helps the customer understand gaps, prevents the customer from being *ambushed*, and permits the customer the opportunity to ensure that the right team members attend. Having a comfortable environment for the event plus a room for the teams to caucus in is a plus. This puts everyone at ease.

Establish agreed-to plans for follow-up to a VoC before both parties depart the co-location. Capture outputs electronically for future use. Remember that during the VoC, some items can, with mutual agreement, be placed in a project action item for resolution outside of the VoC.

4.14 VoC Process Risks

The VoC may drift into discussions about the host-customer relationship. The facilitator will need to close any dialoging; this may be the time to capture that type of issue on a parking-lot chart or action-item chart for later discussion. The host team may also choose to add additional probing questions around this topic.

Prior history around performance, cost, quality, and schedule may have the potential to become key discussion points with the customer. If this happens, a time-out for the host team to reprioritize or add other questions to the process may be needed.

The customer may view all of the intents captured as extremely important. The host panel must decide during the application of the other LPD tools (impact matrix, least-waste-way project plan) what is and what is not important. It is unlikely that all-important intents of the customer can be completely met in the proposal.

In some cases, especially at the proposal stage, the customer may not be available. In this case, creating a surrogate customer using business-development and value-stream director-level staff will at least provide the IPT with a norming insight into the customer. A second VoC post-award with the customer may also prove of great value for project execution.

4.15 Benefits from Using the VoC Process

The use of this type of voice of the customer process has a wide range of benefits. This focused exercise permits the integrated project team to ask specific questions of the customer regarding requirements, issues, and other project lifecycle tasks. It gives the team a rare opportunity to meet with the customer's key decision-makers and be able to listen carefully to them. The result of this listening leads to building trust and openness with the customer. Often, during the VoC, new *gems* of information are revealed and understood. The use of this tool early in the project lifecycle (proposal development) gives the team better understanding of customer needs, requirements, and other issues. The tool is also useful whenever there are gaps in knowledge or potential areas of misunderstanding between you and the customer.

Better and earlier understanding of requirements leads to better project execution. Coupled with risk management, the IPT can have better control over issues that might affect the estimate at completion. Over the past three years of using these tools and processes, ISR Systems has seen a marked improvement in reducing or completely eliminating cost overruns and better maintaining cost performance and schedule performance indices of 1.0.

Discussion Topics

Review the following simulated scenario. Provided is a specification, statement of work, and internal stakeholder inputs. You are now a member of this integrated project team. Based on your review of the provided documents, answer the following:

1. Specification
    a. What requirements exist in this documentation? (Find as many as you think there are, provide rationale why it is a requirement.)
    b. What assumptions exist?
    c. Identify information gaps
    d. Are there any deliverables described? (Please list as many deliverables as you think there are based on the information provided.)
    e. Give four examples of *probing* questions you would want to ask the customer around the theme of "specification."
2. Statement of Work
    a. What requirements exist in this documentation? (Find as many as you think there are, provide rationale why it is a requirement.)
    b. Are there any risks? (Please describe the risk and a possible countermeasure.)
    c. Are there any information gaps? If so, describe how you would handle these as part of the VoC.
    d. Are there deliverables described? Please list.
    e. Do you believe there are missing deliverables? (If so, please list.)
    f. Give four examples of *probing* questions you would want to ask the customer around the theme of SOW.
    g. What other probing question *themes* might the IPT want to create questions for?

Chapter Exercise

Document Review Package for the JMU Sensor for Project X

Note: This simulation is fictitious and does not represent nor is intended to represent a product manufactured by either legacy Goodrich Corporation or United Technologies Corporation. There is no ITAR sensitive data included in this simulation.

DRAFT Specification for JMU Sensor for Project X for Document Review

1. Packaging shall be a single, ruggedized unit as defined in Fig. 4.14.
2. Sensor unit, consisting of two sensors, shall be designed for mild to harsh environmental conditions and the material shall be 300 series stainless steel.

3. Connector shall be of type D38999/29WJ19PC, wall-mount receptacle, series III, olive-drab finish.
4. Sensor shall fit into a 6-×-6-×-6-in space. Integrated into an airborne vehicle (type TBD). In the interest of lowest manufacturing cost, all interface tolerances shall be such that an as-cast part can meet the tolerance without secondary machining if casting used. If casting is used, unobtanium is the desired material.
5. Temperature range: –200 to +200°C. The connector shall be rated to +100°C
6. Electrical
   - Current overload: a continuous applied current of (TBD) shall not damage the sensors
   - Sensor will be supplied with current not to exceed 30 amps
   - Output voltage shall be 28 VDC
7. Mechanical
   - Sensor shall withstand vibration testing from (TBD) hertz to (TBD) hertz, at a maximum acceleration of 10 g's.
   - Pressure: each sensor shall withstand a pressure of 20,000 psig using helium gas as the pressuring medium
   - Visual sensor must be weatherproof
   - Concept of Operations (CONOPS)
   - Sensor shall have a >98% operational readiness capability
   - Sensor shall support day/night missions up to 12 hours
   - Sensor shall be a line replaceable unit (LRU)
   - Sensor shall be repairable at the shop level to the circuit cardboard (CCB) level
   - Secondary visual sensor shall provide a color image in standard video interface format
8. Test
   - Acceptance testing shall consist of demonstrated conformance to the requirements of this specification

Customer Statement of Work for the ISR Project X JMU Sensor
For Document Review Teams

The desired target price for the JMU sensor production units is $12,500 per unit. An incentive will be negotiated to reduce per unit price under $12,500.

Schedule (after receipt of order)

1. The first unit, the engineering quality unit, is due in three months for an initial flight test.
2. The next two units have to be Class-A production units (that means, with released engineering documentation and configuration control), which are needed one month later for a second flight test.

3. Additional units on a per request/per month basis will be requested as part of the final contract under "Option A": up to 20 units for this contract. Follow-on production orders may occur after that over the next five years, depending on unit performance and customer funding.
4. You must meet all contract dates, or you'll be debarred from doing any work for us for three years.
5. This unit will utilize a new design as proposed by the ISR business development and agreed upon as documented in customer-documented Project X Memorandum 2012-011B and final specifications when released.

Customer Mandated Program Reviews

1. VOC (with customer participation) and LPD tool event (to be hosted by ISR)
2. Program kick-off meeting
3. Preliminary design review
4. One technical interchange meeting (location TBD)
5. Critical design review
6. First article inspection review
7. Final factory acceptance test

Contract Document Deliverable Documents

1. Monthly program status reports
2. Preliminary design data package
3. Critical design data package
4. Factory-acceptance test data package
5. Final specification
6. Flight line troubleshooting/repair manual
7. Export license; ITARS documentation

Suggested Discussion for Chapter Exercise

Specification

What requirements exist in this documentation?

Packaging

- Packaging shall be a single, ruggedized unit as defined by Fig. 4.14. This is a vague requirement that needs further definition. Figure 4.14 does not reflect the newly proposed configuration by business development shown in Fig. 4.15, which they proposed without consultation with the project manager. The customer is using an early design concept presented earlier by business development.
- Sensor shall fit into a 6-×-6-×-6-in space. Integrated into airborne vehicle (type TBD). In the interest of lowest manufacturing cost, all interface tolerances

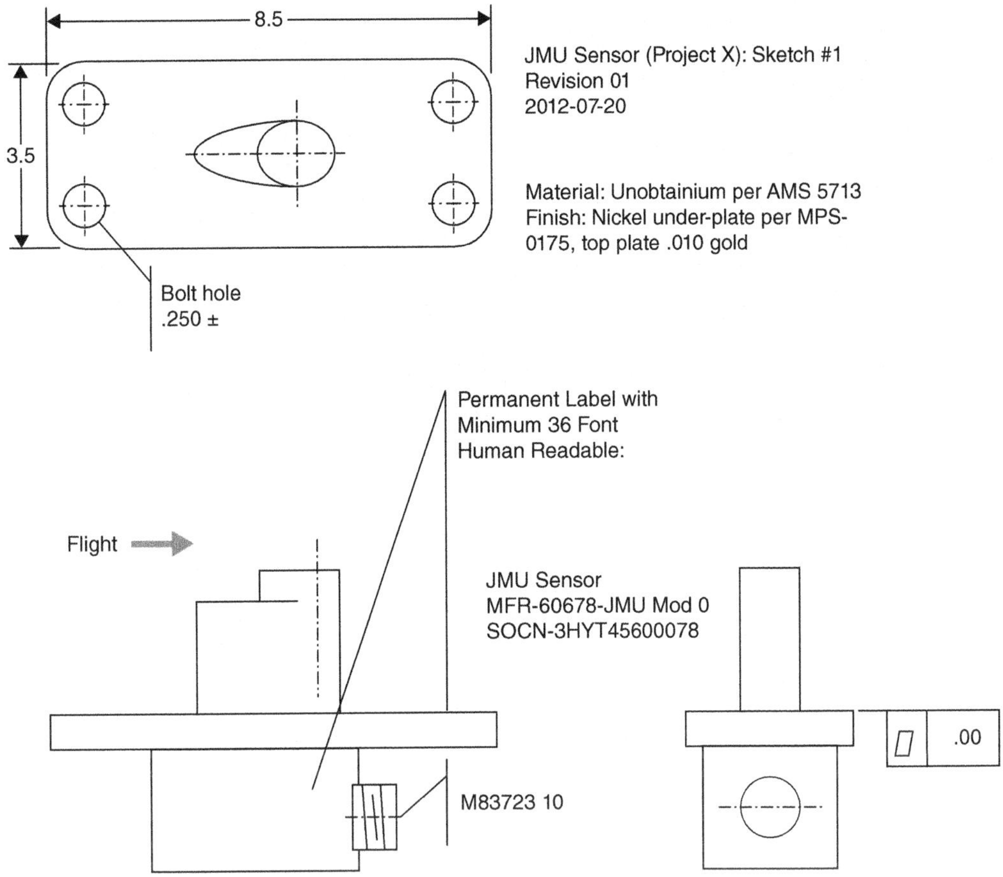

FIGURE 4.14 JMU sensor for project X.

shall be such that an as-cast part can meet the tolerance without secondary machining if casting used. Conflict here. The specification calls for a 6-×-6-×-6-in space, yet the new configuration appears different. Also, the vehicle type is not described. This could have a great affect on packaging requirements. Further definition is needed. Some of the language describes a sensor, other language says "sensors." What is it? Do the requirements apply to both sensors, one sensor?

Environmental

- Sensor shall be designed for mild to harsh environmental conditions and the material shall be 300 series stainless steel. Vague. What is meant by "mild to harsh?" There is a discrepancy in the use of material: 300 series stainless steel versus unobtanium as proposed in the new configuration.
- Connector shall be D38999/29WJ19PC, wall-mount receptacle, series III, olive-drab finish. This looks like a clear requirement. Note the word "shall." But is it really the correct connector? Why did the customer ask for this type?

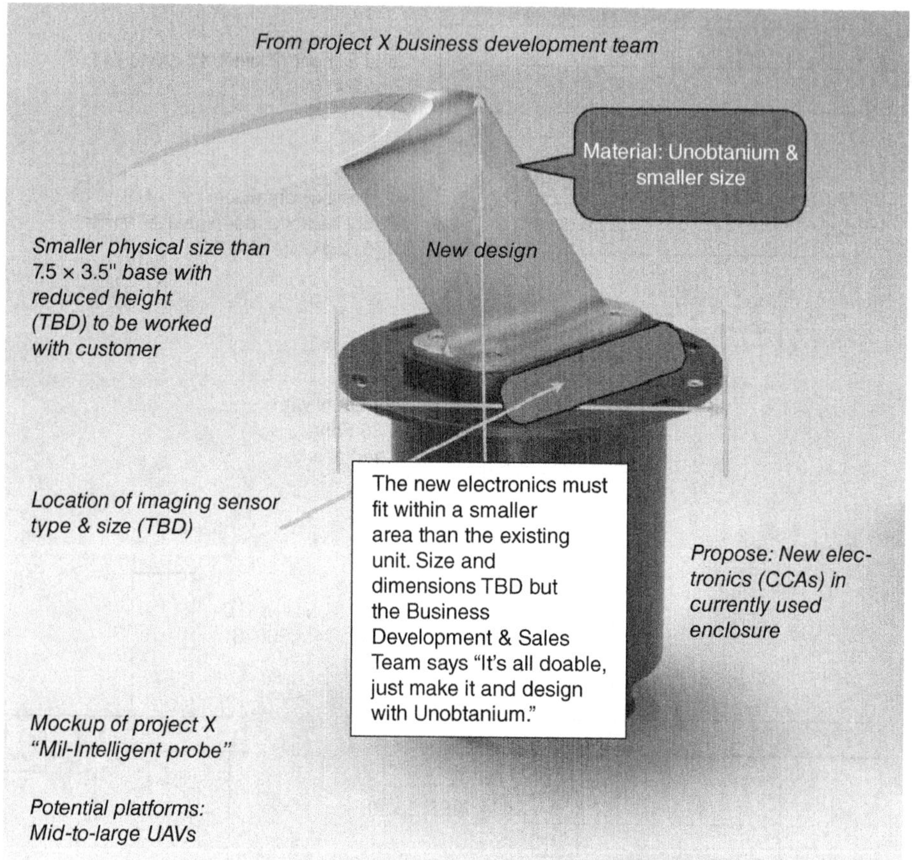

FIGURE 4.15 Business development proposed unit.

- Temperature range: –200 to +200°C. The connector shall be rated to +100°C. On the surface, this appears to be a solid requirement. However, if the vehicle is not yet defined, the concept of operations (how the system will be used) is not defined; the temperature range may change or not be adequate. Further definition will be required.

Electrical

- Current overload: a continuous applied current of (TBD) shall not damage the sensor
- Sensor will be supplied with current not to exceed 30 amps
- Output voltage shall be 28 VDC

Issues: The TBD needs to be defined. It does not appear that all of the electrical requirements are defined. What about EMI, for example?

There are no requirements defined for the addition of the imaging sensor; the circuit card assemblies (CCAs) and the entire electrical design will be affected by this. Any software (embedded or otherwise) as part of this unit?

Mechanical

- Sensor shall withstand vibration testing from (TBD) hertz to (TBD) hertz, at a maximum acceleration of 10 g's
- Pressure: each sensor shall withstand a pressure of 20,000 psig using helium gas as the pressuring medium
- Vibration requirements are incomplete and need further defining. Pressure: the 20,000 PSIG number would need to be confirmed by our engineering team. Might be some questions about packaging here as well.

CONOPS

- Sensor shall have a >98% operational readiness capability; sensor shall support day/night missions up to 12 hours
- Sensor shall be an LRU
- Sensor shall be repairable at the shop level to the circuit cardboard level

The requirements around CONOPS as given could be clarified. There may be several more hidden requirements that would affect design. What kind of resolution is required? Is it a day/night operation? What is the desired field of view (FOV) of the imaging sensor?

Test

- Acceptance testing shall consist of demonstrated conformance to the requirements of this specification.

A requirement would be that the team has additional questions around testing, especially if you do not understand the CONOPS and other environmental requirements just yet.

1. What assumptions exist?
2. Identify information gaps.
3. Are there any deliverables described? Please list internal and external if noted. Give four examples of *probing* questions you would want to ask the customer.

CHAPTER 5

Engineering Project Justification, Financial Aspects, and Return on Investment

The new product development cycle has been undergoing changes that are rapidly accelerating because of the increase in technology and competitive pressures that are rendering new products critical to the financial future of companies and making older products obsolete much faster. In this chapter, we will discuss the financial and economic analysis needed to justify new product development projects. Topics include how to satisfy the concerns of management or outside investors about the potential success of the projects as well as maintaining the company's current financial position.

In the global technology industry, the lifecycle of a typical product is different than that of historical models due to the impact of technological change. The investment in new product development is much higher than that in other industries because the rate of new product development and technological changes is much higher. These development costs tend to be the highest during the start-up phase and lowest in the maturity phase. Costs vary from about 1 to 3 percent of a company's revenues in mature products to more than 25 to 40 percent of revenue in new technology or market segments.

The financial and economic analysis tools necessary for new product development justification will be presented, with examples in design and capital equipment acquisition. In addition, techniques for increasing management confidence in the financial analysis will be discussed using three different methodologies.

The contents of this chapter are as follows:

5.1 *The Business Plan for New Products and Its Potential Impact on the Company's Strategy.* The typical global technology company strategy for maintaining growth based on new products is introduced. The information-gathering stage for new products will be explored with input from marketing. A discussion of various techniques to evaluate risks and returns on the new product development project will be shown.

5.2 *Techniques for Evaluating Projects Based on Economic Analysis*. These include the use of detailed comparative analysis to evaluate alternative projects or to maintain the historical return on investment (ROI) level. Issues of taxes, depreciation, and time value of money are discussed with examples.

5.3 *Capital Equipment Planning and Acquisition Decision*. Decisions based on economic analysis will be discussed as to whether to purchase or design and build the capital equipment in-house. Capacity analysis for new equipment based on technology-advancement issues as well as determining the proper level of estimating equipment use patterns and production requirements will be outlined.

5.4 *Techniques for Increasing Management Confidence in the Economic Analysis*. Several techniques will be discussed, using different methodologies. These analyses will reduce uncertainty in the business plan and financial outcomes for new projects.

5.1 The Business Plan for New Products and Its Potential Impact on the Company's Strategy

New products are developed in companies based on the long-range strategic and business plans for the market segment in which the company wants to operate. The new product strategy plan is dependent on the product/market lifecycle phases discussed in Chap. 2. Whether the product is in the start-up, growth, maturity, or commodity phase will greatly influence the capability, performance, price range, and timing of successive product introductions.

To formulate a new product strategy that is compatible with the rest of the company, it is important to begin with the company goals. The project manager (PM) should outline which business segment is being targeted with the new product as well as the boundaries of that segment. In addition, they should define the competitive advantage of the company: innovation and technology, cost and manufacturing excellence, quality goals, customer satisfaction, and organizational flexibility. The product strategy should include a hierarchy of elements that define the strategy over time:

- Mission Statement: A broad statement for the next five to 20 years of market and product plans.
- Intermediate Range Plan: A more detailed plan outlining goals and actions to be taken for the next three to five years.
- Product Plan: New products (performance and price) to be introduced for the next one to three years.
- Tactical Plan: Action plan for the short range of six to 18 months of product introduction(s) to accomplish long-term objectives.

The new product development process is varied with different companies depending on market requirements, competitive pressure, and the internal company strategy and methodology. Most companies have "go/no-go" decision points at various stages of the concept through the development phases, with checkpoint meetings and targets to be met and revisited. A more detailed discussion on the development phases will be given in Chap. 8.

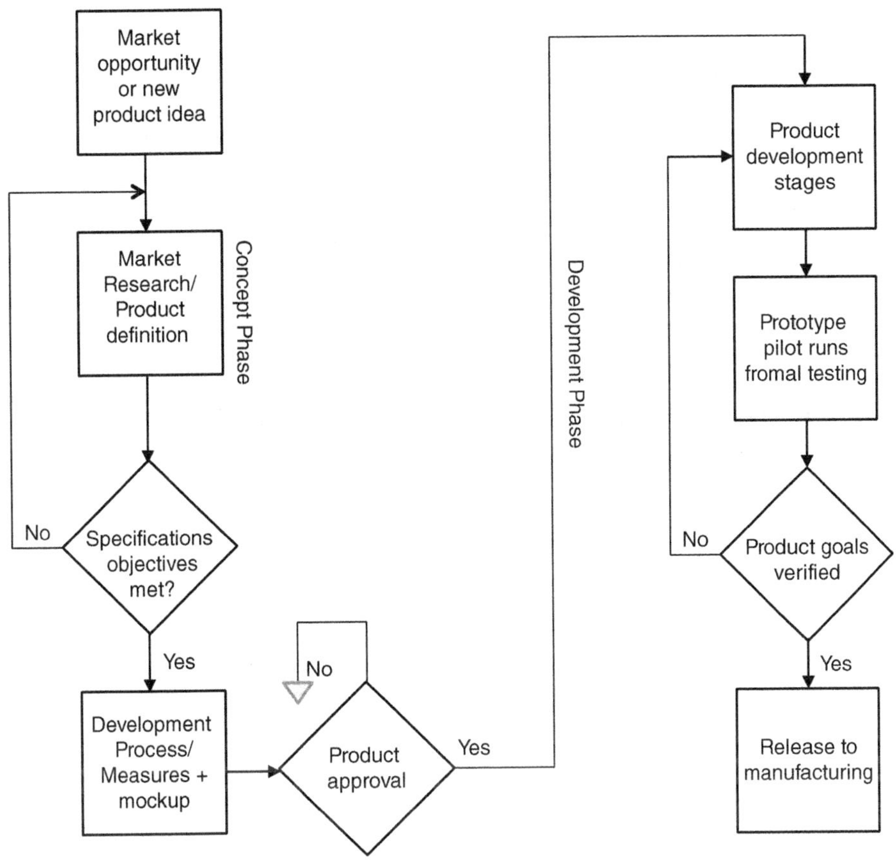

FIGURE 5.1 New product development process.

A simplified flow diagram of the new product development process is shown in Fig. 5.1. It divides the product creation process into two major phases: concept and development. For each phase, a milestone review is conducted to make sure that market and technical performance goals are met. For the concept phase, a review is made to ensure the new product specifications meet market demand through structured methods such as quality function deployment (QFD). For the development stage, a review is made after verification testing to ensure that the product meets its design goals before release to manufacturing.

An in-depth financial review should be made when deciding on go/no-go from the concept to development phase to ensure that the new product development cost and projected profit meet the company's financial goals. At this decision meeting with the company management or investors, a full business plan is needed to evaluate the new product justification at strategic, financial, and technical levels. In addition, the financial analysis should be reexamined at each of the milestone meetings during concept and development phases to ensure the product plan continues to meet the company's financial goals.

The concept stage begins with the identification or the creation of a specific project, product team, and a PM. The product idea can develop from several sources: competitive evaluations, marketing and customer surveys, and, most importantly, from a product champion. The product champion, whether he or she might be a high-ranking executive or a staff engineer on the bench, plays an important part in the introduction of new products. The champion has the vision, the focus, and the determination to carry through his or her new product ideas to fruition. The new product risks are high, but the rewards can be great. If the product champion's ideas are not implemented, he or she might leave the company to start up new ventures. An example is when Apple Computer was developed. Two engineers (Steve Jobs and Steve Wozniak) could not convince the management of Hewlett Packard Corporation to act on their ideas for personal computers.

5.1.1 New Product Opportunities in Technology Companies

These occur when the company management, marketing, or product champion discovers an opportunity in applying the company's core technology in a different product or market segment. Care must be exercised to make sure the company's product development cycle for the opportunity product is compatible with the new marketplace. In many cases, the company is operating in a regulated industry (medical or defense) or in supplying products to the research community where the development cycle is long and may find a product opportunity in the commercial market. In these instances, when technology is used from one market segment to another, the company might have trouble adapting from design to customer (DTC) to design to market (DTM), with many of these issues discussed in the previous chapters. They include the following:

- Identifying proper new product specifications
- Determining new product price
- Developing the proper level of technology adoption in the new product
- Estimating new product sales and preparing for large sales volumes if the product is widely accepted or will be sold in high-volume consumer sectors such as computer, telecommunications, or auto industries
- Managing the needed decrease of new product development time and readiness to counteract competitive measures
- Transitioning to faster new product introductions and the obsolescence of older products
- Releasing new products to customers who do not tolerate quality defects or reliability issues such as recalls and rework, even if a high-technology adoption level is used

In many cases, the company might spin off the new product opportunity organization into a separate business to better focus on the new market environment, rather than keep two separate product cultures and organizations in-house. It might also decide to be acquired by a company more skilled at operating in the new product opportunity market or, inversely, to acquire a company in that market to learn from its experience. The company has to carefully investigate the potential financial returns in the new product opportunity and make sure it is compatible with its core business in profits and specified market segments.

There are many of these examples of technology product opportunities, and a review of some of these can be found in many technology companies growth history, thus:

- A vacuum-leak sniffing technology using nitrogen detection in the commercial market can be used in medical applications to measure lung oxygen absorption efficiency.
- A radar microwave technology for military use can be used in new automobile industry applications such as active braking or automatic parallel parking.
- A large and instant data-acquisition system for particle research can be used for Internet diagnostic applications.

5.1.2 Collecting Data for the Business Plan

Once the new product opportunity is identified, an iterative process begins to take shape in order to further refine, clarify, and specify the product definition. This phase is enhanced by market research, surveys, and economic justifications. Once the process is completed, the development and performance measures of the product are identified in terms of product specifications, potential revenues, costs, product lifecycle, and impact on existing company products.

The results of this iterative process are sometimes built into a business plan. The plan is the blueprint for developing, manufacturing, and selling a new product to the marketplace. It is imperative for all companies to develop business plans for all new products.

Elements of the business plan are:

- The market analysis for the market segment targeted by the product, in terms of market development stage, competitive analysis, and potential sales volume
- The marketing strategy in penetrating the target market, whether to compete on price, features, performance, or quality
- An estimate of the total potential customer base, the percentage acceptance, and penetration of the new product into that customer base
- A target sales price based on customer expectation of new product satisfaction and customer productivity increases as well as competitive factors
- The development plans in terms of the chosen technology, architecture, staffing, equipment, tooling, and material requirements
- The manufacturing plan on how and what is required to produce the product: the supplier strategy, fabrication and assembly processes, equipment acquisition, test strategy, and quality plans
- The product support plan in terms of field support and training, product repairs, and warranty strategy and impact on support for existing products
- The financial analysis and projected ROI for the new product: the product development costs, the expected manufacturing and support costs, and the warranty and service levels as well as the economic impact on existing products

In addition to the business plan, the development of an early prototype or mock-up for the product is important to demonstrate the product idea or feasibility to the management and give potential customers a chance to comment on the utility of the product. There are various rapid prototype techniques using physical mock-ups, software for screen generators, and command and transaction modeling. The advent of advanced computer aided design (CAD) with rendering capabilities and rapid prototyping techniques such as three-dimensional (3D) printing and stereo lithography (SL) can produce a 3D image of the product on a computer screen or in a physical shape. 3D printing and SL are called direct digital manufacturing (DDM) and usually offer less dimensional accuracy than machined or plastic-injected parts in the final product.

Although there are no definite rules as to the correct method of ensuring that a product is accepted from concept into development, many factors predominate: the current financial status of the company, whether the management is willing to take a chance at this particular time, the credibility and previous track record of the project team and the PM, and the current competitive position in the industry.

A detailed financial analysis is needed to support the go/no-go decision, especially for product proposals in the architecture/platform or evolutionary phases of the product process lifecycle discussed in Chap. 2. The financial analysis should contain the following details about the new product:

- Market size and requirements defined
- Specifications and definitions of the new product
- Release schedule to timely meet market and customer needs
- Chosen technology and product architecture are carefully weighed and decided
- Technical feasibility demonstrated through a working model or prototype
- Planned levels of price, performance, and reliability investigated and decided
- Adequate project ROI achieved

It is understood that a certain percentage of the products do not go into development at this point, even if the product idea is sound, because of the company's current financial record or competitive situation and sometimes because of poor preparation by the PM. The company might be anticipating a particular ROI, which is in line with the financial conditions in the industry, but would tolerate a lower rate in the hope of landing a stellar performing product in the future.

Once the go-ahead is given from concept to development, another iterative process begins: that of prototype and early production models' documentation, building, testing, and verification. The number of iterations is important to the overall schedule. Some elements such as very large scale integration (VLSI) and plastic-parts tooling can take a long time to fabricate, so that reducing the number of iterations has become an extremely important part of the design process.

The release to manufacturing is usually a highly formalized process of meeting specific internal as well as external specifications, especially in regulated industries such as medical and defense. At the release checkpoint meeting, usually a long list of items is covered to ensure that the soon-to-be-released product meets all specifications and internal and external requirements. More information about engineering project milestones is available in Chap. 8.

5.2 Techniques for Evaluating Projects Based on Economic Analysis

The success of the product strategy will be measured ultimately through the profitability of the new product being developed. In addition to the expected sales volume and the duration of the product lifecycle, two major cost elements determine that degree of success: the cost of the development project and the production cost of each product. Together these elements contain a certain amount of guesswork. In later sections in this chapter and in the following chapters, the techniques for minimizing risk and increasing the management confidence in the project proposals and business plans are discussed.

Several techniques are commonly used to evaluate the economic viability of a project proposal or business plan. They can be used to compare different project proposals against each other in order to proceed with the ones that have the best ROI. This comparison is important, since project proposals can have different development costs, profits per unit sold, and product lifecycles. The project proposals are competing for the development resources, either from the company's limited development budget or from outside investors. They include many different techniques that are normally discussed in the discipline of engineering economy. In technology companies, only a few of these techniques predominate, since they differ from the traditional product companies in the following ways:

- New product project development costs are higher because most technology companies operate in the start-up or growth phase of the product lifecycle
- Product lifecycle is shorter because of need to incorporate more technology adoption in each product cycle
- Product lifecycle is also variable due to unknown competitive product introduction cycles and the high possibly of disruptive products
- Product manufacturing and material costs are unstable due to rapid technology development that can bring costs down (Moore's law)
- Material variances based on unforeseen shortages due to unplanned demand from other competitors or technology sectors, natural disasters or uncertain political climate in supplier countries can greatly affect material prices and availability
- Monolithic suppliers that capture the majority of the market can obsolete parts without prior notice

Technology companies have historically generated a higher rate of return than do traditional companies and therefore have historically maintained a higher valuation in the financial markets. When some of these companies are listed on stock exchanges, they maintain a zero or little dividend in order to retain earnings or profit for future development projects rather than distribute these earnings to their shareholders. They also encourage or partially fund their employees into the company's stock purchase plans. The message to the stakeholders is that they can get better financial rewards by investing in the company's future.

It is important for these companies to maintain these stakeholder expectations by continuing to develop products that maintain their financial position. This can be achieved in carefully balancing product development projects by funding a combination of revolutionary as well as evolutionary products. The evolutionary products can be the *cash cow* that generates level and steady income, as in an experienced baseball player that hits mostly singles. Many revolutionary products can be funded, with the expectation that only a few can succeed, generating a much higher rate of return. Using the baseball analogy, the company would like a few home runs as well as many singles.

The techniques for economic evaluation of technology products tend to be less structured and more simplified. They include the following:

1. Return Factor (RF): A simple comparison technique that rates a project on expected return versus development investment; it is also called benefit cost ratio or (B/C)
2. Payback Period: A time-dependent analysis that is not discounted to calculate the number of years it takes to pay back or recover the development investment
3. Internal Rate of Return (IRR), aka Return on Investment (ROI): This is to indicate a time-dependent and discounted analysis based on time value of money as well as tax consequences of the product sales profits

These techniques have been historically used to measure project development investments for two major reasons: to ensure the investments have an adequate rate of return in line with the traditional return for the business segment the company is operating within and to select from competing projects due to the finite financial resources of the company.

5.2.1 Return Factor or Benefit/Cost Ratio Calculations

The RF calculations are simple:

$$\text{RF} = \frac{\text{Lifecycle Product Operating Profit}}{\text{New Product Development Costs}}$$

The top number is the expected lifecycle profit generated from selling the new product, which is based on the expected income from selling price of all products into the future versus the new product development cost.

Most new technology products have a low percentage of manufacturing costs versus selling price. The top number is basically an estimate of the number of units sold multiplied by the profit of each unit. In certain situations, the company might decide on a loss-based selling price of the new product in order to leverage this new product for profit generated from other income streams. Examples are Polaroid cameras that sold at or below cost to facilitate the selling of Polaroid film or the Amazon Kindle being sold for lower price to encourage the online selling of e-books.

At this level of new project cost estimation, the new product profit is based on several assumptions:

- The product selling price will remain constant during the product lifecycle.
- The impact of the new product on other products in the same family is undetermined.
- The lifecycle duration is assumed typical of the product family based on technology adoption.

- The product cost is usually estimated from a simple multiplier of material costs, since material costs are easier to estimate than labor costs, especially when the product is in the planning stage.

The new product development costs include the overall cost of the engineering development project for the new product:

- Salaries of the project teams and PMs
- Direct project charges and expenses such as tooling (software or hardware)
- Expenses for prototype parts and the model shop project expenses
- Material purchase and direct labor expended in building prototypes; some of these costs could be recovered in the future since prototypes can be used for non-customer sales such as field demonstration units or training beds for manufacturing or sales support

The RF factor is a simplified first-cut trial to determine the viability of a new product proposal. It should not be used to compare alternative project proposals, as it does not account for variances due to lifecycle time durations of the alternative projects. Technology companies have different levels of expected RF, some with 10 as the minimum needed to approve a new project. The RF factor should be recalculated at each development phase completion to ensure that the project remains economically viable. If the estimated RF falls below the company's minimum value, then the project should be reexamined and project objectives altered. Otherwise, the project maybe cancelled, even if the company has sunk major funding into the project to date. Each financial decision has to stand alone at the time the decision is made, regardless of project history.

5.2.2 Payback Period Calculations

These are used to determine the time needed to recover initial development investment. These calculations are also simple:

$$\text{Payback Period} = \frac{\text{New Product Development Costs}}{\text{Lifecycle Product Operating Profit Per Year}}$$

The major difference with the RF method is that the time element is present in this calculation, so that alternative project proposals can be compared. There is a similar floor applied by companies to the payback period of new projects. In some technology companies, the time period might have to be measured in months as opposed to years. This calculation is not discounted, indicating that difference in the value of money today or into the future is not calculated. This difference is due to inflation expectations based on the current level of interest rates.

5.2.3 Internal Rate of Return (aka Return on Investment)

This is the most accurate of the three economic measures discussed above. It compares competing projects based on a generated theoretical interest rate (ROI) that best approximates the return of the product development investment. It is similar to the two preceding measures in that it can provide a guide to the PM to indicate the minimum

ROI needed for their new project proposal. It can also be used as a tool to select alternative project proposals competing for the development budget. ROI is a measure used to find an interest rate that equates the cash flow for the lifecycle of the product versus its development costs.

One of the important concepts for ROI calculations is the time value of money. Since year after year, inflation erodes the value of money, future dollars are reduced in terms of today's dollar-purchasing value. Alternately, dollars spent in the past have more value or purchasing power than dollars spent today. The amount of money devaluation for the future or evaluation from the past is measured by the interest rate at the time of the project proposal in a simple formula:

$$FV = PV * (1 + i)^n$$

where, FV = Future Value
PV = Present Value
i = Interest Rate
n = Number of Years for the Product Lifecycle

alternatively,

$$PV = FV/(1 + i)^n$$

The terms used in these formulae are labeled differently in engineering economy literature. Present value (PV) is sometimes called present worth (PW), or net present value (NPV), which factors in yearly costs as reducing yearly income from sales. The i in the formulae is sometimes referred to as investment interest rate or cost of capital rate, as compared with the IRR. These terms are interchangeable in broad definitions, although narrower definitions could be obtained by researching the literature.

Another term used in engineering economy is the annualized form of present or future values. This allows for comparing projects in present (P), future (F), or in annual (A) terms. In addition, annual terms can help in visualizing the project economic analysis in terms of arrows pointing up for positive yearly cash flows (profits from sales) versus arrows pointing down for negative yearly cash flows such as losses or project expenses. Present or future single investments can be shown as arrows at the time of occurrence, pointing down since investments are negative cash flows. The formulas for the relationship between annualized and present/future terms take into account the ROI (i), and the number of years (n) the project will yield positive returns.

The conversion from/to these terms is shown as P given A or P/A. There are six (6) sets of conversion units: P/A, F/A, P/F, F/P, A/P, and A/F. They are available as tables in the index of this book.

The ROI of any investment can then be shown as the point where the PV of the costs = PV of the benefits. While there are many applications on the Internet or commercial programs to calculate ROI, they involve a procedure using the following steps:

1. Estimate the development or investment costs and the potential returns (benefits) from this investment in yearly or other time period events
2. Begin with an initial ROI i value and determine if the annualized (A) returns match the initial investment
3. Continue to iterate the value of i until the PV of the investment = PV of the annualized returns

In some cases, the tax consequences of the investment decision might have to be taken into account. Investments such as capital equipment are considered as depreciable assets and hence can be depreciated as they are used up, with different schedules. The depreciation can reduce corporate taxes on benefits and therefore increase the rate of return of the equipment. Example II is given below to show the tax consequences in project economic analysis.

If the investment has a residual value, such as a book value of capital equipment needed to keep the equipment active on the books for accounting reasons, then the future value of the equipment has to be brought from the FV into a PV and included into the ROI calculations. Example III is given below to show tax consequences and book values in project economic analysis

Example I. Non depreciation PV Analysis
A company is considering designing a new rotary counterbalance system for windows instead of the traditional linear spring for $5 profit per rotary spring window. The company needs a $1,000,000 (1M$) investment to develop the rotary counterbalance window system. The company then assumes a rate of return of $i\% = 5\%$. It estimates the market potential at 200,000 units/year for 10 years. Its options are: (1) get a 1M$ loan immediately, develop the rotary spring technology in one year, and then sell the units at $5 profit/unit and repay the loan one year after sales start for five years at $250,000/year; (2) get a major spring manufacturer to license the technology now, sell rotary spring windows immediately, and then pay them royalties of $1 per unit for 10 years after one year. Assume no tax/depreciation consequences.

Solution for Option 1, Example I Plot the timing diagram for investment and benefits for option 1: Get a 1M$ loan immediately, develop the rotary spring technology in one year, then sell the units at $5 profit/unit and repay the loan one year after sales start for five years at $250,000/year.

The arrows indicate the net of the period (years) in annualized benefits (arrows up) or expenses (arrows down) at the completion of the period. The initial investments are shown at the beginning of the project cycle. If the investment is from a loan, then the loan is immediately spent at the end of the period and is shown as a zero net value.

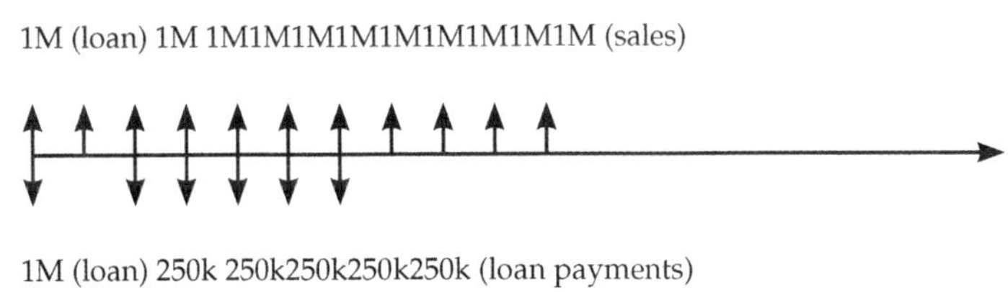

The PV Calculations Option 1

The annualized return will have to be converted back to the present (*P/A*) using discounted cash methods. Since the benefits occur some year(s) after the initial loan and last for several years hence, they have to be calculated for full extent of the project then brought back by subtracting the initial one year of no benefits spent on developing the product. The 10-year benefit cycle starts 10 years after loan start and therefore is calculated at 11 years minus the first year. The *P/A* formula only works as a continuous calculation starting from the first year to year (n). Similarly, the loan repayment starts one year after completion of the project and two years after the initial loan start of the project cycle. The loan payments start on the second year and end at the seventh year. The ROI calculations can be shown as:

PV = ($1,000,000, P/A, 5%, 11) – (1,000,000, P/A, 5%, 1) –250,000 (P/A, 5%, 7) + 250,000 (P/A, 5%, 2)
= 1,000,000 * 8.306 – 1,000,000 * 0.952 – 250,000 * 5.786 + 250,000 * 1.8594
= 8,306,000 – 952,000 – 1,446,500 + 464,850 = $6,372,350

Solution for Option 2, Example I Plot the timing diagram for investment and benefits for option 2. Get a major spring manufacturer to license the technology now, sell rotary spring windows immediately, and then pay them royalties of $1 per unit for 10 years after one year.

1M 1M1M1M1M1M1M1M1M1M (sales)

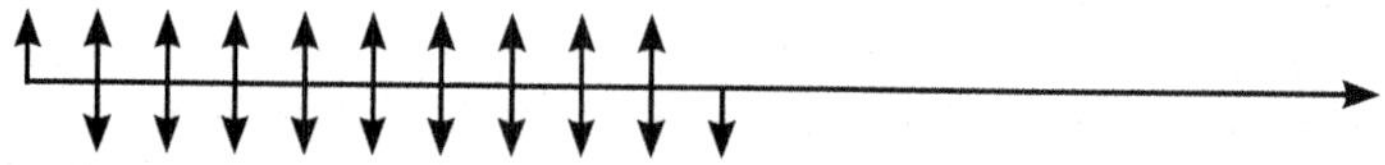

200k 200k200k200k200k200k200k200k200k200k (royalties)

The PV calculations for option 2 are similar. Care must be exercised to maintain the yearly benefits and loan timing. The cash flow is $1 million profit for the first year, $800,000 profit for the next nine years, and $200,000 loan payment for the tenth year.

$$PV = 1{,}000{,}000 + (\$800{,}000,\ P/A,\ 5\%,\ 10) - (\$800{,}000,\ P/A,\ 5\%,\ 1) - (\$200{,}000,\ P/F,\ 5\%,\ 10)\ PV$$
$$= 1{,}000{,}000 + 800{,}000 * 7.722 - 800{,}000 * 0.952 - 200{,}000 * 0.6139$$
$$= 1{,}000{,}000 + 6{,}177{,}600 - 761{,}600 - 122{,}780 = \$6{,}293{,}220$$

Option 1 (loan) is a better choice for maximizing revenue.

Example II. Rate of Return Analysis with Depreciation and Tax Considerations
A company is making consumer (high volume) and industrial (low volume) lines of microwave ovens using the same technology and controls. The company decided to sell the industrial microwave line to focus on consumer products. Your company (IN-microwave) is considering buying the line for $500,000. IN-microwave company depreciation method is straight-line, its tax rate is 20 percent, and marketing thinks it can sell 12 units per year at a profit of $10,000/unit for five years. Assume no book value remaining after five years. Minimum starting ROI for IN-microwave, being an industrial business, is 4 percent.

Perform a complete economic analysis using the three methods for economic evaluation of projects.

Solution to Example II:

1. $$RF = \frac{\text{Lifecycle Product Operating Profit}}{\text{New Product Development Costs}}$$

   $RF = 12 * 10{,}000 * 5/500{,}000 = 600{,}000/500{,}000 = 1.2$

2. $$\text{Payback Period} = \frac{\text{New Product Development Costs}}{\text{Lifecycle Product Operating Profit Per Year}}$$

   Payback Period $= 500{,}000/120{,}000 = 4.167$ years (payback period is not discounted)

3. ROI Analysis with Depreciation and Tax Consequences

In this case, the purchased equipment can depreciate as an asset. The depreciation amount is considered a loss and therefore can be deducted from profits before paying taxes. Many methods are available to calculate a depreciation schedule, the simplest being a straight-line depreciation, which is a division of the value of the asset over the lifecycle of the asset use.

$$\text{Straight-Line Depreciation} = \frac{\text{Value of the of the Asset (\$)}}{\text{Lifecycle of the Asset (Years)}}$$

Year End	Before Tax Cash Flow	Depreciation	Taxable Income	Taxes (20%)	After Tax Cash Flow
0	–500,000				
1	120,000	–100,000	20,000	4000	116,000
2	120,000	–100,000	20,000	4000	116,000
3	120,000	–100,000	20,000	4000	116,000
4	120,000	–100,000	20,000	4000	116,000
5	120,000	–100,000	20,000	4000	116,000

TABLE 5.1 ROI Analysis for Example II

The income or operating profit can be reduced by the amount of the depreciation and is called taxable income. Taxes are then deducted, and the ROI economic analysis can be performed based on after tax income.

Taxable Income = Cash Flow (Operating Profit or Income) – Depreciation
Taxes = Taxable Income * Current Level of Corporate Income Tax Rate
After Tax Income = Taxable Income –Taxes

For current company operating procedures for depreciation schedules and income tax rates, the PM should consult the finance department in the company. For the ROI of this problem, assume a corporate tax rate of 20 percent and a straight-line depreciation method.

The ROI analysis usually involves a year-by-year project investment versus income analysis arrayed in a table. Once the table is arranged as shown in Table 5.1, an iterative process begins with an initial ROI estimate and then proceeds until the PV of the investment minus the PV of After tax income for the lifecycle of the product is equal to zero. The ROI analysis is given below:

Start with $i = 5\% \rightarrow PV = -500{,}000 + 116{,}000 * (P/A, 5\%, 5)$

$= -500{,}000 + 116{,}000 * 4.329 = \$2164$

$i = 6\% \rightarrow PV = -500{,}000 + 116{,}000 * (P/A, 6\%, 5)$

$= -500{,}000 + 116{,}000 * 4.212 = -\$11{,}408$

The i% for this project is somewhere between 5 and 6 percent. The rate of 5 percent is generating a higher income in present-day dollars to offset the initial investment, while the 6 percent does not generate enough income in today's dollars. In other words, if the assumed rate is 6 percent, then the annualized value is worth less in today's dollars than 5 percent rate. Using a prorated average:

$2164/(2164 + 11,408) = 0.16\%$

Then $i\,\% = 5.16\%$ ROI

Alternately, the P/A ratio that results in PV = 0 can be calculated:

$500{,}000/116{,}000 = 4.31$, which is in the middle of 4.329 and 4.212

Extrapolating into the P/A ratios of 5 and 6%;

$5\% + (4.329 - 4.31)/(4.329 - 4.212) = 5 + 0.019/0.117 = 5.16\%$

Example III. Rate of Return Analysis with Depreciation, Including Book Value and Tax Considerations A company is making oxygen (O_2) gas-monitoring equipment for operating rooms to ensure that patients undergoing operations under anesthesia are breathing normally. It was suggested that perhaps developing CO_2 gas-monitoring equipment with new CO_2 sensors would be much less expensive and therefore result in higher sales and profits. The market is 400 major hospitals in the United States and 600 other hospitals worldwide, each with five operating rooms on average. Marketing expects this product to capture 25 percent of the operating rooms during its lifetime of five years, selling equally for each of the five years. You are assigned as the PM for the CO_2 sensor machine development project. The PM estimates the senor machine project total development costs to be $1M, with a profit of $1000 per unit for a five-year life. At the end of the product lifetime after five years, the machine will have a book value of 10 percent of the original machine cost of $1M. Show ROI analysis with straight-line depreciation and remaining book value, using $i = 5\%$, and the corporate tax rate is 20 percent.

Solution to Example III In this case, the depreciation schedule will be reduced by the book value. This is required by the IRS in order to maintain the remaining value of the operating machines active in the company records. The book value will have the effect of altering the depreciation calculations. In the economic analysis, the book value will have to be brought back to the PV through the P/F factor conversion.

$$\text{Straight-Line Depreciation} = \frac{\text{Value of the Asset (\$)} - \text{Book Value (\$)}}{\text{Lifecycle of the Asset (years)}}$$

Sales Estimate Based on Marketing Forecast = (1000 Hospitals * 5 Operating Rooms Average * 0.25)/5
= 250 Sales/Year

Profits (Income)/Year = 250 * $1000 = $250,000 Profits/Year

1. RF = 1,250,000 ($250,000 * 5)/1,000,000 = 1.25 (RF Not Discounted)
2. Payback Period = 1,000,000/250,000 = 4 Years (Payback Period Is Not Discounted)

Yearly Depreciation = ($1,000,000 – 100,000)/5 = $180,000

3. ROI Table for Example III is Shown in Table 5.2

Start with $i = 5\% \rightarrow PV = -1{,}000{,}000 + (236{,}000, P/A, 5\%, 5) + 100{,}000\ (P/F, 5\%, 5) = \text{set to} = 0$

$i = 5\% \rightarrow -1{,}000{,}000 + 236{,}000 * 4.329 + 100{,}000 * 0.7835$

$= -1{,}000{,}000 + 1{,}021{,}644 + 78{,}350 = \$99{,}994$

$i = 6\% \rightarrow PV = -1{,}000{,}000 + (236{,}000, P/A, 6\%, 5) + 100{,}000\ (P/F, 6\%, 5) = \text{set to} = 0$

$= -1{,}000{,}000 + 236{,}000 * 4.212 + 100{,}000 * 0.747$

$= -1{,}000{,}000 + 994{,}032 + 74{,}700 = \$68{,}732$

Year End	Before Tax Cash Flow	Depreciation	Taxable Income	Taxes (20%)	After Tax Cash Flow
0	–1,000,000				
1	250,000	–180,000	70,000	14,000	236,000
2	250,000	–180,000	70,000	14,000	236,000
3	250,000	–180,000	70,000	14,000	236,000
4	250,000	–180,000	70,000	14,000	236,000
5	250,000	–180,000	70,000	14,000	236,000
Book Value	100,000				

TABLE 5.2 ROI Analysis for Example III

$i = 7\% \rightarrow PV = -1{,}000{,}000 + (236{,}000, P/A, 7\%, 5) + 100{,}000\ (P/F, 7\%, 5) = \text{set to} = 0$

$= -1{,}000{,}000 + 236{,}000 * 4.1 + 100{,}000 * 0.713$

$= -1{,}000{,}000 + 967{,}600 + 71{,}300 = \$38{,}900$

$i = 8\% \rightarrow PV = -1{,}000{,}000 + (236{,}000, P/A, 8\%, 5) + 100{,}000\ (P/F, 8\%, 5) = \text{set to} = 0$

$= -1{,}000{,}000 + 236{,}000 * 3.993 + 100{,}000 * 0.68$

$= -1{,}000{,}000 + 942{,}348 + 68{,}000 = \$10{,}348$

$i = 9\% \rightarrow PV = -1{,}000{,}000 + (236{,}000, P/A, 9\%, 5) + 100{,}000\ (P/F, 9\%, 5) = \text{set to} = 0$

$= -1{,}000{,}000 + 236{,}000 * 3.89 + 100{,}000 * 0.65$

$= -1{,}000{,}000 + 918{,}040 + 65{,}000 = \$38{,}900 = -\$16{,}960$

The ROI is between 8 and 9 percent. Extrapolating the answers from 8 and 9 percent:

$$10{,}348/(10{,}348 + 16{,}960) = 0.38$$
$$ROI = 8.38\%$$

The alternate method for calculating another ROI using the P/A ratio in the previous example is difficult because of the presence of the book value.

5.3 Capital Equipment Planning and Acquisition Decision Based on Economic Analysis

In technology companies, capital equipment purchases should always be tempered by analyzing the purchase versus the subcontracting option. This analysis should be undertaken in the case of purchasing specialized capital equipment for manufacturing the new product versus designing the equipment in-house. In addition, economic analysis should be used when deciding whether to manufacture the product in-house by purchasing standard capital equipment or subcontracting the manufacturing or assembly work. In this chapter, only the economic analysis will be performed. Other considerations of the build in-house-versus-subcontracting decision will be discussed in the next Chap. 6.

The capital equipment purchase decision is difficult to make due to the many different alternatives of capability and performance of the equipment. Decision-analysis tools such as QFD or weighted criteria, which were discussed in Chap. 3, can be used in the equipment-purchasing decision. The important steps to consider when making a purchase-versus-in-house decision on capital equipment could be summarized as follows:

- Use decision-analysis tools to select and reconcile different equipment, capabilities, and performance as well as different process methodologies
- Consider the purchase of capital equipment for increased in-house manufacturing capability that can be shared with other products in the business units of the company
- Technology impact could reduce the useful lifecycle and increase obsolescence risks of new capital equipment; consider leasing or a trade-in clause in capital equipment acquisition
- Consider using specialized capital equipment as a competitive factor in reducing product cost

- Compare resources spent on proposed capital equipment versus resources that can be spent on new development projects; subcontracting costs are expended as sales are generated, without the need for initial large capital expenditures

5.3.1 Capacity Planning for Capital Equipment

Planning the capacity for new plants or equipment to satisfy new product manufacturing requirements is an important concern for PMs in technology companies, given the short lifecycle of their products. The issues of technology obsolescence, regulatory changes, product type, and customer changes over time are important considerations when building new plants or purchasing capital equipment. In addition, planning for incremental capacity increases to match longer-term expected demand is necessary when designing an equipment-procurement strategy over time. The PM has to consider all of these elements in order to maximize the ROI to benefit the company for capital equipment purchases.

The first step in the project-planning stages for acquiring a new plant or equipment is to contact marketing to obtain the long-term forecast for the product mix and new product generations being planned during the lifecycle of the equipment. The forecast has to be broken down based on the information given. If the forecast is in sales dollars, it has to be translated into number and types of new products to be sold. A multiplier or *driver* of fabricated or assembled parts per sales dollar, or number of future products to be sold, can be developed to plan for future capacity needs. This would be followed by evaluating the technology requirements and advances to be adopted into the new product that can alter the capacity profile. For example, in the electronics industry, it has been shown that the need for electronic assemblies and components do not grow in a linear fashion with the sales, due to increased package densities in printed circuit board (PCBs) and very large scale integrated (VLSI) components. Mechanical parts or assemblies could be substituted with electro-mechanical parts in the future, changing the capacity needs of these parts.

Once these factors are considered, an initial capacity target can be identified, as indicated in Fig. 5.2, capacity planning over time. Figure 5.2 shows a progression of capacity acquisitions over time. An initial purchase of plant and equipment can be made, and then second and successive machines can be acquired to incrementally increase capacity.

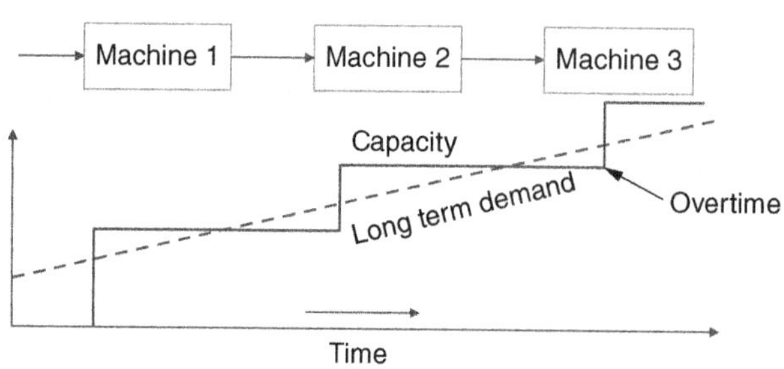

FIGURE 5.2 Capacity planning over time.

The timing of the machine purchases can be modified using overtime to cover the gaps in the demands-versus-capacity curve.

In most cases, capacity planning can be modified by identifying a single machine in a production line that is the limiting item for capacity. In addition, doubling capacity for plants or equipment does not necessarily mean doubling the cost. Regulatory or environmental issues such as treating the plant's industrial-waste water or the exhaust of plant equipment into the atmosphere require treatment plants for approval. Equipment should be sized for the ultimate-demand capacity. Time should be allowed to investigate and develop treatment options and obtain all of the necessary permits. Capacity planning requires the following steps:

- Obtain long-term business plan for the company or long-term marketing forecast
- Convert long-term forecast from sales dollars into product mix of new product types and product generations
- Attach a *driver* of fabricated or assembled parts per sales dollar or number of future products to be sold
- Evaluate the technology requirements and advances to be adopted into the new product
- Care must be taken to ensure that the *driver* relationship between future product sales and fabrication or assembly needed to make that product is not necessarily linear and can change depending on technology advances
- Mechanical parts or assemblies could be substituted with electro-mechanical parts for future products
- Capacity can be increased in timely increments and planned with new equipment purchases over time
- Capacity squeezes can be handled with overtime in the short term
- Usually a single machine in a manufacturing line or cluster determines the capacity of the line
- Doubling capacity does not mean doubling the cost of the plant or equipment
- Regulatory or environmental issues have to be considered for the time needed to research alternative-treatment technologies; permits have to be obtained, and the treatment process has to be scoped for the ultimate plant and equipment capacity

Another important element of the capacity planning project is to maximize the manufacturing-process operation over the capital equipment acquisition. By carefully planning capital equipment acquisition and utilizations, several optimum manufacturing cost points at several levels of production volumes can be achieved. This careful planning can prevent under utilization of equipment leading to cost increases at low production volume. Figure 5.3 shows the different strategies of capital equipment acquisition, with the process cost per component ($/c) optimally designed at three points of production (V_1, V_2, and V_3) corresponding to three levels of cost (X_1, X_2, and X_3) in the figure at right. If only one process design point was selected, then there would be an area of increase of the product cost due to under utilization of the capital equipment, as shown in the figure at left.

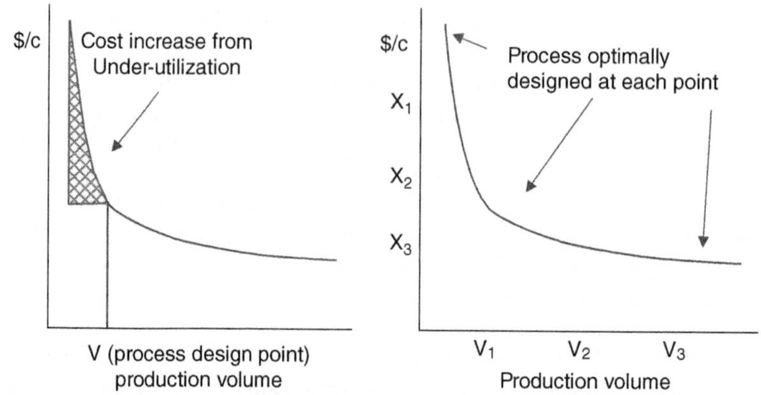

FIGURE 5.3 Different strategies of capital equipment acquisition for manufacturing.

Example IV Capital Equipment Planning Over Time

A company is considering buying a plastic injection mold tool and has two options: a two-cavity mold at $40,000 or a four-cavity mold at $70,000. It is expected that each mold will last 100,000 shots and will have to be replaced with no book value. The company is expected to sell 40,000 parts/year at $0.25 profit per piece. Use ROI techniques to determine which mold the company should buy, assuming an interest rate of 5 percent, a tax rate of 33 percent, and straight-line depreciation for the life of the mold.

Assume that the company decided to select one of these options: Option 1, buy one two-cavity mold now and a second two-cavity mold five years from now or, Option 2, buy one four-cavity mold. Show the ROI of sales from the two molds and suggest the best financial option.

Solution for Option 1, Example IV The two-cavity mold will make 200,000 parts lifetime and will last five years and make $10,000 profit/year. The ROI analysis is shown in Table 5.3.

Year End	Before Tax Cash Flow	Depreciation	Taxable Income	Taxes	After Tax Cash Flow
A	B	C	D = B + C	E = –0.333 * D	F = B + E
0	–40,000				
1	10,000	–8000	2000	–666	9334
2	10,000	–8000	2000	–666	9334
3	10,000	–8000	2000	–666	9334
4	10,000	–8000	2000	–666	9334
5	10,000	–8000	2000	–666	9334
5	-40,000				
6	10,000	–8000	2000	–666	9334
7	10,000	–8000	2000	–666	9334
8	10,000	–8000	2000	–666	9334
9	10,000	–8000	2000	–666	9334
10	10,000	–8000	2000	–666	9334

TABLE 5.3 ROI Analysis, Option 1, Example VI

PW try 1; at $i = 4\% \rightarrow -40{,}000 - 40{,}000\ (P/F, 4\%, 5) + [9334(P/A, 4\ \%,10)]$
$= -40{,}000 - 31{,}340 + 9334\ (8.11) \cong \$ + 4358$

PW try 2; at $i = 5\% \rightarrow -40{,}000 - 40{,}000(P/F, 5\%, 5) + [9334(P/A, 5\%,10)]$
$= -40{,}000 - 31{,}340 + 9334(7.72) \cong \$ + 718$

PW try 3; at $i = 6\% \rightarrow -40{,}000 - 40{,}000\ (P/F, 6\%, 5) + [9334\ (P/A, 6\%,10)]$
$= -40{,}000 - 31{,}340 + 9334\ (7.36) \cong \$ - 2642$

ROI = 5.4% (prorated between 5 and 6%)

Solution for Option 2, Example IV The four-cavity mold will make 400,000 parts and last 10 years and make $10,000 profit/year

$$0 = PW\ (-70{,}000) + 10{,}000\ (P/A, i\%, 10) = -70{,}000 + 10{,}000 * 7 \rightarrow i = 4.9\%.$$

The ROI analysis is shown in Table 5.4.

PW try 1; at $i = 4\% \rightarrow -70{,}000 + [9001(P/A, 4\%, 10)] = -70{,}000 + 9001(8.11) \cong \$ + 2998$

PW try 2; at $i = 5\% \rightarrow -70{,}000 + [9001(P/A, 5\%, 10)] = -70{,}000 + 9001(7.72) \cong \$ - 512$

ROI = 4.9%

Therefore, the two-cavity machine is a better economic choice.

5.3.2 Capacity Planning for Capital Equipment in the Electronics Industry

The electronics industry presents a challenge in capital equipment planning and acquisition. The marketing forecast is a tool for calculating future capacity of capital equipment. It can be represented in dollars or new product sales. However, it will

Year End	Before Tax Cash Flow	Depreciation	Taxable Income	Taxes	After Tax Cash Flow
A	B	C	D = B + C	E = −0.333 * D	F = B + E
0	−70,000				
1	10,000	−7000	3000	−999	9001
2	10,000	−7000	3000	−999	9001
3	10,000	−7000	3000	−999	9001
4	10,000	−7000	3000	−999	9001
5	10,000	−7000	3000	−999	9001
6	10,000	−7000	3000	−999	9001
7	10,000	−7000	3000	−999	9001
8	10,000	−7000	3000	−999	9001
9	10,000	−7000	3000	−999	9001
10	10,000	−7000	3000	−999	9001

Table 5.4 ROI Analysis Table, Option 2, Example VI

have to be translated into either PCBs or components used in manufacturing. The use of drivers to translate future sales into number of components or square footage of PCBs is difficult, given the technology advancement in component density and integration over time. In addition, newer technology might require extra capability or additional capital equipment to be purchased. The issues for electronics capital equipment acquisitions could be:

- Use of components and/or PCBs does not increase in a linear fashion with sales prediction
- Technology impact has to be estimated
  - Increased use of custom integrated circuits (ICs) or software versus hardware use in the product
  - Newer technologies might require extra capability (i.e., flip-chip requiring underfill and chip-on-chip might require thermal profiling)
  - Reduce the space need for test pins if acquiring x-ray inspection technology
- Advanced packaging and interconnections schemes and standards available for new products
  - Use of flex circuits to reduce connectors, altering the PCB space

Example V Electronics Surface Mount Technology Production Line Acquisition Project
In this example, a company is considering the acquisition of a surface mount technology (SMT) production line versus subcontracting the assembly to a supplier. Only the economic analysis based on ROI is performed next. Other issues of the contracting decision and supplier selection are dealt with in Chap. 6.

Solution to Example V The PM decided to adopt a gradual increase in SMT manufacturing capacity and capability. Normally the SMT manufacturing process consists of three operations:

1. Applying solder paste to the bare PCBs with a screen-printing operation
2. Component placement through a pick-and-place machine
3. Reflow the solder paste and permanently attach the components through a reflow oven

The PM proposed an introductory line followed by two additional capacity increase steps by acquiring additional equipment, shown in Fig. 5.4. The additional equipment will also reduce cost using automation in the loading and unloading of PCBs.

- Introduction Line: semi-automatic screen printer, flexible component placement (pick and place) machine, and pass-through reflow oven
- 2nd Step: Double capacity by adding chip shooter
- 3rd Step: Increase capacity and labor reduction
  - Exchange screen printer for fully automatic one
  - Add automatic PCB stackers for automatic loading and unloading

The PM made certain assumptions for the planning of the SMT introduction line:

1. Placement machine determines SMT line capacity
2. All calculations are based on components count, not PCBs
3. All machines are de-rated from their maximum advertised run rate
   - 50 percent machine efficiency for low-volume PCBs/components
   - 70 percent machine efficiency for high-volume PCBs/components

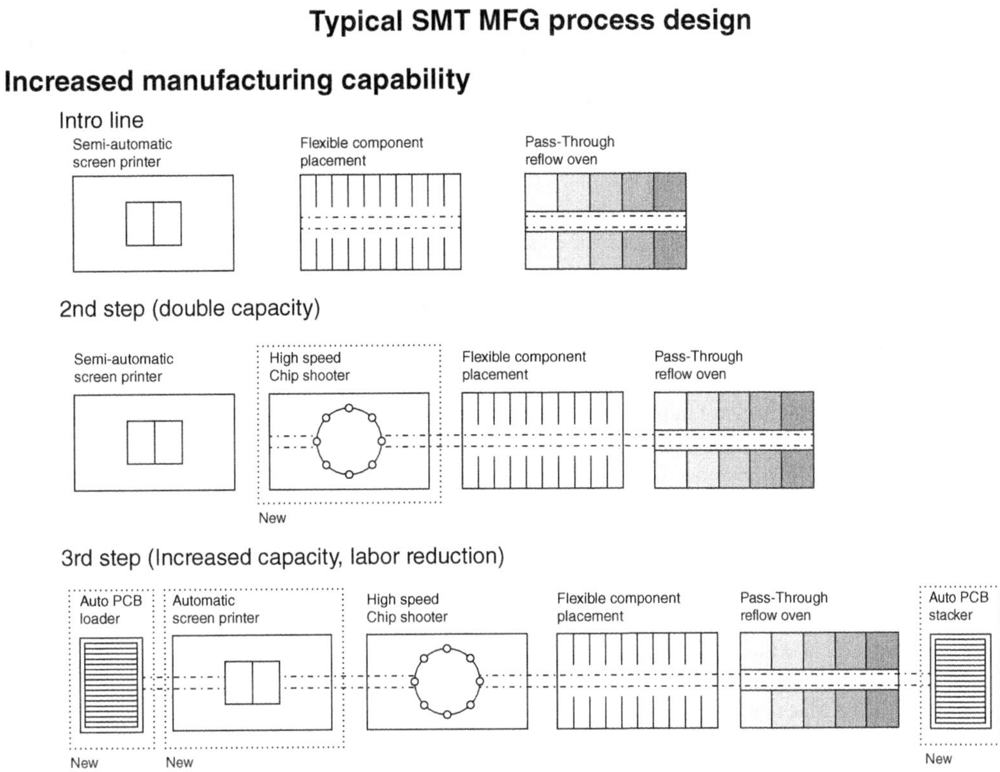

FIGURE 5.4 SMT production capacity and automation plan.

- Use 70 percent estimate of total components as high-volume PCB components; this distinction is important since high-volume PCBs require less setup for each PCB batch, thus costing less per component

4. Minimum labor and support required, based on one shift initially
   - Three operators for each function (screen, pick/place, reflow/rework)
   - One team leader
   - One process engineer
5. Compare capital equipment acquisition to outsourcing, low and high volume
6. The PM investigated the initial forecasted capacity for the introduction line as well as the cost of outsourcing the work to an outside supplier
   - Forecasted demand for the company's products was 2500 PCBs/month, 1750 high volume and 750 low volume, with 4,000,000 (4M) components.
   - 70 percent of components in high-volume PCBs = 2.8M components (C)
   - 30 percent of components in low-volume PCBs = 1.2M components (C)
   - Quotes from outside suppliers resulted in the following estimates
     - Low volume: $0.15 – 0.21/C (1.2M/year)
     - High volume: $0.04 – 0.010/C (2.8M/year)
   - It was decided to use an incremental cost of $0.125/component for low volume and $0.015/component for high volume. The costs per subcontracting are higher since there is

no need to purchase and maintain capital equipment, and the supplier makes a profit for their business efforts. The cost is higher for low-volume components since the contractor is using its equipment inefficiently for low volume.

7. The PM estimated the equipment acquisition investment at $750,000 for the introduction line and calculated the ROI (after consulting with the finance department) with a corporate income tax rate of 33 percent and five-year lifecycle of the equipment with a 10 percent remaining book value after five (5) years.

7.1 Introduction Line Labor Costs (at 40 percent overhead rate)

Employees	#	Labor Rate	Total	Total Loaded
Direct labor	3	$12/hr	$36	$50.40
Team leader	1	$15/hr	$15	$21
Process Engineer	1	$30/hr	$30	$42
Total Labor Cost				
				$113.40/hr

7.2 Introduction-Line Pick-and-Place Machine Placement Rate, Utilization, and Costs

Average Placement Rate	3000 c/hr
@ 50% Efficiency	Low Volume 1500 c/hr
@ 70% Efficiency	High Volume 2100 c/hr
In-House Cost Per Placement	$113.40/1500 = $ 0.076/c Low Volume
	$113.40/2100 = $ 0.054/c High Volume

7.3 Run Time, Based on Component Placements

2.8M High-Volume Components at 2100 c/hr	= 1333 hr, High Volume
1.2M Low-Volume Components at 1500 c/hr	= 800hr, Low Volume
Total	= 2133 hrs

7.4 Machine Utilization

2133 run hours/2088 hr/shift = 102% of one shift

7.5 Outsourcing Average Placement Costs

Low Volume 1.2M/Year * $0.125/c= $150,000
High Volume 2.8M/Year * $0.015/c = $42,000
Total Estimated Savings/year = $192,000/year

The ROI analysis for Example V is given in Table 5.5

Year End	Before Tax Cash Flow	Depreciation	Taxable Income	Taxes	After Tax Cash Flow
A	B	C	D = B + C	E = –.333*D	F = B + E
0	–750,000				
1	192,000	–135,000	57,000	≅–19,000	173,000
2	192,000	–135,000	57,000	≅–19,000	173,000
3	192,000	–135,000	57,000	≅–19,000	173,000
4	192,000	–135,000	57,000	≅–19,000	173,000
5	192,000	–135,000	57,000	≅–19,000	173,000
5 Book value = $75,000					

TABLE 5.5 ROI Analysis Table, Example V

Year End	Before Tax Cash Flow	Depreciation	Taxable Income	Taxes	After Tax Cash Flow
A	B	C	D = B + C	.333 * D	F = B + E
0	–750,000				
1	288,000	–135,000	153,000	–51,000	237,000
2	288,000	–135,000	153,000	–51,000	237,000
3	288,000	–135,000	153,000	–51,000	237,000
4	288,000	–135,000	153,000	–51,000	237,000
5	288,000	–135,000	153,000	–51,000	237,000
5 Book value = $75,000					

TABLE 5.6 ROI Analysis for 1.5 Work Shifts, Example V

Start at an initial ROI of $i = 10\%$
PW = –750,000 + 173,000 (P/A, 10%, 5) + 75,000 (P/F, 10%, 5)
= –750,000 + 173,000 (3.7908) + 75,000 * .6209 = $ – 47,624
ROI < 10%; (PW = –8217 @ 8%, PW = 12,808 @ 7%), ROI = 7.5%
Payback Period = 4.3 years

Since the ROI @ 7.5% is less than the minimum required 10 percent for supporting projects in the company, the PM contacted another business unit in the company and offered to internally supply the SMT production for that unit. The additional work will increase the production volume by 50 percent or 1.5 shifts. The yearly income, or cash flow before taxes, is now $192,000 * 1.5 = $288,000. The resulting ROI was much higher than the minimum required by the company financial department, as shown in the ROI analysis for 1.5 shifts in Table 5.6.

Start at an initial ROI of $i = 18\%$
PW = –750,000 + 237,000 (P/A, 18%, 5) + 75,000 (P/F, 18%, 5)
= –750,000 + 237,000 (3.12) + 75,000 * .437 = –750,000 + 772,215 = $22,215

ROI of $i = 20\%$; –750,000 + 237,000 (P/A, 20%, 5) + 75,000 (P/F, 20%, 5)

–750,000 + 237,000 (3) + 75,000 * .4 = –750,000 + 711,000 + 30,000 = –$9000

ROI > 18% and Payback Period = 3.2 years

The PM suggested the company purchases the equipment, with agreements to supply other parts of the company with the excess production capacity.

5.3.3 Issues with Manufacturing Machines ROI Calculations

Example V illustrates many of the pitfalls in acquiring capital equipment and using their equipment specification as a basis for economic analysis. A PM has to carefully assess the new capital equipment specifications and compare the benefits from acquiring them to subcontracting. Issues to consider in the ROI for capital equipment include:

- Advertised machine rates are never achieved, especially in low-volume high-mix capacity or JIT scheduling
- Machine run-time ratio equals 45 to 70 percent advertised maximum rates
- Analysis should consider nonproduction time for maintenance and breakdowns
- Quality and machine yields should be estimated by asking other machine users

- Target a quality yield level based on the design quality of new products
- Other costs (NRE costs) that should be considered:
  - Tooling required for each product type
  - Maintenance; assume costs for contracts or local kits
  - Operational costs such floor space, utilities etc.
- Schedule when to buy additional capacity based on expected customer demand
- Start with line of maximum flexibility at low volume and high mix
- Add production lines (high speed, lower mix) as necessary
- Additional capacity should be purchased when production lines are running at 125 percent of one shift capacity
- Standardize on one set of equipment for each additional line
- Never mix processes in any single production line; danger of contamination could be high

5.4 Techniques for Increasing Management Confidence in the Economic Analysis

Since many of the economic-analysis examples performed in the previous sections are based on marketing forecasts and assumptions, the PM must be prepared to convince the management team or outside investors with the soundness of the new product business and financial plans. ROI should be used as well as additional analysis to better understand the risks and probabilities of the business plan assumptions. Several techniques can be used to reduce uncertainty in the business plan and financial outcomes for the new projects. All of these analyses should be performed with time-discounted cash flows to render them more conservative. They include:

1. Breakeven Analysis. In this analysis, the breakeven new product sales volume to defray the costs of the product development equipment is determined. New product sales volume greater than the breakeven point will bring more revenue than was spent on the initial investment. The sales volume less than breakeven point will not defray the initial costs of the investment. For the breakeven analysis, either PV or annualized value (A) of income returns can be used for determining the breakeven point.
2. Sensitivity Analysis. In this analysis, the factors that constitute the economic analysis are varied within a range of high and low values for each factor, comprising the high and low range of estimate of each factor. The expected outcome is shown as a result of these various combinations of factor variations. This analysis is similar to the design of experiments (DoE) techniques discussed in Chap. 1. It can show which factor is more critical than others in the analysis and which factors can be varied with less impact on the economic analysis.
3. Probabilistic Analysis. In this analysis, a probability is assigned to each factor value that constitutes the economic analysis. The analysis uses a combined probability present value for the final outcome of the economic analysis. A more

simplistic method is to use probability estimates as conservative and optimistic limits to the estimation, with 50 percent (optimistic outcome) and 90 percent (conservative outcome) probability estimates being the most common.

Example VI Increasing the Confidence of Example IV Economic Analysis

In this example, the economic analysis of Example IV is expanded to include the three types of confidence-building measures explained in the previous section. They are intended to show the management different aspects of the economic analysis to help in increasing their confidence in the proposed project by exploring *what-if* conditions.

In Example IV, a company is considering buying a plastic injection mold tool and has two options: a two-cavity mold at \$40,000 or a four-cavity mold at \$70,000. It is expected that each mold will last 100,000 shots and will have to be replaced at no book value. The company is expected to sell 40,000 parts/year at \$0.25 profit per piece. ROI techniques were used to determine which mold the company should buy, assuming an interest rate of 5 percent, a tax rate of 33 percent, and straight-line depreciation for the life of the mold.

Solution to Example VI In this example, three different post ROI analyses are used to further examine the purchase decision. To make each analysis simpler, assume there is no depreciation or tax consequence of the purchase decision.

1. Breakeven Analysis. A more realistic situation is that the two machines have different costs for electricity, floor space, supervision, and engineering support. Let's assume that the cost per piece of operating expense is 4 cents/piece for the four-cavity mold and 21 cents/piece for the two-cavity mold. Show a breakeven analysis for the economic use of the two molds in the first five years in terms of annual value benefits minus costs with profits at 25 cents per piece and assuming $i = 5\%$.

For large mold (four cavity), benefits – costs/year = \$0.25 – \$0.04 = \$0.21/part
For small mold (two cavity), benefits – costs/year = \$0.25 – \$0.21 = \$0.04/part

Solution using annual value AV analysis:
AV for option 1= $AV_1 = +N * (\$0.21) - \$70{,}000\ (A/P[.231], 5\%, 5) = 0.21N - \$16170$
AV for option 2 = $AV_2 = +N * (\$0.04) - \$40{,}000\ (A/P[.231], 5\%, 5) = 0.04N - \$9240$
Breakeven is when $AV_1 = AV_2$ or $0.21N - 16170 = 0.04N - 9240 \rightarrow 0.17N = 6930$
$N = 40{,}765$ volume needed to reach breakeven on the project

This analysis can also be performed in time (days) to breakeven

40,760/40,000 = 1.019 years of machine use
= 1.019 * 365 = 372 calendar days to reach the breakeven point

Solution using PV analysis:
PV for option 1= $PV_1 = +N * (\$0.21) * (P/A[4.329], 5\%, 5) - \$70{,}000 = \$0.909N - \$70{,}000$
PV for option 2 = $PV_2 = +N * (\$0.04) * (P/A[4.329], 5\%, 5) - \$40{,}000 = 0.17316N - \$40{,}000$
Breakeven is when $PV_1 = PV_2$ or $0.909N - 70{,}000 = 0.17316\,N - 40{,}000 \rightarrow 0.73584N = 30{,}000$
$N = 40{,}770$

This PV solution is slightly different than the AV solution given the slight difference in the factor accuracy (to three decimal points) from the ROI tables.

The breakeven analysis using PV is shown in Fig. 5.5. While the breakeven point occurs at 40,000 parts, the investments in the molds have not paid off yet, using discounted returns. For the large mold (four cavity), the investment is paid off after making 77,000 parts (= 70,000/0.909), while the small tool (two cavity) is paid off after making 231,000 parts (= 40,000/0.17316), using discounted returns.

The PM can assure the management of the soundness of his decision. Assuming the worst case of lower sales, the decision made is the right one after approximately one year of sales (40,000), while the ROI was based on 10 years of sales.

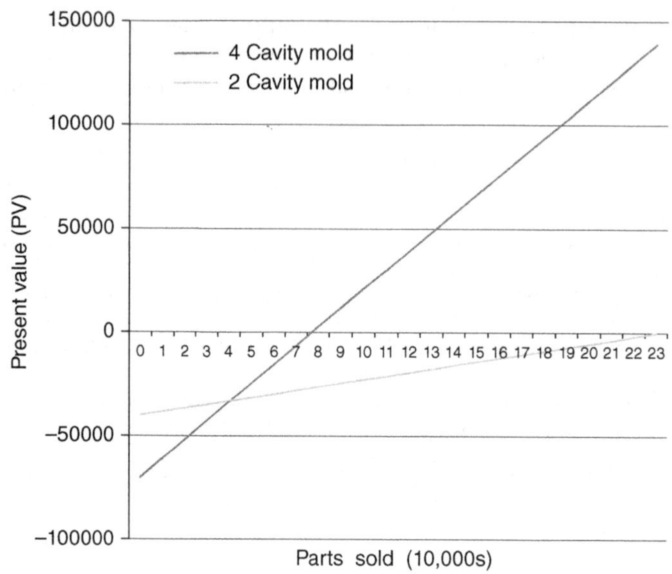

Figure 5.5 Breakeven analysis of Example VI using present value.

2. Perform a sensitivity analysis on the two-cavity mold machine, assuming $i = 5\%$ and changes of ±10% on machine costs and on profits (that is nine different outcome calculations based on current, high, and low values for each factor). Plot the annual value (AV) results versus the machine costs and benefit variations (that is six plotted points)

For two-cavity mold annual worth (AW) or AV using initial assumptions of the analysis = + 10,000 + 10% − 40,000 + 10% (A/P[.231], 5%, 5) = the results are nine (9) values corresponding to two factors with three variations levels. The mean of all nine values is $760 nominal value, shown in Table 5.7.

% Change Benefits	Resulting Benefits	% Change Cost	Resulting Costs	AV Outcome ($)
−10	9000	−10	36,000	684
−10	9000	0	40,000	−240
−10	9000	+10	44,000	-1164
0	10,000	−10	36,000	1684
0	10,000	0	40,000	760 (Nominal Value)
0	10,000	+10	44,000	−164
+10	11,000	−10	36,000	2684
+10	11,000	0	40,000	1760
+10	11,000	+10	44,000	836

Table 5.7 Sensitivity Analysis of Example VI Using Annual Value

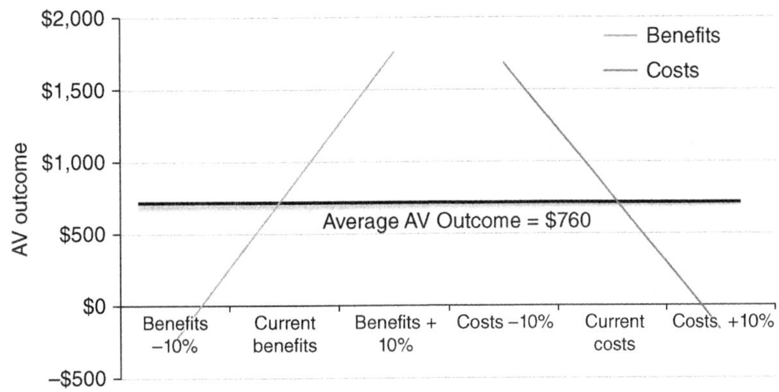

FIGURE 5.6 Sensitivity analysis of Example VI using annual value.

Benefits averages for the three combinations (–10, 0 + 10) = –240, 760, 1760 as follows:
For the 9000 benefit, the annual value is the average of (684, –240, –1164) = –720/3 = –240
For the 10,000 benefit, the annual value is the average of (1684, 760, –164) = 2280/3 = 760
For the 11,000 benefit, the annual value is the average of (2684, 1760, 836) = 5280/3 = 1760
Note that the average is always 760 for benefits and costs because analysis is symmetrical with ± 10% changes in value.
Cost averages for the three combinations (–10, 0 +10) = 1684, 760, –164
For the 36,000 cost, the annual value is the average of (684, 1684, 2684) = 5052/3 = 1684
For the 40,000 cost, the annual value is the average of (–240, 760, 1760) = 2280/3 = 760
For the 44,000 cost, the annual value is the average of (–1164, –164, 836) = –492/3 = –164
Benefit is slightly more important (greater span) than cost.

A plot of different levels of AV versus the changes in benefits and costs is shown in Fig. 5.6. There are three points on each line, corresponding with the effect of changing the level by –10%, 0, and +10% of each factor (benefit and cost). It can be seen that if we vary the assumption of benefits and costs by as much as ±10%, then there is a significant ratio of > 2 * PV in the analysis outcome. Both factors are important (significant) in altering the outcome. This analysis shows a great sensitivity of the outcome to small changes in the assumptions of the economic analysis.

3. Perform a probability analysis on the two-cavity machine producing 40,000 parts/year and assuming different probabilities for benefits and numbers of years of use. Derive the compound present value (PV), using $i = 5\%$ assuming probabilities for three benefits/part ($0.25, $0.30, and $0.20) and three years of service (four, five, and six). A combined probability of benefits (B) * number of years (n) is shown as B^*n. Note that the total combined probability should be equal to one.

   $\sum B^*n = 1$. Probability of benefits are: P(Benefit = $0.25) = 0.20; P(Benefit = $0.30) = 0.40; P(Benefit = $0.20) = 0.40 and probability of number of years, n: $P(n = 5) = 0.30$; $P(n = 4) = 0.40$; $P(n = 6) = 0.30$.
   PV (each level of benefit and years) = –40,000 + (Benefits = 8000/10,000/12,000) * P/A (3.5460/4.3295/5.0757, 5%, n (4/5/6)

   $$\text{PV compound} = \sum[\text{PV} * \text{P}(B^*n)]$$

The probability analysis is shown in Table 5.8. The total PV compound number is a sum of all of the probabilities of present values assumed for the analysis. It represents a more balanced number based on the probabilities than single PV calculations.

Benefit/Part	P(Benefit)	Actual	Years	P(years)	P (B*n)	P/A, 5%, n	PV Formula	PV Each Level B, n	PV Compound
$0.20	0.4	$8,000	4	0.4	0.16	3.5460	–40K + 8k*3.5460	–$11632.00	-$1861.12
0.20	0.4	8000	5	0.3	0.12	4.3295	–40K + 8k*4.3295	–5364.00	–643.68
0.20	0.4	8000	6	0.3	0.12	5.0757	–40K + 8k*5.0757	605.60	72.67
0.25	0.2	10,000	4	0.4	0.08	3.5460	–40K + 10k*3.546	–4540.00	–363.20
0.25	0.2	10,000	5	0.3	0.06	4.3295	–40K + 10k*4.3295	3295.00	197.70
0.25	0.2	10,000	6	0.3	0.06	5.0757	–40K + 10k*5.0757	10757.00	645.42
0.30	0.4	12,000	4	0.4	0.16	3.5460	–40K + 12k*3.546	2552.00	408.32
0.30	0.4	12,000	5	0.3	0.12	4.3295	–40K + 12k*4.3295	11954.00	1434.48
0.30	0.4	12,000	6	0.3	0.12	5.0757	–40K + 12k*5.0757	20908.40	2509.01
Total	3			3	1				$2399.60

Table 5.8 Probability Analysis of Example VI Using Compound Present Value

5.5 Conclusions

This chapter presented the basis for project consideration through economic analysis, using three types of tools to justify new project or equipment investment. It is important for the PM to get involved in the project justification and thus greatly influence the analysis by applying his or her great knowledge about the new project, product, or equipment. The chapter introduced a methodology of applying marketing and sales data into the project analysis and consideration for lifecycle issues for development project and equipment acquisition. Issues of obsolescence, technology impact, and time-dependent activities in increased project returns were also presented.

References and Bibliography

Baker, J., *CMOS: Circuit Design, Layout, and Simulation*, 3rd ed., 1174, Wiley-IEEE, New York, 2010.

Boardman, A., D. Greenberg, A. Vining, and D. Weimer, *Cost Benefit Analysis: Concepts and Practice*, Prentice Hall, New Jersey, 1996.

Cary, D., "Calculating NPV and IRR with Sub-Annual and Seasonal Cash Flow Patterns," *Journal of American Academy of Business*, 1(16): Volume 16 Number 1, 2010.

Chen, W., *The VLSI Handbook*, 2nd ed., CRC Press, Boca Raton, 2006.

DeGarmo, E., W. Sullivan, and J. Canada, *Engineering Economy*, 10th ed., New Jersey, Prentice Hall, 1996.

Duggal, V., *CADD Primer*, Mail Max Publishing, New York, 2000.

Feibel, B., *Investment Performance Measurement*, Wiley, New York, 2003.

Fries, C., *Mathematical Finance: Theory, Modeling, Implementation*, Wiley Publishing, New York, 2007.

Groover, M., *Fundamentals of Modern Manufacturing: Materials, Processes, and Systems*, 4th ed. Wiley & Sons, Inc., Danvers, MA, 2010.

Hartman, J., "The Relevant Internal Rate of Return," *The Engineering Economist*, 139(158): Volume 49, Issue 2, 2004.

Hazen, G., "A New Perspective on Multiple Internal Rates of Return," *Engineering Economist*, 31(51): Volume 48, issue 1, 2003.

Mian, M., *Project Economics and Decision Analysis, Volume I: Deterministic Models*, 2nd ed., 269, PennWell Books, Tulsa, OK, 2011.

Moore, G., and D. Brock, *Understanding Moore's Law: Four Decades of Innovation*, Chemical Heritage Press, Philadelphia, 2006.

Narayan, L., M. Rao, and M. Sarcar, *Computer Aided Design and Manufacturing*, Prentice-Hall of India, 2008.

Pillers, M., "MCAD Renaissance of the 90s," *Cadence Magazine* 3: 1998.

Tang, S., "The Variable Financial Indicator IRR and the Constant Economic Indicator," *Engineering Economist*, 69(78), Volume 48, issue 1, 2003.

Weste, N., and D. Harris, *CMOS VLSI Design: A Circuits and Systems Perspective*, 4th ed., 840, Pearson/Addison-Wesley, Boston, 2010.

Williams, J. *Financial And Managerial Accounting*, McGraw-Hill, New York, 2012.

Websites

http://en.wikipedia.org/wiki/Rule_of_72
http://money.msn.com/home-loans/mortgage-calculator.aspx
www.ifitbreaks.com/table.htm
www.hughchou.org/calc/formula.html
www.mlcalc.com/#mortgage-300000-0-30-5-0-0-0-2-2013-year
www.mortgagecalculator.org/
www.google.com/ig/directory?type=gadgets&url=hosting.gmodules.com/ig/gadgets/file/104685095803626246633/mortgage-gadget.xml

Discussion Topics

1. Growth technology companies typically do not issue dividends to their shareholders. Their argument is that their IRR is much higher than the historical Wall Street growth indices, and their earning are better invested into new products. If we assume that share prices grow at the same rate as a company's IRR, how many years would it take to double your share price if you invested in a growth company's shares? Each participant should have a unique set of IRR (i%) and number of years (n) values to enter into the discussion board for doubling your investment.
2. One of the major concerns for middle-aged folks is when to retire and start receiving social security payments. The Social Security Administration (SSA) gives you the following guidelines:
   a. Retire at age 62. SSA payments are reduced by 25 percent, and you can only earn less than $14,000/year before further reduction of your payments.
   b. Retire at 66 (full retirement age). You can also continue to work and earn as much as you want with no penalties. Typical yearly SSA payments for engineers are $20,000 to $30,000/year.
   c. Retire at 70, and SSA payments will increase by 25 percent.

   The decision is obviously dependent on the life expectancy of the person and the expected future interest (inflation) rate. Each participant should use a different inflation rate (i%) or consumer purchasing index or (CPI) to advise whether to retire at 62, 66, or 70 based on the person's life expectancy. Indicate the age where there is a breakeven between any two of the three choices. You can choose 0 percent, low or hyper-inflation i % rates, or half-percent increments not in the tables, depending on your outlook for the economy into the future. Hint: treat the SSA problems posed here as independent items from your income. SSA is an annuity where the decision when to take it rests on the merits of the three options above, nothing to do with your salary.
3. The ROI tables are based on simple compounding of interest, looking into the present (P), future (F), or into annual (A) equivalents. Can you use them to calculate the mortgage monthly payments used for purchasing a home? Without getting too bogged down in formulas, explain how mortgage monthly rates are calculated and how they are different than the ROI tables provided in this book. Hint: use the websites cited in this chapter.

4. Show how the monthly payments for a $300,000 home-mortgage fixed-interest loan for 30 years would look like using the mortgage industry tables. You can use online mortgage calculators. Calculate the present value of the $300,000 mortgage monthly payments. Use the websites cited in this chapter.

Problems

5.1. An engineering student will graduate this year with a BS degree at the age of 26. He expects to make $60,000/year for the rest of his career. He is considering his options for further study for an MS degree, where he can make an additional $10,000 per year. He is also thinking of retiring at the age of 55 or 66 years.

Option 1. Enroll in a master's degree program and pay full tuition at $40,000 and graduate in one year.

Option 2. Enroll in a master's degree program with a teaching assistantship, pay no tuition at all, and graduate in two years.

Option 3. Enroll in a master's degree program and pay full tuition at $40,000 and graduate in one year; borrow the money with a long-term student loan to be paid out in 10 years at $5000 per year starting at the end of the first year after he receives his MS degree.

Make a decision table based on present worth analysis (PW = PV). The table should include the baseline (no master's) and the three master's options. Assume an inflation rate of 5 percent. Make a snapshot of the four alternatives after five years and after a lifetime of work retiring at either 55 or 66 years. Assume no tax consequences. Please show all of your calculations in details. Put your results in a matrix of three rows of years of work and four columns (baseline and choices 1, 2, and 3)

5.2. A company is considering buying a plastic injection mold tool and has two options: a two-cavity mold at $45,000 or a four-cavity mold at $80,000. It is expected that each mold will last 100,000 shots and will have to be replaced at no book value. The company is expected to sell 40,000 parts/year at $0.25 profit per piece. Use ROI analysis techniques to determine which mold the company should buy, assuming a tax rate of 33 percent straight-line depreciation for the life of the machine. Do not include a replacement for the two-cavity machine after five years. Use hand calculations (no software) for ROI determination.

5.3. A company decided to develop low-cost photo-voltaic solar-energy cells. It decided on two development choices: a fully automatic line for $800,000 capable of producing 200,000 cells/year for a profit of $1/cell, or a semi-automatic line for $500,000 capable of producing 120,000 cells/year for a profit $1/cell. It is estimated that both lines will last for five years and will have a book value at the end of the five years of 10 percent. Assume both lines will be working to 100 percent capacity. You are the PM for development. Please make all calculations with or without taxes.

Show the ROI calculations for both machines, based on five-year life, 10 percent book value, and 33 percent taxes when applicable.

a. Fully automatic machine = assume 33% taxes, salvage value 10%, and five-year life
b. Fully automatic machine = assume no taxes, no salvage value, and five-year life
c. Semi-automatic machine = assume 33 percent taxes, salvage value 10%, and five-year life
d. Semi-automatic machine = assume no taxes, no salvage value, and five-year life
e. Conclusion: show which plan is better

With taxes and remaining book value: fully or semi-automatic?
Without taxes and remaining book value: fully or semi-automatic?

f. Show fixed and variable profit (including operating costs) breakeven point based on yearly volume for the first five years for both machines. Assuming $i = 5\%$, 10 cents/cell cost to operate the automatic and 40 cents/cell to operate the semi-automatic machine. Selling price for both is \$1/unit. Please make a plot of the breakeven sales point.

g. Assuming an interest rate of 5% and life of $n = 5$ years with no tax consequences or book value remaining, show probability analysis for PV for \$500,000 semi-automatic @ 120,000 cells/year machine assuming two probabilities of profit per cell and useful number of years (n) for the machine. P (profit = \$0.80) is 0.8 and P (profit = \$1) is .20; and P ($n = 5$) is 0.4 and P ($n = 4$) is 0.6

h. Assuming an interest rate of 5% and life of $n = 5$ years with no tax consequences or book value remaining, show sensitivity analysis for PV for semi-automatic machine @ 120,000 cells sold/year @ \$1 profit per cell machine, assuming ± 20% variations in the profit per cell and useful life. Plot the PV results versus the profit and useful life ± 20% variations.

5.4. For this problem, use the following abbreviations:

- Project organization to manage project: silo (S); liaison (L); matrix (M); self-direct teams (T)
- Product as needing resources: research (R); advanced development (A); and development (D)
- Describing the product as revolutionary (R); evolutionary (E); or platform/architecture (P)
- Describing the process as job shop (J); batch (B); assembly line (L); flow line (F)
- Estimate R&D investment in terms of percent of potential sales as low (L); medium (M); high (H)

Please provide your answer with the most appropriate selection given the scant details in the questions.

A company is designing a new inductive proximity technology for touch screens in smartphones. The company needs a \$1,000,000 (\$1M) investment for a project to develop the product. It assumes a rate $i\% = 8\%$ and estimates the market potential at 500,000 units/year for the future up to eight years. Their options are:

1. Get a \$1M loan, develop the technology in-house for one year, sell 500,000 units for eight years at \$1/unit with income to the company starting after one year of loan approval, and repay the loan after one year for five years at \$240,000/year.
2. Get an existing company to licensee a less-capable technology, plan to sell the phones immediately for only five years, since the technology will becomes obsolete earlier, and then pay them royalties of \$0.25 per unit for five years. Assume no tax/depreciation consequences.

a. Describe the project characteristics:
Organization ____, resources needed____, product description____, process ____ R&D%____

b. What is the best choice for the company? Which tool would you use: ROI or PV?
Which choice: borrow or license?

c. Show calculations (both formula and actual #) for Option 1. Get a \$1M loan, develop the technology in-house for one year, and sell 500,000 units/year for eight years at \$1/unit with income to the company starting after one year of loan approval, and repay the loan after one year for five years at \$240,000/year.

d. Show calculations (both formula and actual #) for Option 2. Get an existing company to licensee a less-capable technology, plan to sell the phones immediately for only five years, since the technology becomes obsolete and then pay them royalties of $0.25 per unit for five years.

5.5. A company is making consumer (high volume) and industrial (low volume) lines of microwave ovens using the same technology and controls. The company decided to sell the industrial microwave line to focus on consumer products. Your company (IN-microwave) is considering buying the line for $500,000. IN-microwave company depreciation method is straight-line, its tax rate is 20 percent, and marketing thinks it can sell 12 units per year at a profit of $10,000/unit for five years. A project manager was selected to manage the industrial microwave business. Assume no book value remaining after five years. Hint: minimum starting ROI for IN-microwave being an industrial business is 4 percent.

1. Describe the project characteristics:
   Organization ____, resources needed____, product description____, process ____ R&D%____
2. Show ROI calculations: ROI (%), payback period (years), and return factor

5.6. The PM from problem 5.5 proposed an alternative track. To make the calculation easier, the PM assumed no tax and no depreciation consequences.

Scenario A: Buy the line for $500,000. Marketing thinks it can sell 12 units per year at a profit of $10,000/unit for five years.

Scenario B: Develop the line in-house at $300,000.Marketing thinks it can sell 12 units per year at a profit of $5000/unit for five years.

1. Show the breakeven point *in yearly sales* between the two scenarios (buy or make in-house), using an $i = 7\%$b. Plot a simple graph to show your result
2. For Scenario B, develop the line in-house at 300,000.Marketing thinks it can sell 12 units per year at a profit of $5000/unit for five years. Assume $i = 7\%$.
3. Perform sensitivity analysis for scenario B if the in-house project cost estimate ($300,000) varies by ± 20%, Marketing thinks it can sell 12 units per year at a profit of $5000/unit for five years.
4. Perform probability analysis if the probabilities of the project cost were as follows: Marketing thinks it can sell 12 units per year at a profit of $5000/unit for five years.

P($200,000) = 30%; P($300,000) = 40%; P($400,000) = 30%

CHAPTER 6

Make or Buy: Subcontracting and Managing the Supply Chain

The increase in outsourcing and global contract manufacturing has led to significant impacts on the development of new products and their manufacturing operations conducted around the world. This chapter outlines some of the outcomes of this trend in terms of changes to the product development process, manufacturing strategy, financial, legal, and business issues in global outsourcing. A strategy for new product realization and its manufacturing with the supply chain is presented with emphasis in selecting, managing, and negotiating with suppliers in order to minimize adverse consequences of contract manufacturing.

Over the last 30 years, the world's major original equipment manufacturers (OEMs) have transformed themselves using a business model where the majority of their manufacturing, as well as design, is subcontracted to an industry design and manufacturing segment known as contract manufacturers (CM), or external manufacturing suppliers (EMS), also known as the supply chain.

The trend of moving to the supply chain began with the concept of core competency, where an OEM company is focused on its own niche or its particular knowledge of the market segment it serves. The OEM's focus should be on enhancing the propriety technology it develops into its products. Once the OEM identifies and builds on its core competency, it can have the noncompetency operations subcontracted out to the supply chain.

EMS companies have grown in size, diversity and complexity and have developed their own particular competency in the industry segments they serve. Many EMS companies collaborate to offer a variety of services ranging from design in different engineering and software disciplines as well as component engineering and services to supplement or even alternatively replace the OEM design team. Large EMS companies offer the full slate of new product realization menu of design, analysis, and prototyping, along with new product introduction as well as manufacturing, customer service and warranty of new OEM products.

OEM companies must have effective means to introduce and support new products that are reliable and delivered in a timely and cost-effective manner. This is especially important, given the rapid technology changes in today's competitive environment, with knowledgeable and demanding customers. This chapter will deal with proven supply-chain strategies and methodologies in order to plan successful product launches in rapidly changing high-technology improvements in design and manufacturing techniques. An OEM company may elect to introduce, build, and support new products in-house or partner with the supply chain for some or most of the new product realization efforts. The pros and cons of subcontracting will be discussed with tools, techniques, and procedures outlined that can help the OEM as well as the EMS companies make a reasoned subcontracting decision on the supply-chain role in their overall business plan and operations. Examples and case studies of subcontracting will be provided to show the successes and failures of contracting.

This chapter will include the following topics:

6.1 *The Lean Enterprise Concept and the Supply Chain.* The issues of identifying core competencies and balancing the company's competency with the increased dependency on the supply chain will be discussed.

6.2 *The Outsourcing Strategy To Be Considered and the Associated Pitfalls.* This section includes the business concerns in subcontracting and planning for the impact of outsourcing on the OEM operations.

6.3 *The Changes to the Product Realization Process and Communications with the Supply Chain.* In this section, the changes that have to be implemented in the product realization process will be outlined for successful new product realization and collaboration with the supply chain.

6.4 *The Supplier Selection Process.* This section includes how to prepare for management approval for subcontracting as well as legal and liability issues to be considered when subcontracting.

6.5 *Chapter Summary and Case Studies of Subcontracting.* This section will illustrate successful methodologies and potential problems.

6.1 The Lean Enterprise Concept and the Supply Chain

The focus on core competency and the increase in outsourcing and supply-chain growth has resulted in having many companies discard their manufacturing capability and, hence, become lean organizations dependent on outside suppliers for design and manufacturing resources, as shown in Fig. 6.1. At the same time, the cost of acquiring expensive modern manufacturing equipment has become prohibitive, and the pace of

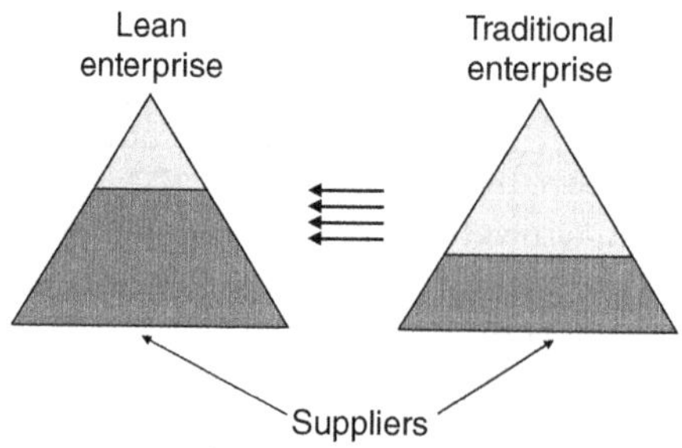

FIGURE 6.1 Transition of the traditional enterprise into a lean enterprise.

new manufacturing technology has quickened. This rendered OEMs reluctant to invest in their manufacturing plants, lest those plants become obsolete in a short time. In addition, the advent of global standards and competition for high-quality, low-cost, and shorter new product introductions has increased the need for new product design teams to incorporate design and manufacturing feedback through early supplier involvement (ESI), design for manufacturing (DFM), and concurrent engineering (CE) into their design process. These new tools of new product realization have to be incorporated into the relationship of an OEM with its supply chain.

The outsourcing strategy of an OEM should begin with identifying the core competency of the company. The company should determine its identity and the uniqueness inherent in its technology, products, and services. It should also determine which operations or elements of the design and manufacturing of new products need not be unique. For example, in design, there is probably no need to develop new connectors or cables or even power supplies or display devices. These components could be purchased or their design and manufacturing could be subcontracted out. Only if a design or manufacturing process offers a competitive advantage, it should be substituted with industry-standard components or designs. It is best for the company to participate in industry-wide forums or standards committees in order to ensure the selections of standardized components or processes that are beneficial to their strategy. For the electronics industry, there are industry-based organizations such as the International Electronics Manufacturing Initiative (iNEMI) that collaborate on technology roadmaps for new designs. Many industry-supported organizations such as Association Connecting Electronic Industries (IPC) and Surface Mount Technology Association (SMTA) allow competitive companies to collaborate in developing new standards and new processes in manufacturing and quality.

Some OEMs will retain competency for special reasons and continuously invest in design or manufacturing expertise in core areas. Examples would be Fisher Controls in cast valves, Gillette in plastics molding, Frunderburg-NOK in auto gaskets, and Hewlett Packard in inkjet printing. Other companies retain competency in the market segment they operate because of regulatory oversight such as:

1. FDA regulations require extensive testing and specialized (clean room) manufacturing environment for medical instruments and monitors. Extensive knowledge of FDA regulations, medical standards, and good manufacturing practices (GMP) is also required.
2. Defense contractors are regulated by military specifications and standards (MIL-STDs) in materials and processes, design specification and specialized project management and reporting.
3. Aerospace companies are regulated by long design cycles and FAA testing and certifications.

An OEM company needs to select an appropriate algorithm for subcontracting, based on its business plan and the cost structure of the competition. The tactical issues of subcontracting involve the changing of the product creation process within the organizational boundary and communications with the suppliers. A supplier selection, qualification, and management process has to be in place to facilitate a smooth transition to subcontracting.

Sector	# Interviewed	Design Outsourcing		MFG Outsourcing	
		Design	Analysis	Cost Driven	Value Added
Communication	2	35%	20%	0%	100%
Computer	4	25%	10%	40%	40%
Consumer	2	15%	25%	50%	0%
Industrial	5	10%	10%	25%	0%
Military/Medical	4	0%	5%	10%	0%
Design Services	3	N/A	N/A	50%	50%
MFG Services	10	10%	0%	N/A	N/A

Table 6.1 Status of Companies Outsourcing Hardware Design and Manufacturing Capabilities

6.1.1 Development of Outsourcing

This trend toward outsourcing selected portions of design and manufacturing competency has been occurring at different rates, depending on the industry sector and the maturity of the product offerings in that sector. Table 6.1 is a summary of data collected from 30 companies and 50 interviews conducted by the author.

Table 6.1 shows that manufacturing and design outsourcing correlates strongly with time-to-market pressures in the particular industry sector. Military program development tends to be long range and dependent on the use of proven technology. This is contrasted with increased outsourcing in the communications and computer (C/E) industries, which are under greater pressure to reduce time-to-market and are early adopters of new technology. The C/E industries are also heavy users of value-added manufacturing outsourcing, where a primary manufacturing service provider delivers supply-chain management for the total product material procurement, assembly, and test. Consumer and industrial sectors rely on their manufacturing competency in their own plants, which is difficult to outsource, since they are specific to each product and not typically enclosed electronic boxes found in C/E products.

Manufacturing service providers are moving up and down the supply chain to provide more value to their customers. While most suppliers are cost-driven, focusing on one type of manufacturing competency, many are adding more value to their customers either by providing design services, or managing other suppliers in the supply chain. A plastic supplier said to the author, "We can turn a 25-cent plastic part into a five-dollar assembly with the additions of electronics and cabling."

Contract design companies are also leveraging specialized core competencies in order to offer design and manufacturing services to their customers, such as engineering analysis and access to tooling and manufacturing outsourcing in low labor cost countries. In many cases, they can provide complete design and manufacturing resources for specialized subcomponents such as printers, motors, and electronic box packaging.

These rapid changes, combined with the growth of the global economy, are resulting in globally competitive companies, operating in many countries with design and manufacturing sites, and simultaneously launching worldwide products over a wide spectrum of countries and customers. These companies are partnering with global suppliers to achieve the best strategy for worldwide design and manufacturing optimization of

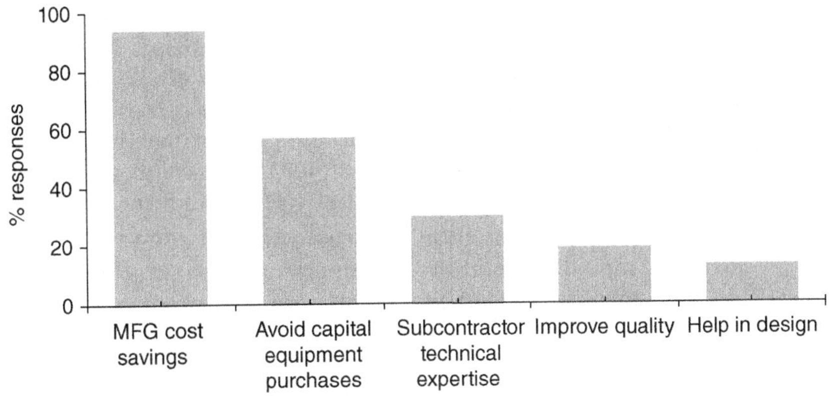

FIGURE 6.2 OEMs outsourcing reasons.

their operations. An example of consolidation of suppliers into few global ones is the auto industry. The auto-supply industry will shrink from 1000 first-tier supply companies today into around 25 well-financed global suppliers in the future.

The reasons for outsourcing also differ based on the business goals of the company. Fig. 6.2 shows a ranked survey of OEM reasons for outsourcing. The reasons are as follows:

1. Manufacturing Cost Savings. This is the most obvious reason, especially when using developing countries' labor rates. Unskilled labor cost in some of these countries is less than 10 percent of U.S. labor cost, with little or no benefits such as health insurance. However, several issues have to be considered:
   - Unstable infrastructure as well as political or socio-ethnic situations in the subcontracting region can result in unforeseen problems, such as natural or political disasters, resulting in work stoppage or workers suddenly demanding significantly higher wages.
   - The local workforce can sometimes act as a cohesive extended family unit, which might hinder the normal operations or proposed changes to the work conditions, counteracting the benefits of lower wages.
   - Low wages and skills in the contracting region may be offset by the higher productivity and skills of the U.S. worker. With a lower-skill workforce, the subcontractor might need excess capacity in case of a longer time necessary to train the workers and repair or maintain expensive and complicated production equipment.
   - In case of technical problems and equipment failures, some OEM engineers might have to travel to the contracting site to repair equipment, update processes, or conduct training. The cost of frequent travel has to be factored into the labor cost equation.
   - The labor cost of high-technology products, especially those in the start-up phase of the product life cycle is very low, less than 5 percent of the sales price. Labor-cost benefits in subcontracting might not be of great importance in these conditions.

2. Material Cost Reduction. This is an important issue not mentioned in the survey and of greater importance in the electronics industry. The majority costs of materials in a high-technology product are in the specialized parts used specifically in the product. However, many products use a plurality of standard components. The contractor can achieve significant material savings through the ability of leveraging large purchases of standard components over individual OEMs. This can be seen in the case study of subcontracting later in this chapter.
3. Reduced Capital Expenditures. The benefits to the OEM in subcontracting by reducing capital expenditures are important. The contractor can use economies of scale to purchase large-capacity automatic production machines, much more than the OEM can hope to achieve individually. In addition, the OEM can also reduce the impact of other ancillary issues such as floor space, equipment maintenance, and obsolescence. Two developments in the capital equipment topics need to be highlighted:
   - Contrary to popular perception, the capital equipment in low-wage contracting regions are more up to date than U.S. home-based OEM facilities. This is due to the newer installations in the contracting regions and, in many cases, local government support for acquiring expensive capital equipment.
   - Capital equipment manufacturers are sensitive to the equipment obsolescence issue and will offer trade-in clauses with the purchase of their equipment for newer-technology equipment into the future. This trade-in issue should also be part of the OEM subcontracting decision.
4. Leveraging the Subcontractor Expertise. In a specific industry sector, contractors can develop expertise on their own to help OEMs reach efficiency in design and manufacturing by undertaking the cost of development themselves. They can work together with one OEM customer to develop a new process or design and then offer that process or design to other OEMs. Examples in different industries include the auto sector, where the subcontractors are driving the technology enhancements with features such as active suspension and comfort and safety enhancements such as rich sound speakers, stability systems, collision avoiding radars, and rearview cameras. Examples of these innovations are common in many industry and product segments:
   - In the electronics industry, the advent of denser packaging such as chip-on-chip designs allows OEMs to further integrate a design into smaller packaging.
   - In the plastics industry, the use of specialty molds able to perform mechanical actions (close a container) in the molding machines.
   - Modern packaging design services using tight tolerances, decorative trim, and automatic threaded inserts are helping to reduce parts' costs and design flexibility.

   An OEM should be concerned about some of the issues of using the subcontractor expertise in the following areas:

   - Subcontractors might decide to make the transition from a supplier to a competitor of the OEM. Once they gain the expertise in manufacturing a certain product family, or supplying the components, they might begin to develop competing products. One example is Samsung products that compete with Apple's. While Samsung is a large supplier of ICs to Apple, Apple and Samsung have sued each other, which made for strange bedfellows.

- Patent infringement might occur if a supplier offers a technology feature co-developed with one OEM to other OEMs in the same industry sector. The OEM should carefully review the patent history of the new technology or component from its suppliers before making a technology adoption or incorporating a new part. One example is the touchpad attachment into laptops, where Hitachi held the patent, while an Acer offshoot company incorporated it into many of the major OEM laptops it built for other OEMs. Hitachi sued the Acer supplier in the International Trade Commission (ITC) in 2010, and the suit was settled after major litigations.

5. Improving the Quality of the Product. Suppliers are usually sensitive to meeting quality and safety standards and can help navigate the maze of many different standards and regulations, removing the requirement that the OEM has to keep abreast of developments in these standards. Examples of these standards include ISO 9000/QS 9000 series for design and manufacturing quality for the electronics and auto industries, ISO 14000 for environmental issues, and the EU RoHS (lead-free) standards for restrictions on hazardous substance content in products shipped into EU countries.
6. Using the Subcontractor to Help in Design. Increasingly, suppliers are moving up the new product realization chain into design services, by completely providing the project management and design and development of new products. More on this issue in the next section of this chapter.

6.1.2 Competency versus Dependency

As companies rush to outsource design and manufacturing, they are concerned with several issues. Does outsourcing really save money? What to outsource and when? What to do with the remaining competencies in-house? How to avoid some of the dependency problems of outsourcing?

One of the engineers for a major OEM interviewed by the author explained, "I do not see how design outsourcing makes any sense, we normally have two-thirds MEs (mechanical engineers) designing the box. . . now it takes two-thirds MEs just to manage the design contractor! "Another manufacturing engineer said, "My company has decided to contract our manufacturing outside. We want to send out our oldest products . . . and no contractor wants to do it at a reasonable savings to us!"

Both these quotes highlight a common problem with outsourcing. Both cases mentioned above highlight the tendency of OEMs to begin outsourcing either older products or completed designs. This would leave new designs and products within the company. The difficulty is in the transfer of large amounts of knowledge about these older products, together with their nonstandard methodologies and operations. It is best to begin outsourcing at the beginning of a new product cycle, so that the competencies of an OEM and its suppliers are maximized.

An important issue for the PM of a new product is the balance of retaining the project team competency in noncore elements to be outsourced. This decision might be

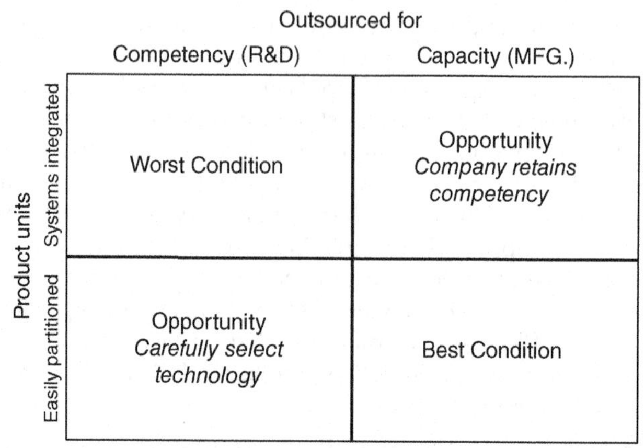

Figure 6.3 Outsourcing competency versus dependency.

attractive in retaining the ability to better manage new product design and react quickly to manufacturing crises, versus losing the competency and increasing their dependency on the supply chain. Many companies opt initially for the middle of these two options to act as a safety valve in case of supplier failures. They maintained prototype shops in manufacturing and certain expertise in design of noncore competency technologies, such as power supplies and motors. The reality is that once the contracting decision is made, these safety-valve organizations are usually standing idle at a great cost to the project. The proper decision is to study the new product architecture and then determine the best strategy for outsourcing.

Figure 6.3 illustrates the choices available to the PM on the issue of what to subcontract. If the new product is designed with well-partitioned system architecture, then it can be easily subcontracted for manufacturing capacity for all or some of the production volume, such as printed circuit boards (PCBs) in electronics products, or a suspension system for automobiles. In addition, there is an opportunity for subcontracting noncore competency parts of the design. Poorly designed or integrated systems with proprietary processes, such as nonstandard cables and special controls and displays suited for harsh environments in medical or military applications, might be difficult to partition. These systems can only be partially contracted for manufacturing of certain components such as cables. In these conditions, the OEM will retain competency, since the supplier is only manufacturing a small part of the overall system.

Outsourcing should be implemented in various steps and according to the company's needs. Table 6.2 is a summary of the common issues in outsourcing, ranked by importance, both in terms of what to outsource and what are the necessary and remaining minimum versus questionable competencies in the company for managing the supply chain.

In summary, outsourcing for design or manufacturing occurs for two basic reasons: capacity and competency. If the design of the product or system is well-partitioned, then outsourcing for capacity is easily accomplished with minimum effects on the company retaining its competency. Outsourcing for competency should be carefully selected, so that the company retains its desired or core competencies.

What to Outsource?	Minimum Competencies Needed to Select and Manage Supply Chains
Commodity product/assembly with standard interfaces, manufacturing cost, and quality requirements	Identify and select qualified suppliers
	Write appropriate system specifications
Product/assembly with well-defined interfaces	Evaluate incoming bids and select best bidder
Product/assembly requiring multidisciplines	Validate deliverables meet specifications
Product/assembly containing new technology	**Questionable In-house Competencies**
Product/assembly and system integration	Improve submitted bids
Basic core competency product or assembly	Help supplier technically in design/manufacturing
Basic new product with many new characteristics needing evaluation, justification and ROI	Help supplier operationally and in training and rework
	Improve deliverables after receipt

TABLE 6.2 Common Issues in Selecting Outsourced Products and Competencies

The advantages of outsourcing are many:

1. The enterprise does not have to keep abreast of noncore manufacturing or development technologies, saving engineering resources, capital equipment, and large payrolls.
2. The enterprise can easily expand and contract due to business conditions and product introductions, without the burden of hiring or firing.
3. The enterprise is free to seek the lowest-cost contractors, especially those that can leverage their size into lower-cost material procurements.

As the CEO of a new networking company said, "We can grow to $1 billion and never have to spend a cent to expand a plant or upgrade a computer."

Outsourcing can take form in many different directions, depending on the company's willingness to increase its dependency on its suppliers, starting with manufacturing then moving up into development and design. As each outsourcing scenario is embraced by the enterprise, the concern to maintain competency and reduce dependency has resulted in maintaining some unnecessary competencies in-house.

In the initial stages of manufacturing outsourcing, the OEMs were concerned that unused competency will disappear. While direct labor was contracted away, more skilled resources, such as manufacturing and process engineers, were kept on to manage in-house prototype shops or outside supply chains. Lamented one manufacturing manager to the author in a telecommunication company, "We used to design our manufacturing process for our products to last for 40 years . . . the life cycle of new products has shrunk now to months instead of years!"

In some companies, as more manufacturing was outsourced, there was not a comparable reduction in the manufacturing staffing and expertise within the company, reducing the benefits of outsourcing. This should not be the norm. The author interviewed many companies that are successfully outsourcing without the benefit of internal competency in manufacturing processes, relying on early supplier involvement (ESI) for DFM feedback and rapid prototype services.

A similar effect is also occurring in design outsourcing. OEM design engineers will perform all of the system aspects of the design, such as design specifications, connectivity, testing, and analysis required for the outsourced product component or assembly. They will specify the design space of the outsourced component and subcontract the component development to a design supplier. The supplier can model and analyze the internals and study the mechanical interference and performance specification in relation to the overall product. The supplier can also design the mechanical interface and electrical connection to the rest of the product. In addition, suppliers also can perform validation for the design they developed in collaboration with the OEM design engineers.

The OEM could achieve more efficiency by allowing suppliers to increase their contribution to the design. A mechanical engineering (ME) manager from a consumer company who contracted the design of a motor system mentioned to the author, "We defined the envelope, the design house can decide on the geometry inside the envelope, what size of pins, connectors, etc. . . . They send us back an FEA analysis based on their design. Then they can go ahead and tool up and produce qualifications and validations reports for their designs. Based on those activities, we can approve the supplier design."

The competencies shown in Table 6.2 are divided into those that must be retained in order to select and manage the supply chain and those that some OEMs might continue to support internally, when in fact they are not needed. Examples of those questionable competencies abound. Many OEM companies maintained prototype shops or small manufacturing cells in case the supplier did not deliver the product on time and within specifications, requiring rework. The auto industry, at the beginning of the supply-chain conversion, created an organization called American Supplier Institute (ASI) that took on the training of the suppliers. Most of these questionable competencies can be turned over to a well-qualified supplier with internal competencies maintained with rigorous training.

6.2 The Outsourcing Strategy to be Considered and the Associated Pitfalls

Business concerns should be paramount in developing a good outsourcing strategy. There should be synergy between the business model of the company and its outsourcing efforts. Some of the issues of proper outsourcing strategy and the selection of a supply chain can be summed up as follows:

1. The types of customers in the industry that the company operates in. Issues such as customer expectation in cost, quality, reliability, life cycle, support, deployment, and speed of delivery are paramount in selecting the proper supplier that is focused on these issues and may be supplying other companies in the same market.
2. The type of market environment, including other competitors supplying similar products. This would include typical financial metrics prevalent in the particular OEM industry such as profit margins, material overhead, and efficiencies from contracting. Other financial and operational indicators for the

industry include cost of goods sold, product turnover timelines, development and technology investments, workforce skill-level required, and timely cost-improvement efforts.

3. Prevailing method of product operations, such as the expected levels of reliability, cycle time, system assembly and configurations, revision control, and upgrade policies as well as any applicable design, manufacturing, or quality standards.
4. Business concerns have to be addressed in terms of forecasting the impact of outsourcing on the business indicators for the company. These indicators include absorption ratios of overhead, materials, direct labor, and cost of goods sold ratios. Other indicators include material and labor efficiencies from supplier material procurement and material and capital equipment leverage as well as adequate planning and reporting of cost-improvement projects.
5. Legal issues should be considered carefully to eliminate potential liability. These include the instructions to bidders, subcontractor documentation (such as their quality plan), certificates of compliance, warranty for delivered parts and products, subcontractor management, delivery schedules, liability for late delivery, and conditions of forecast demand and how to manage changes to the forecast.

A total plan to formulate the outsourcing strategy can be divided into three parts: (1) an outsourcing competency matrix; (2) a selection process; and (3) a communications plan.

1. An outsourcing competency matrix summary is presented in Fig. 6.4 for an electronics high-technology company. The matrix shows the organizational competencies currently needed for manufacturing the product. The shaded areas in Fig.6.4 indicate the competencies that will be affected by subcontracting. The content and resources for each affected competency should be examined as outsourcing decisions are made: whether to outsource a particular competency partially or fully, the impact of outsourcing on staffing levels for each competency, and the needs to support and manage outsourced competencies.

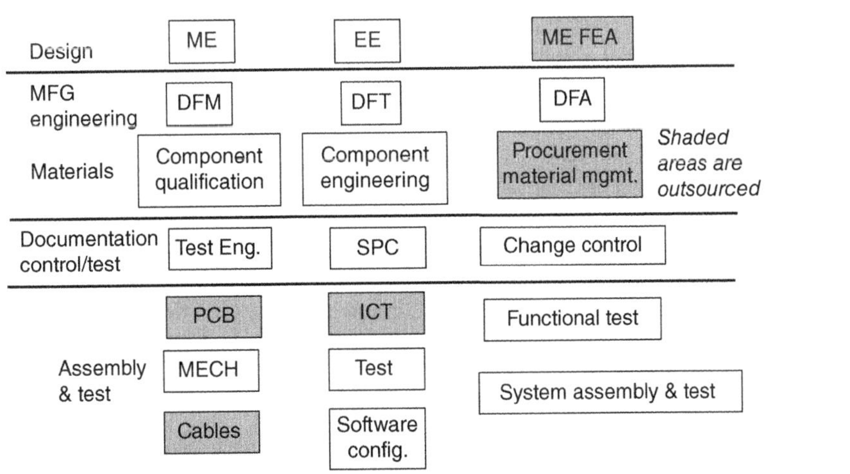

FIGURE 6.4 Outsourcing competency matrix.

2. The selection of the supply chain should occur at the concept stage of the product. The selection and bidding processes should be based on historical relationships. Suppliers could be initially qualified and then bid their cost model for the design and manufacture of the product.

   The supply-chain qualification process is dependent on industry standards in cost, quality, and activity timeliness. Cost should be well-quantified through a cost model based on activity-based costing (ABC). Quality should be quantified by standards such as ISO 9000 and Six Sigma, which outline comprehensive methodologies for controlling, maintaining, and implementing quality programs. Timeliness should be measured in terms of turnaround time of orders and historical delivery performance of the supplier and include any incentives for early delivery. The supply chain should also provide the company with its technology and equipment acquisition plans as well as plans for reducing cost over time from more efficient process operations.

3. The interface between the supply chain and the remaining company's core elements should be the same as in-house capabilities. The supply chain should provide for ESI and design guidelines to their manufacturing capabilities and constraints early into the product realization process. The communication links between the supply chains and the OEM should be available as easily as the in-house accessibility. Regularly scheduled meetings such as design reviews, as well as on-demand meetings to discuss problems and their resolutions, should be planned accordingly. The communications should be instantaneous as well as simultaneous. As one design house manager put it, "Meet your consulting engineer anytime you desire." The best plan is to have a single point of contact, the product manager (PM) or designee at the OEM, and a PM at the supplier assigned to the OEM.

6.2.1 Operational Issues When Outsourcing at Different Levels of the Product Realization Process

Before committing to an outsourcing strategy, the PM should examine the operating procedures at the OEM and consider the changes that will have to be addressed as well as the concerns to be met when design and/or manufacturing operations are outsourced. Steps include:

1. Marketing forecast and satisfying customer orders on time. The marketing forecast is critical to the success of the supply chain. Once a contract is signed with the supplier, strict guidelines are put into place as to the time horizon of the forecast for the contracted product, with penalties for schedule changes based on a time horizon. This is necessary for the supplier to protect its material procurement process and schedule the product build slots into its production queue. This requires better planning for the OEM marketing and the sales force in order to level the customer demands and eliminate the "end of the month" or "end of the quarter" syndrome. Otherwise the OEM will pay strict penalties for changing the schedule for contracted product build.

2. Operations and capacity planning. At some OEM manufacturing operations, whenever there is an "end of the quarter" crunch, when sales orders finally arrive to meet forecasted sales quotas, the operations department goes to work

and miraculously ships the customer orders in the last few days of the quarter. These are false heroics since the OEM operations department probably keeps excess capacity to meet these *crunch* needs. When a product with "end of the quarter syndrome" is outsourced, severe penalties will occur. The supplier operates on a more realistic capacity than the OEM and will protect their operations by contractual means for their schedule stability. The OEM operations are faced with the loss of control and flexibility when outsourcing the product and have to make sure that the planning and material procurement is level with the contractual forecast. This loss of control can also extend to engineering changes (ECOs) and their impact on inventory procurement and obsolescence.

3. Financial department and maintaining forecasted revenue. When a product is outsourced, the supplier has to maintain the shipping profile as forecasted; otherwise, there will be a severe impact on the profit outlook for the OEM. The supply contract will usually have rewards/penalty clauses pertaining to shipping on time as well as early or late delivery.

4. Design teams have to maintain close communications with design outsourcing suppliers and integrate the different functions of design, analysis, and prototyping. It is important to identify the interfaces of the different design elements. For example, a design team can allocate a certain space for a motor that is designed by an outside supplier and ask the supplier to engineer the motor to design specifications within that space. If the supplier needs more space for the motor in the product, a lengthy discussion can ensue about the reasons and the consequence of changing that dedicated space in the new product and the potential adverse consequences.

5. General management concerns are based on the impact of outsourcing on the bottom line of the OEM's financial statements. Will outsourcing have a significant impact on the financial reporting numbers and thus cause the financial markets to investigate the reasons behind those changes? Some concerns the management team might have with outsourcing include the following:

   - Absorption or the need for resolving the capacity of direct labor and overhead impacted by subcontracting: how many direct labor skills need to be retrained or maintained and how many overhead resources such as materials, manufacturing, and quality engineers are affected? The outsourcing competency matrix, shown in Fig. 6.4, and a skills-distribution matrix might have to be generated by the PM to address these concerns.
   - Cost of goods sold (CGS), which contains the overhead placed on labor and/or materials, including the materials overhead, which is normally about 5 to 12 percent for medium-size technology companies. This number is a measure of labor and general administration costs. Will this number change significantly with subcontracting, since there will be less material to defray the overhead costs?
   - What are the material efficiencies from supplier-purchasing leverage as well as labor efficiencies from supplier automation? Will there be adequate periodic reporting for improvementsin these costs?
   - What is the flexibility model with the supplier in terms of changes to the forecast? How will the supplier operations handle the end-of-the-quarter

syndrome? Will the supplier be flexible in its capacity planning in order to handle surge orders or keep up with long-term growth?

- What are the consequences of missed shipments, and how will the supplier guarantee delivery of products on time?
- How to handle non recoverable expenses (NRE), such as tooling costs, and other ancillary product manufacturing issues such as incoming inspection, fault analysis, and software configurations?

The PM should consider researching answers to these issues before presenting subcontracting proposals to the general management of the OEM.

6.2.2 Types and Levels of Outsourcing

As mentioned earlier, there are levels of outsourcing based on the competency versus dependency matrix (Fig. 6.3). They are:

1. Cost-Driven Manufacturing Supplier. This level requires the supplier to have a reputation for high-quality and low-cost operations, with focus on a single technology such as PCBs for the electronics industry. The supplier has to exhibit great agility in reacting quickly to changes in process requirements and have the capability for rapid turnaround. The supplier has to demonstrate competency in world-class manufacturing techniques in quality and productivity and exhibit a strong commitment to DFM for cost reduction. The supplier has to be able to set up clear communication channels with the OEM with focal point managers at each site. The supplier has to maintain good schedule tracking in its facilities. For example, the supplier can point to an OEM subcontracted job location and progress on its production floor anytime it is requested. The supplier must be able to use standardized documentation and meet high-quality expectations and standards.
2. Value Added Manufacturing Supplier. This level of supplier should encompass all of the elements of the world-class manufacturing capability of cost-driven suppliers. The supplier must demonstrate additional multi-technology capability such as electronics, sheet metal, plastics, box assembly, and software configurations, along with mechanism fabrication, assembly, and test.
3. Design Services. This level of supplier should demonstrate a multidisciplinary capability of electronics, mechanical, and software design skills. The supplier should have most of the major design and analysis tools in electronics and mechanical assembly such as PCB, ASIC design and layout, CAD and finite element analysis (FEA), computational fluid dynamics (CFD), and software development and documentation. The supplier should also be familiar with rapid prototyping techniques and regulatory and international standards and should have connectivity to major manufacturing subcontractors and suppliers.
4. OEM Design Contract Manufacturer. This level of supplier specializes in specific market segments such as office automation (printers and displays). It can provide turnkey services by performing the entire product realization cycle and produce its own products as well as contracting for other OEMs. These suppliers are usually the technology leaders in the market segments they serve.
5. Product Development Services. This is the highest level of supplier that provides design optimization and implementation from specifications.

They possess multidiscipline competencies such as mechanisms, optics, control, and software. They are familiar with design analysis tools and methods. They have extensive access to research labs and universities. They use the latest concepts and control software in project management. They also have intensive experience in manufacturing processes such as electronics assembly and fabrication, plastics, metal turning, and sheet metal. They can also offer industrial marketing and design in order to develop a consistent look and feel of the product that is in synergy with the OEM corporate and product identity and planning. They have in-depth experience in ergonomics and user interface, including human factors engineering, simulation and testing, and multicultural global needs. They can offer marketing skills and tools such as customer surveys, product validation, and market share analysis as well as modeling and rapid prototyping services and domestic and international regulatory reviews.

6.3 The Changes to the Product Realization Process and Communications with the Supply Chain

The product realization process has undergone several changes with the advent of concurrent engineering. The change from a serial process of product development to a more parallel process has resulted in the need for new paradigms.

Traditional product development required a top-down control of the various activities of product creation. Very formal organizational structures were developed and managed with a phase review process. Plans and milestones had to be completed at the end of each phase of product development and were subject to several levels of management reviews. After each review, the project was allowed to proceed and be funded until the next review.

The pressure toward shorter project time frames, global teams, and design and manufacturing outsourcing has resulted in significant changes in the relationship between company personnel and their suppliers with more frequent communications occurring earlier in the product development cycle. The changes can be summed up as follows:

1. Less-frequent formal milestones in the development process, but many smaller informal meetings, most of which are one-on-one engineering interactions. Included are those with the supply-chain engineers, discussing the merits of design functionality, engineering analysis, design validation, and manufacturability feedback (DFM and ESI). There is a need to quickly have these meetings available to all interested engineers without having to hop on the next plane and meet somewhere in the middle. As one design manager explained, "OK, I have a problem . . . let's put an agenda together, and let's get together (on the Internet)."
2. The trend to faster time-to-market has necessitated earlier than usual release of hard tooling. Although investing in rapid prototyping and soft-tooling methodologies can minimize risks, production tooling is being launched much earlier than before. One manager in a plastics supplier said, "Steel is being cut for the tooling before design reviews are completed."

3. This early release of tooling has resulted in the supplier taking on more of the risk for tooling being reworked or scrapped. In addition, the reimbursement profile for the supplier is changing since more work, such as tooling and prototyping, is performed earlier in the design cycle.
4. The selection of the potential supplier(s) has to occur early in the design stage, without the benefits of being able to select suppliers from a bidding process. This is difficult since a complete set of product documentation has not been finalized. Therefore, the selection process will depend on intangible issues such as the history and financial position of the suppliers, their communication methodologies with their customers, and their demonstrated quality levels and cost models. This has led to many new methods for supplier selection, including bidding on manufacturing a particular commodity design, such as PCBs, using their cost models and not through actually examining the physical product before bidding on it.
5. Though supplier exchange networks have sprung up in many industries, their focus is on commodity items, not on outsourced designs.
6. The OEMs are using various methods to obtain manufacturability feedback prior to awarding contracts, such as inviting a selected list of potential suppliers to provide ESI information before contracts are awarded. These additional ESI costs will have to be negotiated between the suppliers and the OEMs or become part of the operating overhead structure of the suppliers.
7. In some OEMs, the PM will invite potential suppliers, which were qualified beforehand, for a closed session where the initial design details are shown to each supplier individually. DFM/ESI critiques are solicited from each supplier in order to input the DFM data into the design even if a supplier is not selected. This allows for proceeding further in the design process before selecting the final supplier. A modified version of this scenario is to send initial design drawings via electronic means, with the details time-stamped to expire in a certain period to solicit DFM feedback. In either situation, a nondisclosure agreement (NDA) is signed by both parties to ensure proprietary information protection.
8. The trend toward lower manufacturing costs and DFM is changing new designs into fewer but more complex major parts. The pressure toward faster product development is also increasing the amount of information available on engineering drawings. Engineers prefer a smaller number of drawings with the maximum amount of information attached to them, such as showing as many parts as possible in one drawing, as well as adding assembly information and special instructions. Tooling and NRE expenses per part will increase, but the total number of parts will decrease.
9. As a result of the items above, the need of increased communications between distributed project teams, design analysis and support experts, and manufacturing resources in the company and its supply chain is expanding. There have been many technological developments in communications among the various stakeholders of the design process, which will be discussed in Chap. 10.

6.3.1 Supply Chain Development

The use of the supply chain is changing as the manufacturing services sector continues to grow. The major supply companies have mimicked the OEMs' reach by distributing their manufacturing centers globally, to be near their customers' sites. In this manner, supply companies can service global OEMs. Based on interviews highlighted in Table 6.1, the issues of global supply chain can be summed up as follows:

1. The suppliers are focused on their customers' issues. Time-to-market is one of the most important goals for the communications and electronics industries. The supplier companies have set up plants around the world to react quickly to their customers' demands. Flextronics' CEO said, "In all of our facilities around the world, with the exception of burned-in products . . . nothing takes longer than a day to build."
2. The location of the supply companies is essential for their ability to provide increased service during the prototype phase of the product development. One manager from a major supplier said that, "Locating my plant near my customer can save two days of Fed-Ex time for prototypes."
3. The OEMs are forcing their suppliers to conform to their design specifications. For example, most OEMs will specify that the design and manufacturing documentation from suppliers will conform exactly to their in-house computer aided engineering (CAE)/CAD systems, including the system type and model number. This is forcing their suppliers to maintain several CAE/CAD systems with operators knowledgeable in more than one system, since OEMs in the same industry sector use several design systems.
4. The OEMs are also asking their supplier for final testing, including troubleshooting of their products and systems. In these cases, suppliers are absorbing the cost of the training programs for test technicians. To make matters worse, these technicians are recruited by competitive suppliers, and even by the OEMs themselves. This is resulting in wage competition, and raising labor costs for skilled supplier personnel in the supply chain.
5. Increased dependence on supplier quality and lower-cost goals have resulted in eliminating incoming inspection for parts, making companies vulnerable to spurious quality problems in the supply chain.
6. The trend toward increasing the links in the supply chain by further subcontracting to achieve even lower-cost manufacturing has resulted in low-technology suppliers getting into the manufacturing cycle of high-technology products. These suppliers do not have the sophisticated technology or the controls in place to make sure that all necessary specifications are inspected. Serious variances in quality are not promptly reported up the supply chain. For an OEM, a poorly managed supply chain is vulnerable to quality problems if changes are made in the subcontractor chain without their approval or notification. It is recommended that the supply chain does not extend beyond three levels down from the final assembly.

6.4 The Supplier Selection Process

A selection process consisting of two steps should be applied: (1) qualifying potential suppliers who meet a minimum set of requirements for financial, operational, and technology issues, and (2) comparing the qualified suppliers through a supplier matrix to

	Weight	Max	Alternative Suppliers			
			A	B	C	D
Quality	30%	96	91/27.3	90/27	88/26.4	48/14.4
Process	15%	120	110/16.5	110/16.5	93/14	68/10.2
Service	10%	58	44/4.4	47/4.7	53/5.3	39/3.9
Delivery	20%	63	49/9.8	50/10	44/8.8	29/5.8
Cost	25%	57	43/10.8	39/9.8	45/11.3	35/8.8
Total	100%	78.5	68.8	68.0	65.8	43.1
% of possible maximum			87.6%	86.6%	83.8%	54.9%

TABLE 6.3 Weighted Criteria for Supplier Selection Matrix

determine the final supplier selection, as shown in Table 6.3, for the selection of a PCB supplier for an information technology product. The matrix is based on a criterion rating system of comparing alternatives. The max is the maximum value that can be attained by a supplier for a particular criterion, if it has met all expectations. The types of criteria and weight assigned for each is determined by the PM and the supplier selection team. For suppliers, their score is multiplied by the weight of each criterion, under the supplier column. Each criterion is composed of many sub-criteria in order to render a complete analysis of the decision. Examples of sub-criteria matrices are shown in Table 6.3. All data from these tables were drawn from actual case studies of electronic assembly subcontracting performed by the author.

The PM and the supplier selection team should select the suppliers who can best develop, operate, and control their processes. They should plan to develop partnerships with the most capable and proven suppliers. In addition, they should use standardized processes in design and manufacturing to certify suppliers' new products and processes in order to obtain best value from the suppliers. Before deciding on subcontracting, a new product has to be targeted for outsourcing and the following steps taken to ensure the proper selection of the final qualified supplier:

a. Define the new project or product to be subcontracted, preferably early in the design stage

b. Select potential suppliers from word-of-mouth or industry supplier databases

c. Narrow number of potential suppliers to three to five alternatives based on synergy of operations and products

d. Establish contact with alternative suppliers using phone or verbal screening

e. Prepare technical data package (TDP) for subcontracting and send to each alternative supplier

f. Send request for quotation (RFQ) to each alternative supplier

g. The selection team should plan for supplier on-site visits for technology, capability, and quality assessment.

h. Final selection of approved suppliers should be based on the criteria rating matrix methods similar to Table 6.3.

6.4.1 Criteria for the Supplier Selection Process

In this section, the suppliers are investigated through a prescribed set of criteria, which are then ranked in rating matrices as shown in Tables 6.3 and 6.4. The criteria should be agreed on before the site visits and filled out after each visit. A sheet for each alternative supplier should be completed with information gleaned from the public domain as well as querying each supplier and investigating the supplier data from published records as well as observed operations. The individual criteria shown in Table 6.3 each have their own subset of criteria and weights. They were divided into six categories with the criteria as shown below. The first category is used for the initial selection of qualified suppliers for business and operational soundness, and the remaining five categories are used for final selection of the suppliers in the weighted criteria matrix shown in Table 6.3.

1. Business and Technology. Some of these criteria can be obtained from formal business records of the suppliers, while others can be ascertained by interviewing the supplier management team during the onsite visit.
    - Dunn & Bradstreet ratings as well as other financial rating services.
    - Supplier business profile, historical development, and growth pattern; whether they are a stand-alone entity or part of a larger enterprise.
    - List of customers, business volume, and commitments; there should be no dominant OEM customer for more than 40 percent of the supplier's business.
    - Deep commitment to compliance with relevant industry standards, such as ISO 9002 certification, GMP, Underwriters Laboratories (UL), Canadian Standards Association (CSA), Technischer Überwachungs-Verein, English: Technical Inspection Association (TUV) listings, specialized department of defense (DoD) MIL-STDs, FDA, FAA, and electronics industry standards (IPC, SMTA), depending on the industry sector.

	Alternative Suppliers				
Criteria	**Max**	**A**	**B**	**C**	**D**
ISO 9000 Certified	10	10	10	10	3
Manufacturing Standards	10	10	10	8	8
Statistical Process Control	10	8	9	10	1
Quality Diagrams Available	10	10	10	10	5
Test Failure Reporting	10	8	7	10	5
Incoming Inspection	10	10	10	10	4
Defect Analysis	8	8	8	5	8
Obsolete Material	8	7	6	8	4
Design for Manufacture	10	10	10	7	5
Continuous Quality Improvements	10	10	10	10	5
Total	96	91	90	88	48

TABLE 6.4 Weighted Quality Criteria for Supplier Selection Matrix

- Internal controls for financial and process operations.
- Supplier's financial terms and contractual obligations and any financial burdens that can adversely affect the supplier operations such as patent infringement lawsuits and obsolete or written-down inventory.
- Supplier materials, workmanship, and product discrepancy outcomes.
- Supplier design capability and systems compatibility, including CAD (ME, EE) and design analysis if available.
- Supplier DFM/ESI guidelines, kept up to date and strictly enforced.
- Supplier staff of direct and indirect labor force including operators, technicians, process development and test, and their supporting engineers and managers. This should include information about staff experience (years employed) in and outside the supplier and wage rates and benefits. The ratio of direct to supervisory and professional staff should be noted, and the ranks of the top management as well as and their historical experiences should be investigated. Employee turnover should also be noted.
  - Supplier training profile and whether the supplier is maintaining current training and skills distributions matrices.
- Supplier customer base, including names and sizes of customers, from largest to smallest. The supplier selection team should interview some of the customers for more insight into the supplier performance and operations.
- Supplier customer service operation and methods of communications with the OEM.

2. Quality and Reliability. These criteria were selected by the PM and the supplier selection team. They consist of the following criteria:
   - ISO 9002 certification, which is given a higher priority than other standards
   - Other manufacturing standards complied with (IPC III), British Approvals Board for Telecommunications (BABT), UL, CSA, TUV listed
   - Statistical process control (SPC) and control charts prominently displayed; check that they are current
   - Assembly and test yields
   - Pareto of defects and root-cause analysis
   - Test failure database and corrective action
   - Incoming inspection equipment and capability
   - Defect analysis laboratory and capability
   - Disposition of nonconforming, obsolete materials, and returns
   - Continuous flow lines
   - DFM/DFA/DFT/ESI guidelines and manuals

   The summary of the supplier selection team quality matrix is given in Table 6.4 for a typical supplier selection process.
3. Process and Materials Capability. These consist of the following criteria:
   - Sheet-metal capability
   - Cabling system design and build and whether it is subcontracted

- Ability to handle various electronic component technologies: through hole (TH), surface mount technology (SMT), fine pitch, ball grid array (BGA), flip chip, chip scale, chip on chip, etc.
- Separate stockroom for each customer
- Availability of standard components
- Long lead and material shortage reports
- Ability to locate jobs in progress and logistics tracking
- Inventory turnovers/year (four to 10 typical)
- Stock pulls and replenishment processes
- Is there a separate prototype line?
- Final product and subassembly testing capability
- Repair capability and types of repair equipment and processes

4. Service and Support
   - Single point of contact
   - Supplier team assigned to customer
   - Joint technical problem resolution process
   - Communication facilities: email, phone, fax, and electronic data transfer
   - Collaboration on DFM/DFA/DFT/ESI
   - Openness and sharing of process and technology roadmaps, schedules, and test results
   - Continuous process improvements
   - Check references for adequacy of service and support
   - Attention level given by supplier during evaluation period
   - Supplier organizational and management stability
5. Delivery Performance
   - On-time delivery profile: (–3, 0) days
   - Lead time, including:
     - Compliance to committed lead-time
     - Competitive lead-time (kit complete to delivery, four to 10 days)
   - Locked and loaded (schedule freeze) lead-time (seven to 30 days)
   - Current production capacity
   - Upside capacity limit and plans for capacity expansion
   - Capacity trigger when supplier acquires additional capacity (at 70 percent loaded capacity is preferred)
   - Rework capacity
   - Check on incentives and fines for early or late delivery
   - Check on the supplier contractual language

6. Contracted Product Cost Structure and Bid Analysis. Many suppliers, as well as OEMs, have a cost model for estimating costs at the early design stages. Some of these cost model methodologies will be explained in Chap. 11. These costs models use ABC techniques to quickly provide for an estimate to the bid proposal. Most of the cost models are simple and are based on the material cost for the product as well as the NRE associated with manufacturing tooling and testing. The topics to be agreed on by the selection team included the following:
   - Cost estimate of sample product or assembly to be subcontracted
   - NRE for sample product or assembly
   - Material vendor warehousing
   - Purchasing leverage from the entire supplier customers' demand to highlight possible materials savings
   - Cost models and hourly rates for assembly, in process and final tests
   - ECO and rework hourly cost rates
   - Frequency of cost reductions
   - Historical yearly cost reductions achieved in the past (range 4 to 10 percent per year)
   - Supplier materials overhead
   - Supplier warranty of workmanship (type and duration)
   - Forecast downside liability (30 days typical)

Table 6.5 is a summary of the cost structure for an electronic manufacturing services (EMS) supplier cost estimates for PCBs. Relevant data was collected for each of three potential suppliers, in addition to the internal manufacturing department at the OEM.

It can be seen that the cost differential between the three alternative suppliers is small, ranging less than 4 percent from the highest to the lowest. This is a testament to

Metrics: Suppliers	1	2	3	Internal
Cost of Sample Board, $	700	728.52	703.00	738.91
Materials/Sample Board, $	N/A	531.60	521.06	518.67
Materials Divisor	N/A	0.7297	0.7412	0.7200
NRE for Sample Board, $	N/A	15,924.00	$6,325.00	19,289
Material Vendor Warehousing	Yes	Yes	Majority	Minority
Purchasing Leverage, $M	225M	500M	15M	120M
Frequency of Cost Reductions	Quarterly	Quarterly	Quarterly	Quarterly
Typical Reductions Achieved	4 to 6%	4 to 5%	N/A	10%
Workmanship Warranty	12 months	30 days	12 months	Negotiable

TABLE 6.5 Supplier Cost Structure for PCB Assembly

the convergence of manufacturing processes for a mature sector such as PCBs. Most suppliers provided their cost model, which is a multiplier of the materials cost (materials divisor) as well as NRE expenses. Additional information solicited from the suppliers included the materials purchasing leverage, which is the total materials purchasing power from all of their customers. In the case of suppliers 1 and 2, they exceeded the internal OEM material leverage. Cost-reduction frequency and scale were also noted, which derives from the reduction in the learning curve, which is similar for all alternatives. Finally, the workmanship warranty was typically set for 12 months.

Given the cost data from Table 6.5, it would seem that the best alternative was supplier 1. The additional work performed by the PM and the supply selection team (summarized in Table 6.3) also indicated that supplier 1 was the best choice for the total selection of criteria used.

On-site visits to the suppliers were used to generate some of the criteria rating in all of the different attributes. The visits were an opportunity to audit each supplier for the six types of criteria categories as follows:

- Each potential supplier was provided with the criteria categories so they are ready to answer the questions posed by the PM and the selection team.
- Schedule at least a half-day and lunch for on-site supplier audit.
- Verify all major concerns with introductory meetings and plant tour.
- Meet supplier PM and team to be assigned to the OEM.
- Ask for quality data and cost models.
- Meet with all suppliers' functional managers.
- Examine procedure books, wall charts, test documents, and job locating tags for clues of supplier efficiency and process control.
- Talk to all levels of employees to gauge supplier culture.

6.4.2 Presenting the Subcontracting Plan to Management

After the supplier selection team has made the final supplier selection, the team should plan to make a thorough and thoughtful presentation to management. The team should have a game plan for the presentation and invite the internal manufacturing and operations management to the meeting. The team should plan to:

- Address all management concerns as mentioned earlier in this chapter.
  - Absorption, CGS, capacity, delivery, impact on schedules
- Show where the OEM's own competencies are lacking
  - Use core competency map shown in Fig. 6.4
  - Be careful not to downgrade OEM competency too much
  - Focus on new product requirements
  - Show the advantages of standardized processes
- Be prepared to battle in the matrix details
  - Make sure that supporting metrics and documentation are available for each score in the matrix
  - Have the matrix available from all of the suppliers' plant tours

    - Discuss all points with own OEM operations staff
    - Show all data, details as well as summary
- Keep discussions at high levels
    - Use of graphs is a must
    - Be ready to focus on details if necessary
- Focus on your product needs
    - Competencies required for your product
    - Mismatch of your needs versus available supplier competency; be prepared to discuss intangibles such as DFM, quality, marketing forecast, and test strategy
- Keep away from the minutia of technical details that could slow down the meeting; don't get bogged down with analysis paralysis.

6.4.3 Issues to Address Before Signing a Contract with a Supplier

Before the contract is signed with the selected supplier, these issues should be addressed with the relevant parts of the OEM organization:

1. Legal and Liability Issues. These should include the standard boilerplate used in the OEM bidding process:
    - Contractual clauses
    - Bidding instructions
2. OEM Standard Terms and Conditions, including:
    - Instructions to bidders
        a. Quality and 100 percent conformance to specifications
        b. Commitment to supply defect-free parts
        c. Supplier quality plan, including quality objectives, responsibilities of each party and testing, quality control, inspection and audit programs
        d. Supplier warranty, including workmanship defects; aim for six to 12 months replacement warranty for workmanship and field failures
        e. Buffer stock for replacement parts to be kept by the supplier for the OEM. Define quantity and duration of stock availability
        f. Agreement process for rejects and returns between the OEM and supplier
            - Liability of reject agreement and determination
            - Consequences of reject and cost-sharing agreements for rejects, if any
        g. Engineering specifications and ECO handling and time horizon
    - Infrastructure to manage the supplier subcontractors, as well as the responsibility of the supplier to audit and rate their own subcontractors. Details of subcontractor supply-chain qualifications and audit plans should be available. No subcontracting, delegation, or changes without written permission from the OEM. The supplier quality plan for its supply chain should be made available to the OEM.

3. Supplier Project Management. Details of critical activities between OEM and supplier:
   - Dates, development, equipment, activities, and costs timelines
   - Procurements activities and manufacturing lead-time
   - Interface teams, duties, and responsibilities for each party
4. How to Handle Variances in the Contracts
   - Supplier to provide monthly material variance reports as well as actual bill of materials (BOM) and invoices
5. Supplier Quality Plan. Supplier should warrant that its goods and services comply with conformance certificates. Other elements of the quality plan include:
   - Performed within industry standards and practices
   - Equipment, process, and tools as well as calibration and maintenance procedures
   - Supplier to be environmentally responsible; its conformance plans adhere to environmental standards
   - Supplier is subject to inspection and approval within specified time
   - Options if supplier goods fail
   - Rework plans if needed, at OEM, supplier, or third party
   - Acceptance tests
     - May occur at any time with or without sampling plans
     - Maximum wait period and notification in writing of defects
     - Acceptance does not relieve supplier from latent defects to the OEM
     - OEM can accept nonconforming material if noted
     - Defective product handling including repairs, returns, and failure analysis conditions as well as no-trouble-found (NTF) determination
   - Batch and Inventory Control
     - Part-numbering system and methods to align the OEM and supplier material management systems
     - Supplier batch and equipment setup control and flags for assembly and test
     - Tooling and NRE costs should be reviewed each quarter for absorption and amortization
6. Termination of the subcontractor should be based on written notice (three to six months)
7. Delivery: liability for late shipment and cost of OEM lost shipments
   - Supplier notification of late shipments
   - Supplier liability and changes with time (from the date of order)
   - Late-delivery penalties as percent of contract price

8. Schedule Management
   - OEM forecast (six to 12 months)
   - Fixed purchase-order period (three to six months)
   - Long lead-time cover (establish rules)
   - Delivery schedule (four-week outlook)
   - Schedule-commit process (establish rules)

6.4.4 Outsourcing Quality Issues

The two guiding principles of the quality practices for the supply chain are (1) do not generate any defects within your span of control, and (2) do not pass on any defects to the next link in the chain. Quality is controlled by the use of adequate tests at different stages of the supply chain. It is generally assumed that the lowest total cost of quality is testing at the lowest level of product assembly possible. Defects are more expensive to find and remove at the higher levels of product assembly.

Given that tests are necessary at the lowest possible level, companies had their suppliers assure their quality by testing their product at the supplier location, deliver the product into their factories, then test them in their incoming inspection department. This is an obvious duplication of effort and, therefore, a quality system for incoming inspection was devised to include the supplier management program and an incoming inspection and test program. The supplier management program is a cornerstone of quality assurance in a supply chain. It consists of many steps, including:

1. Qualifications. The qualification process begins with a list of potential suppliers, which is then audited either by reviewing their documentation (financial, manufacturing, and quality procedures) or visiting the supplier manufacturing facility. A review of their procedures and conformance to international standards of manufacturing and quality is conducted before placing them on the approved list of suppliers.
2. Supplier Ratings. After the initial approval, the supplier is usually placed in an approved status. Purchase orders are placed, and goods are received with consistent incoming inspection and testing procedures.
3. Preferred Status. After a period of time, and with increased communications and confidence in the supplier demonstrating its capabilities and quality in a consistent and continuous manner, a supplier can be placed in a preferred status. This will encourage the company to place orders with this supplier, reducing tests and inspection of incoming products.
4. Partner Status. The next step in the supplier status is the full partnership, sometimes a preferred ranking such as "A preferred" or "partner." In this case, purchase orders might be placed with the partner without bids. Partner parts can go directly into the company's stockroom without incoming inspection tests in a program known as "ship to stock" or "dock to stock."
5. Quality Audits. In the process of qualifying or rating suppliers, quality audits are performed to ensure conformance with industry and OEM company standards. The quality audits could take the form of visits to the suppliers to check on performance or special testing and inspection of supplier materials, beyond normal test and inspection procedures.

The level of incoming inspection is dependent on the status of the suppliers. For a nonqualified or new supplier, there should be an extended incoming inspection and test program. On the other hand, a preferred supplier should be in a position to ship to stock, where its materials are received directly into the stockroom with no incoming tests or inspections. A supplier whose materials are to be altered in production should have a greater level of inspection and test, while a supplier with materials, which will stay intact in the product, can have a reduced incoming inspection program.

Another method of substituting for the incoming inspection function is to ask the supplier to provide certificates of compliance and/or test to specifications. The conformance or test certificates should be mutually agreed upon to include the specifications that are relevant to the product being manufactured.

An important part of the supplier quality system is the resolution of quality and nonconformance problems. The quality system, as described earlier, can provide a measure of protection to the company from spurious quality problems in the supply chain. Many specifications are given for new products, and not all of them are tested in incoming inspection or at final product test. The probability is significant that one of the suppliers will incur some quality problems during the lifetime of the product. Rather than react to such quality problems, it is more desirable to plan ahead and include provisions for resolving legal and liability issues in the contractual agreements signed by the company and its supply chain. Otherwise, nonconformance to specification by a supplier could result in costly litigation.

6.4.5 Legal and Liability Issues in the Instruction To Bidders

In order to provide a clear understanding of the relationship of the suppliers to the company, the following clauses should be included in the instructions to bidders for prospective suppliers:

1. Quality Level. The supplier provides for 100 percent conformance to specifications, with a commitment to supply defect-free parts.
2. Quality Plan. The supplier's quality plan should include quality objectives, responsibilities of the supplier, and a description of their quality, testing, inspection, and audit programs for their own suppliers.
3. Warranty. The supplier should have a warranty of their parts for workmanship and field failures for a given amount of time, usually six to 12 months. The supplier should keep a certain level of buffer stock of parts in case of increased defects and a policy for handling rejects and returns from the field.
4. Project Management. The supplier should provide details on critical activities such as dates and costs of equipment development, procurement of materials, and manufacturing leadtimes. The supplier should also identify the interface team and its plan for managing its subcontractors. The supplier should also indicate monthly material variances and cost sharing agreements, if any.
5. Subcontractor Management. The supplier subcontractors'relationship plan should be outlined, including the subcontractor qualification process and its management for quality and performance. No subcontracting, delegation, or changes in the subcontractor status should be initiated without written notification from the OEM.
6. Delivery. The supplier should identify its liability for late shipments, including the cost to the company of lost or late shipments. Procedures for notification of late shipments as well as a weighted scale for liability-versus-time should be outlined.

7. Forecast. The supplier schedule management should include a forecast horizon, usually six to 12 months. Special provisions should be made for long-lead items, short-term delivery schedules, and the commitment-to-build horizon. The short-term build schedule should be fixed and cannot be changed without penalty to the OEM.
8. Operational Issues. Typical operational issues to be addressed in the contract legal clauses include items such as the part-numbering system for materials, batch control, and flags for assembly and test. They should include the cost and ownership issues of tooling and purchasing, amortization of equipment, and a termination of contract notification period.

6.4.6 Infrastructure to Manage Subcontractors

The supplier management program should be controlled by a single PM responsible for communications with and control of a particular supplier. The program should not be based on a commodity management model as shown in the top of Fig. 6.5. The commodity model involves many lines of communication and could be slow to react to problems with the product design or manufacturing. For example, an engineering change to the product has to propagate through the OEM commodity organization, rippling through multiple points of contacts and increasing the chances of miscommunications. In the decentralized supplier management model, at the bottom of Fig. 6.5, design and manufacturing information and decisions flow through the supply chain, as each tier of the supply chain passes the relevant information down the chain. In Fig. 6.5, the point(s) of contact, shown as a circle with a letter c inside have been significantly reduced.

The steps involved in good supplier management involve the following:

- Select PM and their team and assign tasks and responsibilities
- Identify resource requirements and technical personnel for the supplier management team
- Establish an interface link to the subcontractor and agree to the frequency of regular contract

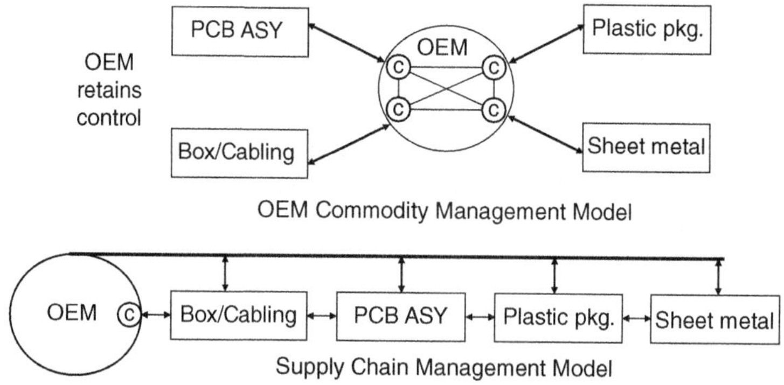

FIGURE 6.5 Supplier management models for electronic products.

- Identify key project criteria and goals for the contract and the managers at the OEM and supplier
- Ensure proper targets are set and reviewed regularly in quality, cost, and delivery

6.5 Summary and Case Studies of Subcontracting

In order to illustrate successful methodologies and potential problems of contracting, several case studies will be presented based on the author's experience in research, consulting, and expert-witness litigations of the supply chain.

Subcontracting is an important part for the global high-technology industry. It has taken on different methodologies depending on the industry sector (electronics, automobiles, aeronautics, defense, plastics, etc.), yet remains surprisingly uniform in each industry sector. The important issues for successful subcontracting discussed in this chapter are as follows:

1. Examining the financial aspect of subcontracting and performing the necessary economic analysis for sound decisions on the amount and level of subcontracting, including the impact on the companies' financial indicators and reporting.
2. Performing due diligence on following a robust process for potential supplier identification and selection, including a thorough analysis of adverse consequence if the supplier fails in quality, delivery, and cost improvements.
3. Enshrining these efforts in an appropriate set of legal documents such as instructions to bidders and standard terms and conditions in the final contract negotiations.
4. Establishing a proper structure to manage the suppliers for accurate communications and rapid reaction to problems and crises.

Examples of subcontracting successes and failures shown below are direct results of either applying the principles above or failing in one or more of these categories:

I. Subcontracting Case Study I

Blood Monitor for Consumer Market in a Medical Company Engineering Subcontracting

1. OEM Resources and Plans
    - Product Description: 38 custom parts, including PCBs, mechanics and kinematics, interface and cable, with assembly documents for each
    - Sales estimated to be in the 200,000 to 1,000,000 range per year. Prototype runs of 100 to 300 for FDA approval
    - OEM Product Design Team: 10 mechanical engineers, 45 electronics and software engineers
    - Mechanical Design Software Tools: 14 ProE™ 3D workstations for design team and suppliers
    - Several AutoCAD™ stations for 2D manufacturing support
    - All final products assembled and tested at the OEM (250 associates)

2. Supply Chain
    - Total: 11 design suppliers used for pumps, valves, with additional ANSYS™ design analysis supplier at $8K/day
    - Design Cycle: six month concept phase to negotiate with suppliers; 18 to 24 months total design cycle for the product
    - Communications: two conference calls/week, one email/day (with faxes); video (deemed to be poor); face-to-face meetings one to two times per month; local suppliers visited more frequent than transnational suppliers
3. Example Outsource Design (Pump)
    - 15 parts, OEM modeled internals, specified output envelope; interferences and specified operating conditions of environmental, temp, flow rate, and power consumption
    - Supplier negotiated more power in the early design phase; OEM required quality certification and complete engineering analysis and testing of first pieces of production

II. Subcontracting Case Study II

High-Technology Communications Company Engineering and Manufacturing Subcontracting

1. OEM Resources and Plans
    - Product Description: electronics box with multiple PCBs and connectivity to the outside.
    - Sales estimated to be 10,000s to 100,000s range.
    - Major customer demanded fast-track development for an enhanced existing product. Led to conscious decision of PM to speed up development for time-to-market and handle manufacturing cost and quality problems with ECOs after product release.
    - OEM Product Design Team: distributed at two locations in the U.S. 30 percent MEs, 70 percent EE and software, CAD modeling and analysis completed in-house.
    - All parts assembled and tested at the supply chain.
2. Supply Chain
    - Total (3): one design and two manufacturing suppliers at global locations
    - Start with five design concepts; narrow down based on ESI/DFM from manufacturing suppliers input to three concepts
    - Communications: weekly local OEM team meetings; OEM design team visits (any two sets of three supplier locations), every two weeks; four to six OEM engineers travel for each meeting
    - Design Reviews: video conferencing for design reviews; supplier comments by email attachments before meetings; Q&A summary published one week before meeting
3. Consequences of PM Development Speed-up Decision
    - The PM decision to speed up development resulted in many ECOs after production release. Decision originated with customer request.

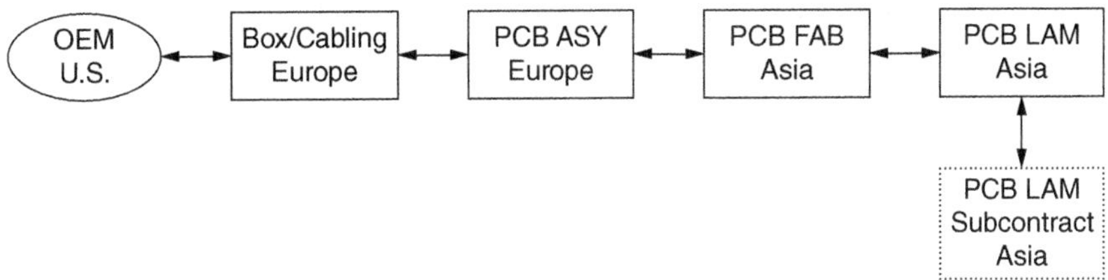

Figure 6.6 Subcontracting Case Study III: Failure at a Computer Company OEM.

- ECOs were executed for costs and quality improvements.
- Products rushed to market before they were ready.
- Multiple service complaints from customers leading to potential loss of future orders.

III. Subcontracting Case Study III

Supply Chain Management Failure at a Computer Company OEM.

This case study shows the risks of having a long (more than three levels) supply chain and the difficulties encountered, as shown in Fig. 6.6.

The OEM decided to initiate a supply-chain structure for its large system designs. A major design supplier was contracted, tasked with the establishment of a supply chain it managed, and delivered the product to the OEMs customers. The following scenario ensued and caused major product quality issues.

- The box and cabling supplier designed the systems and set up the supply chain to provide PCBs to its system-assembly facility. The supplier chose a PCB assembly supplier in Europe, who chose a raw PCB fabricator in Asia, who in turn chose a subcontractor for PCB lamination (LAM) in Asia.
- The system design was validated from pilot units provided throughout the chain with full functional testing of all specifications to certify all manufacturing processes in the supply chain.
- The box and cabling design contractor provided all test equipment for the supply chain to test production units. Only major specifications were tested, based on economic return analysis. It would be impractical to test all specifications that were verified in the pilot run, since the cost would be prohibitive.
- Product and delivery ran smoothly for a period of time.
- The PCB lamination supplier in Asia decided to subcontract the lamination process because of capacity issues. They chose another Asia subcontractor for lamination without prior approval from the supply-chain management.
- The additional PCB lamination subcontractor in Asia did not conform to specifications in the design. It resulted in spurious quality and data communications problems in the field for the OEM systems.
- This problem was not detected by the in-house test equipment and potentially defective systems were shipped.
- Problem was eventually detected at customer sites.

- Since the quality problem occurred in lamination, the potentially defective PCBs were not repairable. Those already shipped had to be replaced. Large amounts of inventory throughout the chain were unusable.
- Financial issues had to be resolved by litigation.

This case shows the typical risks when not completing due diligence in supplier selection and searching for the lowest-cost supplier as well as too many levels in the supply chain. In addition, poor management of the supply chain and improper notification and prior approval of subcontractor changes compounded the problem. Another issue in this case study is the great damage that a small, low-technology supplier can inflict on a high-technology product. In resolving this case in court, the supply-chain contract clauses were the main documents used by the judge and attorneys in negotiating the financial settlement between the parties in the supply chain.

References and Bibliography

Blanchard, D., *Supply Chain Management Best Practices*, 2nd ed., Wiley & Sons, New York, 2010.

Cohen, S., and J. Roussel, *Strategic Supply Chain Management: The Five Disciplines for Top Performance*, McGraw-Hill, New York, 2005.

Dowlatshahi, S., "Implementing Early Supplier Involvement: A Conceptual Framework," *International Journal of Operations & Production Management*, 18(2):143–167, 1998.

Jacoby, D., *Guide to Supply Chain Management: How Getting it Right Boosts Corporate Performance*, Bloomberg Press, New York, 2009.

Kusiak, A., *Concurrent Engineering: Automation, Tools, and Techniques*, John Wiley &Sons, New York, 1991.

Lambert, D., *Supply Chain Management: Processes, Partnerships, Performance*, 3rd ed., Supply Chain Management Institute, Sarasota FL, 2008.

Lavassani, K., B. Movahedi, and V. Kumar, "Developments in Theories of Supply Chain Management: The Case of B2B Electronic Marketplace Adoption," *International Journal of Knowledge, Culture and Change Management*, 9(6):85–98, 2009.

Leonard-Barton, D., "Core Capabilities and Core Rigidities: A Paradox in Managing New Product Development," *Strategic Management Journal*, 13(S1):111–125, 1992.

Ma, Y., G. Chen, and G. Thimm, "Paradigm Shift: Unified and Associative Feature-based Concurrent Engineering and Collaborative Engineering," *Journal of Intelligent Manufacturing* (2008).

Mclvor, R., and P. Humphreys, "Early Supplier Involvement in the Design Process: Lessons From the Electronics Industry," *Omega*, 32(3):179–199, 2004.

Mentzer, J., "Defining Supply Chain Management," *Journal of Business Logistics*, 22(2):1–25, 2001.

O'Neal, C., "Concurrent Engineering with Early Supplier Involvement: A Cross Functional Challenge," *Journal of Supply Chain Management*, 29(2):2–9, 1993.

Prahalad, C., and G. Hamel, "The Core Competence of the Corporation," *Harvard Business Review*, (68)3, 79–91, 1990.

Simchi-Levi, D., and P. Kaminsky, *Designing and Managing the Supply Chain*, 3rd ed., McGraw-Hill, 2007.

Discussion Topics

1. Discuss and rank the top three changes in the product realization process due to outsourcing. Select an industry (communications, industrial, and regulated such as medical/military and consumer) or discuss in general. Please give the reasons for your selections and ranking of the changes in the realization process.
2. Pick an industry type (communications, industrial, and regulated such medical/military and consumer). Show the top four steps in the supplier selection process and indicate reasons for your ranking.
3. You are ready to present your outsourcing recommendation to the CEO and the board of directors of the company. The internally affected operations manager is present at the meeting. Describe the top-three agenda topics you should discuss to allay the CEO fears and how to counteract them. Rank the topics and discuss your reasons for selecting and ranking them.

Problems

6.1. Analyze one of following suppliers below (select only one supplier from any of the outlined industries) or you can pick you own company using the following criteria. Analyze the industry (auto, plastics, electronics, or medical suppliers), but write a specific report on one company that you think is far above the rest in that industry.

- Client industry (OEM) prospects in the next three to five years
- Customer base and business diversification
- Financial position and profit history
- CEO and critical managers' history and prior position at other companies
- Market position and product strategy
- New products and existing product portfolio
- Quality and cost strategy
- Intellectual property issues
- Workforce issues (union contracts, etc.)

Please pick one subcontractor from one of these industries:

a. Tier 1 auto suppliers: www.buyupside.com/sample_portfolios/autosupply.php and www.thestreet.com/story/10949243/1/3-post-bankruptcy-auto-supplier-stocks-for-2011.html Or you choose one the following: American Axle & Manufacturing Holdings, Autoliv, BorgWarner, CLARCOR, Federal-Mogul, Gentex, Johnson Controls, Magna International, Tenneco, TRW Automotive Holdings, WABCO Holdings. There are also two subcontractors that emerged from bankruptcy: Visteon and Lear.

b. Plastics suppliers: www.plastics.com
Focus on local companies in a region, such as the northeast: Sterelite Corp. (Townsend, MA), NYPRO (Clinton, MA) and many others.

c. Electronics MFG services (EMS). Select an international supplier such as Solectron: Milpitas, CA; Sanmina: Huntsville, AL; Celestica: Toronto, CA; Flextronics: San Jose, CA; Jabil Circuits: St. Petersburg, FL; Benchmark: Hudson, NH.

d. Medical design/manufacturing services. Select national or regional; companies such as Quest Medical/Atrion Corporation: Allen TX. Other local and national companies at www.qmed.com/supplier-categories/contract-manufacturing-services.

6.2. A company makes extremely fast data-collection products for research labs around the world using Hadron colliders. Its products use extremely fast triggers for collecting massive amount of data for particle acceleration in real time to be analyzed later off-line. The products are complex, expensive, and sell at low volumes. The products offer a large amount of options and different types of controls and tend to have long lifecycles. The engineering staff is experienced, competent, and highly educated, but the product development cycle is long and tends to function in technical silos. Manufacturing is a job shop organization, with a poor shipping performance. It usually has *end-of-the-month syndrome*, where the shipping profile accelerates toward the end of the month.

Management decided to use the technology to develop products for hot diagnosis of intranets in companies; that is, an intranet can be troubleshot while it is functioning to locate connection problems, without having to turn it off. This would be a good product opportunity and a great advantage to potential customers. The company realized that to manufacture and sell high volumes to the technology market is not within its expertise. The company hired you as a consultant to help with developing this new technology into products to the telecommunication market. You are scheduled to visit with the CEO and make a presentation for how to proceed with applying the technology in a new field, with an organizational structure that can leverage the supply chain for quick product manufacturing.

Discuss organizational and technical issues, use of tools discussed in this and previous chapters, supplier management, changes or proposals, including time horizons, you would like to make to the CEO.

CHAPTER 7

Engineering Project Planning and Execution

In this chapter, the traditional methods of project planning are demonstrated, using techniques that were originally developed and are in common use in construction management. These techniques include listing project tasks, their duration, and who is assigned to each task as well as the interdependencies of these tasks. Once these tasks are identified, their relationships are laid out in sequential fashion, and a critical path is determined to initially calculate the project completion date. Leveling methods of resources and other techniques are used to reduce project cost and duration or selectively shortening activities on the critical path. A plethora of project management software, such as Microsoft Project©, are available, to make the project planning tools much more automated and optimized for project schedule and cost.

The high-technology industry in the past had focused on the technology aspect of their development projects and did not use the construction management approach to plan projects. This is especially true for start-up technology companies, as shown in Fig. 2.1. As global technology products sales increased, and most of these companies began to grow and mature into commodity companies, these project planning techniques became imperative for achieving shorter time to market and increased profits. This is especially true for companies that are sensitive to design to customer (DTC), such as military suppliers.

Once the project planning is completed and actual development begins, the concern is to keep the project schedule on track and synchronized with the project timeline. Many techniques have been developed to keep the project manager (PM) informed of the design team progress and allow for early warning and resolution of potential problems in project execution.

In this chapter, the techniques for engineering project planning using non structured and structured methodologies will be shown. Tools for reducing project costs and shortening project duration will be discussed, with additional emphasis on optimizing techniques for achieving these results. Methodologies for companies to adapt these planning techniques to their special needs will be shown. The project execution focus will be on managing project cost through collecting data from the project's actual expenses and comparing them with the planned project expense profile.

Topics to be discussed in this chapter include:

7.1 *Historical Approaches to Engineering Project Planning.* In this section, the need for more structured approaches to project management is shown, where the emphasis on technology is less important, and the focus on completing the project on time is viewed as a competitive weapon.

7.2 *Project Requirements Definitions.* In this section, the elements of good engineering project planning are shown, together with the requirements for setting the completion date for the new projects, and the negotiations necessary to achieve this completion date.

7.3 *Engineering Project Scheduling Tools.* In this section, the scheduling tools of Gantt and PERT charts are explored, with examples of specific conditions and how to handle them. The critical path method for determining the shortest time to complete the project is shown.

7.4 *Methods and Techniques for Reducing Project Duration and Cost.* These methods are discussed using common sense techniques as well as tools used to level and shorten these projects using the principles of concurrent engineering.

7.5 *The Causes of Engineering Project Execution Problems and How to Mitigate Project Delays.* It has been shown earlier in this book that engineering project delays result in serious consequences in terms of market share and customer satisfaction. In this section, the major reasons for project delays are discussed with methods to mitigate them.

7.6 *Techniques for Monitoring Project Progress and Estimating Project Completion Profile.* These techniques are shown through different methodologies as well as formal and informal communications. A formal system that integrates work scope with schedule and budget resources is shown. Schedule and cost variation are identified and reported to the PM for corrective action. Many of these business best practices are documented in ANSI/EIA Standard 748-98.

7.7 *Successful Project Execution and Lessons Learned.* In this section, details of successful projects that combined financial with communication tools to increase the enterprise awareness of project progress and deliver early warnings of project problems will be examined. The use of increased communications to deliver projects on time and on budget and meeting the original specifications will be discussed.

7.1 Historical Approaches to Engineering Project Planning

The PMs in the early part of the technology revolution in the 1970s matured mostly under the military tutelage of the post World War II and the Cold War generations. The U.S. military funded many technology projects, including the technology innovations produced during the space race with the Soviet Union. The management style of these early PMs was focused more on command and control military style leadership and less on team based and collective project collaboration. The U.S. military won wars through industrial might and superior technology. There was no need to change this winning management style, especially with no other major overseas industrial competition.

Competition with Japanese technology based products in the 1980s resulted in U.S. companies beginning to emulate the Japanese management style in terms of teamwork and collective leadership. The rapid introduction of new products, with emphasis on reducing the number ECOs, resulted in focusing the technology industry on speeding up the time-to-market. Many industries took to the *new products in half the time* idiom, which spawned the need to take a hard look at project planning and execution. PMs became aware of ideas such as, "First to market captures 70 percent of sales." Technology companies were cancelling 50 percent of the new product development projects because they took too long, while technology advances made them obsolete.

The key to successful project planning is to focus on participatory management with the project team. Table 7.1 is a contrast of the command versus team focus management styles. The team focused PM operates in an open style that solicits team inputs and makes collective decisions that are always more reasoned than decisions made solely by the command style manager. Team respect for the manager flows from the manager's proven technical skills and managerial ability and not from a command and control structure. The team-focused PM has a much greater chance in successful project planning and development than a command-focused PM.

7.1.1 Initial Project Planning Steps and Project Statement

The initial project planning begins with the project statement. In design to customer (DTC) projects, the statement flows directly from the specifications provided by the customer. In design to market (DTM) projects, the statement is generated after careful analysis of customer needs as well as the current industry competitive position and the company strategy. Project statements have to consider regulatory and technical issues in terms of technology adoption levels and trade-offs in cost, schedule, and performance and resource constraints of the company.

The project statement should have additional information as to the execution of the project as well as tools and metrics that will be used to develop the product. The statement should be specific such as, "The new product will establish 30 percent market share in the United States, 15 percent globally, at 20 percent gross profit, 1.33 *Cpk* production quality, 100 ppm software defects, less than 5 percent returned units and $100/unit in scrap and rework by the third quarter, 2016."

Style	Command Focus Management	Team Focused Management
Project manager authority	Position based and authority driven	Relationships and credibility based
Project information	Withheld, screened, need to know basis	Open, shared, and available
Management competency	Derived from position or alleged	Technical skills, proven performance
Decision making	By command, individual, specific	Collective and inviting of team input
Sharing of praise, rewards	Individual, selfish	Shared with; acknowledge the team

TABLE 7.1 Team Focus versus Command Focus Management Styles

7.1.2 Development Plans for Design to Customer Projects

The PM should consider these phase steps and milestones before formulating the new product or system development plan for a DTC project:

1. Customer bid review and validations phase: When a bidding opportunity becomes available, the PM should examine this bid in terms of the business strategy of the company and the synergy of the proposal to the company's core competency. After management review of the proposal, a decision is made whether to pursue the opportunity and bid or not bid on the proposal.
2. If the bid is approved, then the PM should begin a resource requirement planning phase as well as a concept and system architecture design phase. This phase should be followed by a concept or preliminary design review milestone.
3. Once the concept is approved, the design and development plan is activated, which ends in the critical design review milestone.
4. The next phase is the design verification and validation phase. It consists of a full performance and specification test for the product as well as environmental, life testing, vibrations, and shipping tests. These tests could be completed using new product prototypes developed in the design laboratories or on a pilot run built with hard tooling in production. Building prototypes with mature tooling and production facilities is a great opportunity to test the tooling and production readiness and obtain design for manufacturing (DFM) feedback. Most companies have opted for production-ready prototypes to remove the extra step of prototype then pilot run phases. This phase is completed with the specifications verification milestone.
5. The final phase is release to production, where all of the hard tooling and documentation for the production phase of the new product is completed. The release to production milestone is a final review of production tooling, documentation, production costs, quality, and reliability estimates.

In some companies, the last phase is combined with the previous phase into one large milestone for design verifications and release to production milestone. More details on this topic are discussed in Chap. 8 on engineering project phases, communications, and control.

7.1.3 Development Plans for DTM Projects

The PM in DTM projects must work with the company business and marketing departments to distill a new product opportunity or a follow-on product proposal into a project statement or future product map. The product map would outline the stepwise adoption of new technology into a family of products for the future. The PM might consider these phase steps and milestones before formulating the new DTM product or system development plan:

1. Customer needs have to be converted into new product requirements via marketing surveys, customer interviews, focus groups, and structured methods such as quality function deployment (QFD).
2. The amount of technology adoption in the new product has to be determined through a product map to outline new product streams for the next three to five year business plan for the company.

3. The company strategy in defining its core competency and whether further investment in the new product features is warranted.
4. Regulatory issues in terms of industry standards and environmental and ergonometric factors have to be considered.
5. Technology risk assessment and using the supply chain to provide additional competency or capacity to the product plan.
6. Trade-off analysis for cost, quality, schedule, and performance and their impact on project completion dates.
7. Project constraints in terms of availability of resources in design and manufacturing and capital equipment acquisition.

Once these issues are analyzed by the PM, then a new product proposal is presented to the management, with a go/no-go decision to proceed into the concept stage, which is similar to step two for the DTC discussion above.

7.2 Project Requirements Definitions

There are several techniques used to define the project duration as well as the delivery date and the plan to complete the project on time. The worst scenario is to have the marketing department or sales force promise the customer delivery of the new product at a certain date, then turning the responsibility over to the PM to design the product. The PM should begin the project-planning effort to develop the new product by completing the following plans and documents:

- Statement or Statement of Work (SOW): This is a narrative of work required, to be determined from customer inputs and agreement by management on the scope of the new project.
- Work Breakdown Structure (WBS): This is the collection of all efforts to be expended by the project, defined by task activities, activity input and output, and duration of each activity and person(s) responsible for each activity.
- Subdivided Work Description (SWD): This is the work authorization containing performance description and hours allowed to work on each activity.
- Master Production Schedule (MPS): This is a list of resources needed in production for meeting the SWD, including personnel, equipment, and facilities.
- The project cost plan and the earned value (EV) plan of the project after completing each activity. These are plans to monitor the financial progress of the project. There is need for a system for tracking project activity cost variances versus the EV (to be discussed later in the chapter).
- Staffing plans, disciplines, and skills needed to accomplish all tasks.
- Supply chain plans to determine which tasks or activities to be subcontracted in terms of competency and capacity.
- Process capability plans to assess process improvement needs and new process capabilities for the new project.
- Capital equipment plans to satisfy any special requirements for the new product that is currently not available.

- Facility and space plans to house the project teams, any special equipment and the manufacturing areas required for in-house production.
- Quality plan to track the quality progress of design and manufacturing and to validate the product specification requirements and the manufacturing quality and reliability estimates.
- Communication plan to negotiate with customers and suppliers and reporting plans to convey progress and status to the management.
- The project metrics plan to extract design effort metrics as compared with company and industry norms.

7.2.1 Task Identification Plans

One of the important elements of project planning is to divide the project SOW into individual tasks to be performed by members of the project team. These tasks will be plotted or listed in PERT or Gantt charts to show the layout of the project and the precedence of each task. It is important to strike a balance so that the tasks are not too small (accounting for less than one day of engineering work) or too large (multi-disciplinary tasks accounting for weeks at a time). Too many tasks will lead to complex charts and relationships, and therefore the overhead for managing the information on the PERT or Gantt charts will be exceedingly high. It will also require many adjustments if the project is late. If the tasks are too large, then the effectiveness of the PERT and Gantt charts will be reduced, since they do not show problems in the execution of large tasks. It is important to keep the number of tasks at a reasonable size, depending on the number of milestones and skill sets required for the project.

Each task should have a special document to identify the purpose, a brief description, the team members assigned, the task activities, and input and output of each task. A typical task definition example is given in Table 7.2.

The listing of all of the tasks and their dependencies can be built into a network such as PERT or a listing (bar chart) such as Gantt to identify the start and end point of a project and the personnel assigned to that project as well as their time commitment to the project and task dependency or network precedence.

7.2.2 Project Planning Methodology

Once the tasks have been identified by the PM, a process of planning for project start and end date begins. The process takes form in the following steps for the team focused PM:

1. Bottoms Up. This is a process by which the task assignee(s) estimate the duration of each task, then the total task dependencies and the duration matrix are loaded up on a PERT/Gantt chart to obtain start and end times. The release-to-manufacturing milestone marks the end time for development projects. Obviously, the team members will be on the conservative side when they make their estimates, since they are being measured on completing the tasks on time. The procedures for bottoms up planning consist of the following:
   - Collation of task definitions and time for each task from each task leader and team member(s)
   - Collection of all tasks' durations and dependencies determines the end point (release-to-manufacturing milestone meeting)
   - Activity duration as estimated by the project team members

1. Task # and Description: System Interface requirements
2. Task Leader: ______________
3. Task Goal: Define the system interface requirements specifications and design concepts
4. Task Summary: Characterize the interface control requirements and associated interface documents
5. Task Team Members: ______________
6. Task Activities:
   a. Work with the systems group to define system requirements
   b. Work with the subsystems team to define interfaces and requirements
7. Task Inputs:
   a. External requirements document
   b. System connectivity guidelines
8. Task Outputs:
   a. System interface requirements specifications
   b. Systems specification tolerances
9. Preceding Task # and Description: ______________
10. Following Task # and Description: ______________
11. Task Duration: ______________
12. List Other Tasks or Processes Affected or Related: ______________

TABLE 7.2 Task Definition Example

2. Top Down. In this case, the PM and the management staff determine the duration of each task. Usually, the management will be more inclined to set shorter completion times for the activities than the team member estimates. Aspects of this type of project planning are:
   - Management determines the project end point
   - Requirements determine task breakdown and schedule
   - Activity duration is estimated by the PM
3. Synchronization. In this phase, the PM and the project team members negotiate to determine a reasonable equilibrium between the team and management estimates for each task. The goal of this phase is to resolve the management deadline set for project release to manufacturing:
   - Trade-offs should be considered in new product scope, performance, and resources for task activities
   - Interrelated and concurrent tasks should be identified in order to reduce project total time
   - A robust risk analysis plan should be considered to manage the trade-offs and concurrent tasks
4. Iteration. In this phase, the project team and the PM continue to iterate the synchronization process. The goal is to find additional efficiencies in processes,

tools, and technologies for reaching the management goals for project completion dates. Before agreeing to the final schedule, a risk management assessment and priority plan has to be generated. Some companies allocate about 10 percent of the project negotiated costs as a reserve for contingency planning.

Once the start and end points of the project planning occur, then the PM can begin to lay out the project schedule using one of the tools outlined in the next section.

7.3 Engineering Project Scheduling Tools

Project planning and identifying a critical path in a network of interconnected activities is part of the historical mathematical innovation known as graph theory. It was first suggested by Leonhard Euler based on solving a problem of trying to cross seven bridges in the city of Königsburg (Kaliningrad) without backtracking in 1736. He introduced the idea of nodes and paths to represent physical entities. Other techniques of optimization of certain activities began to take shape in different types of applications in different industries. By identifying each activity and its attributes, as well as the precedence relationships with other activities, the shortest path through the network can be determined, such as the total distance travelled or the minimum time to complete all activities. Examples of optimization are:

- Design of experiments (DoE) for optimization of product design and manufacturing processes.
- The travelling salesman problem of identifying the various destinations for a salesman, then creating a path for the minimum distance travelled.
- Optimizing the operation of an oil refinery by minimizing crude oil landed costs at the refinery. It includes optimizing input crude mix as well as overall product output mix (gasoline, diesel, heating oil, and jet fuel) and dispatch.
- Factory scheduling techniques, such as:
  - Optimize machine use by studying setup versus run techniques. This is accomplished by looking back at the job queue and selecting the jobs that best match the current machine(s) set up and tooling requirements.
  - Optimize multiple machines in the shop by looking at the job queue and creating the optimum sequence of the job through the different machines in the shop.

With exception of the DoE, all of the above optimization scenarios use linear programming (LP) techniques for finding the optimum operating conditions or activity selection and routings. For a typical high-technology project, the optimization can be performed by trial and error or through running exhaustively through all different path combinations.

An important outcome of the PERT or Gantt charts is the determination of the minimal time required to complete the project. The methodology is called the critical path method (CPM). CPM identifies the critical tasks needed to be completed on time. Tasks that are not in the critical path can have slack or float time, whereby they can start later or finish earlier. These concepts are called earlier start (ES) and earliest finish (EF) as well latest start (LS) and latest finish (LF). Critical tasks have no slack and therefore have to start and finish at the appropriate time(s).

Optimization in PERT or Gantt charts can be planned by leveling the tasks and reassigning the staffing during slack times, assuming the skills sets are interchangeable within the project team. In addition, once the critical tasks are identified, more resources can be added to complete some of the critical tasks in less time to allow for shorter total project time. In the latter case, caution must be exercised to make sure not to shorten the activity too much, lest another path becomes the critical path. In that case, the CPM analysis has to be performed all over again.

7.3.1 Project Planning Tools and Techniques

There are many tools that can be used for project planning and scheduling for tasks and milestones. Two being the most common: Excel© and Microsoft Project©. There is usually a distinction of project size where each tool can be most useful:

1. Spreadsheet based planning:
   - Useful for short-term projects
   - Less than three months total project duration
   - Low number of team members (≤ 5)
   - Less than 25 tasks
2. Project management software planning:
   - Greater than three months project lifecycle
   - More than five team members
   - Greater than 25 tasks

The traditional tools of project planning consist of three elements that are discussed below:

1. PERT Charts (program evaluation and review technique): These are network diagrams that document the relationship of different tasks in order of start and finish times. It will also show the latest start and earliest finish time for each task and can be used to determine the critical path in the project.
2. Gantt Charts: These include a listing of tasks that documents the schedule, events, activities, and responsibilities necessary to complete a project. For each task, the start and finish dates are recorded, with a summary of all tasks for the project. The tasks are drawn from the work breakdown structure (WBS). Some Gantt charts also show the task dependency relationships between tasks. Gantt charts can be used to show current schedule status using different shading or colors to describe completed tasks.
3. Milestones: These include predetermined end points of project development phases or sets of tasks. They are integration points to ensure synchronization of all functions in the project. They serve as information points and make evident to management whether a project is on track. They should be scheduled whenever all preceding tasks are completed. They can be used as management checkpoints to meet with the PM and the project team to review progress and for deciding to continue with the project funding.

7.3.2 PERT Chart Methodology

PERT charts were originally developed to improve management of defense contracts. In particular, they were extensively used for the nuclear submarine construction program managed by Admiral Hyman Rickover in the 1950s. Though they are in wide use in construction management, their use is applicable to projects in any field. They are helpful for planning the duration and cost of a project and can be used for optimizing both. In addition, uncertainty in task duration can also be handled mathematically using probability theory. PERT charts can be used successfully for the following project planning and control methodologies:

- Document or monitor all projects activities and tasks.
- Plan and track costs of all activities/tasks, either individually or in a cluster, such as the project concept phase. The overall current projects costs and variance to the planned values can also be measured.
- Activities/tasks can be clustered to encompass a project design phase, and a project milestone can be set at the end of the cluster. A design revue meeting can be held to ensure the proper completion of all activities in that phase and obtain management approval to transition to the next phase.
- Provide feedback in case of project delays or over budget, and alerts the PM to problems so they can decide when to intervene to keep the project on track. PM intervention could result in changes to the critical path and the project scheduled end date. The PM can do one of the following:
  - Sacrifice original specifications or project goals
  - Delay project completion date
  - Add more funding and or resources to critical project activities
  - Change project implementation plan or dependencies of activities; an example would be to start hard-tooling construction before the detailed design review meeting

7.3.3 Steps in Creating and Implementing a PERT Chart

The most important function in creating a PERT chart is the identification and selection of the scope of each task or activity. A typical task is shown in Table 7.2, which includes information about each activity, input and output, and any other tasks or processes affected by or related to this task. Task selection and planning should consider the following:

1. Task selection:
   - Divide the total project into tasks small enough to be assigned to one team member with an identified deliverable, within a reasonable time.
   - The minimum time should be commensurate with the planned PERT chart update plan. If the PERT is updated every week, then the minimum task time period should be the same.
2. Task duration:
   - Use historical databases to determine the most likely (realistic) task time based on similar activities in the past. Add technology uncertainty and engineering skills variability to specify variations in task times. Engineers could vary as much as 2:1 in completing technical tasks.
   - Temper task duration analysis with a probability level (50 and 90 percent confidence) or use pessimistic or optimistic times. These variations could be used for estimating expected task times in the project PERT chart analysis.

3. Plan for turnover, training, and non project variables. Decide on de-rating engineers' availability times for the project based on their other activities such as:
    - Engineering team members being asked to solve older product problems by supporting production or field problems.
    - The amount of training time allocated to each team member as mandated by the company.
    - The typical turnover rate and what percentage of the team is expected to leave before project completion.
    - How much time do senior members of the team spend attending conferences or mentoring younger engineers.
4. Planning Details:
    - Use fewer levels for small or well defined tasks.
    - Use more details for large, risky, or unproven technology tasks.
    - Number each task and who is responsible.
    - Show task duration based on the discussions above.
    - Determine task dependencies and precedence.
    - Identify predecessor and successor activities; they can be single or multiple tasks.
    - Some activities can be performed in parallel.
5. Build a PERT network showing all various paths from start to finish.
6. Critical path is the minimum time to complete a project; the critical path method (CPM) identifies all of the critical activities along the critical path.
7. Optimize the project by:
    - Using leveling techniques to minimize staffing requirements throughout the project duration and assuming that a skills-set matrix can be used to overlay the leveling plans and propose shifting of staff between slack activities.
    - Using additional resources to shorten (crash) critical activities by substituting higher cost for shorter time for the critical activity. Estimate the impact on the critical path, since another path might become critical.
    - Using fast-tracking of tasks to re-order the sequence and dependencies of tasks in order to perform them in parallel and reduce overall project time. An example would be ordering the tooling or building prototypes before the final design review. Risk assessment should be made as to the impact of the fast-tracking decision.

7.3.4 Example of the Planning of a PERT Chart

In this PERT chart example, the project was to institute a control chart program in an area in the company. The following activities were identified and grouped into three major clusters of tasks, leading to three major paths:

1. Launch the control chart program with communications to all areas and employees, select the control chart team from the designated area (the designated area was selected in the second path), and train the team on the software.
2. Select the area to host the control chart launch, select the quality parameter or process output to be charted, and begin to take production data for the control charts.

Activity	Activity End Node	Activity Path	Duration (Weeks)
Launch Program	A	Start – A	5
Select Area	C	Start – C	5
Oder Software	F	Start – F	20
Select Team	B	A – B	3
Train Team	E	B – E	5
Select Output	D	C – D	5
Take Data	E	D – E	5
Train in Software	E	F – E	10
Plot	Finish (Fi)	E – Finish	2

TABLE 7.3 List of Activities for PERT Chart Example

3. Order the control chart software and train the control-chart facilitator in the software. The facilitator will train the control chart team while he is self-training in the software.

The list of activities and duration of each activity is shown in Table 7.3. The duration of each activity is marked by the number of weeks it takes to complete it. The end of each activity is marked by a node or event and given a letter to identify it. The project comprised of all activities has a start point or node S and finish node Fi (to distinguish the end from node F, the completion of the order software activity).

Note several interesting issues with this example project:

- Activity "Select Team" (which starts at node A, end of "Launch Program," and ends in node B) cannot start in path 1 until the area is selected in path 2. This dependency is strictly a timing one, with no activity between nodes C and A. It is best to present this dependency as a dotted line with a value of zero for duration since there is no work to be performed for this activity. The dotted line is usually called a dummy activity.
- Node E is the end node for three activities: "Train Team" originating in path 1, "Take Data," originating in path 2 and "Train in Software" originating in path 3. Multiple inputs and outputs for each node can be accommodated in the PERT chart.
- Activities "Train Team" in path 1 and "Train in Software" in path 2 can be in parallel, yet they are in different paths. This dependency could be registered in the PERT chart by having a second dummy activity with a duration value of zero between nodes F and B.

Considering all of these activities and their dependencies, a PERT chart is drawn, shown in Fig. 7.1. This example shows all activities as an arrow, including the dummy activity mentioned above. The activity end nodes (designated by letters) are shown as circles, and the start and finish nodes are identified as S and Fi.

The dummy activity is shown in light dotted line arrows, and the critical path activities are shown in heavy line arrows. The critical path is determined when all paths

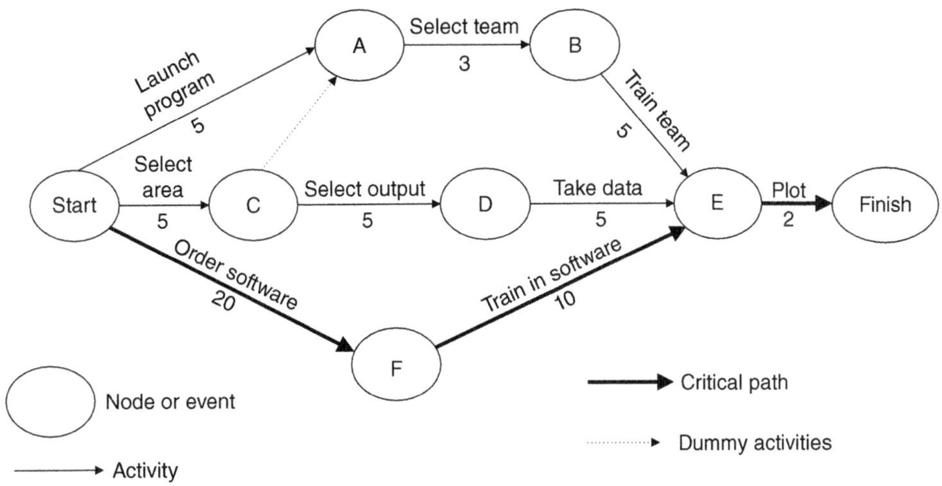

FIGURE 7.1 PERT chart example.

in the PERT network are identified, and the total time measured in each path as shown in Fig. 7.2. In this case, the extra path made by the dummy activity is not included in Fig. 7.2, since it did not fall into the critical path.

The path in Fig. 7.2, Start – F – E – Finish is the longest path in terms of duration; therefore it is deemed the critical path, and activities "Order Software," "Train in Software," and "Plot" are deemed critical activities. Other activities have slack or float time, whereby they can start later or finish earlier as well as start earlier and finish later to a certain extent without affecting the overall project completion. Activities on the critical path have no slack time. The slack time can be calculated for each path by adding the total time for each activity in the path and subtracting that from the critical path times. The slack times are shown in Table 7.4.

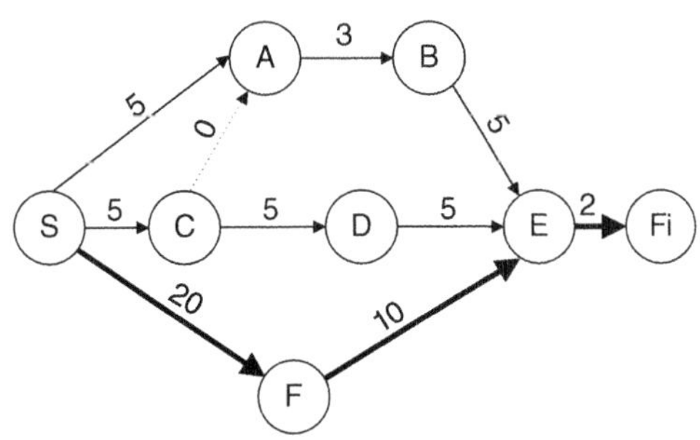

Path	Activities	Path duration(w)
Start-A-B-E-Finish	5 + 3 + 5 + 2	15
Start-C-D-E-Finish	5 + 5 + 5 + 2	17
Start-F-E-Finish	20 + 10 + 2	32

FIGURE 7.2 Critical path determination.

Path	Activities	Path Duration(w)
Start – A – B – E – Finish	5 + 3 + 5 + 2	15 (slack = 17 = 32 – 15)
Start – C – D – E – Finish	5 + 5 + 5 + 2	17 (slack = 15 = 32 – 17)
Start – F – Finish	20 + 10 + 2	32 (slack = 0)

TABLE 7.4 Slack Time for PERT Exercise

7.3.5 Determining Slack (Float) Time Extension

There are two additional analyses that can be performed on the PERT chart. The collection of these does not affect the CPM:

1. Forward-pass analysis determines ES and EF times for each activity in the PERT chart. For each network path, begin at the start node and work forward until all activities are included, and you reach the end node. Place the ES and EF values at the top of each activity arrow in the chart.
2. Backward-pass analysis determines the LS and LF for each activity in the PERT chart. It is the reverse of the forward pass. For each network path, start at the end and work backward until you reach the start node. Place the ES and EF values at the bottom of each activity arrow in the chart.
3. Note the following when the float time extension analysis is completed:
   - For critical activities, ES = LS and EF = LF
   - For all non-critical activities
     a. Activity time = EF – ES = LF – LS
     b. Path slack = LS – ES = LF – EF
     c. Total slack for the PERT chart = largest path slack
4. To make PERT charts consistent over many different definitions and software displays, the following guidelines can be standardized:
   - Activities are presented in the PERT chart either as $\frac{\text{ES}-\text{EF}}{\text{LS}-\text{LF}}$ or with more information as in $\frac{A(\text{ES}-\text{EF})}{t(\text{LS, LF})}$ where

     A = activity identification
     t = activity time
     ES = earliest start time
     EF = earliest finish time
     LS = latest start time
     LF = latest finish time
   - The first activity(ies) in the project has the earliest start time of 0. ES of the first activity can be set to 0, or it can be set to a specified starting time defined for the project.
   - Unless LF of a project is otherwise specified, it is set equal to the LF of all of the activities preceding it
   - LF of project = LF of all activities preceding it. It is also equal to the EF of the project since the project completion node is on the critical path
   - If a desired deadline is specified for the project then: LF of project = specified deadline; the project start date can be determined with backward pass method

- ES for any activity is equal to the maximum of the EF of the immediate predecessor(s) of the activity
  - ES = maximum {immediately preceding EF}
- EF of an activity is the activity's ES plus its activity time: EF = ES + (activity time)
- LF for an activity is equal to the smallest of the LS times of the activity's immediate successors
  - LF = minimum {immediately succeeding LS}
- LS for an activity is the LF minus the activity time
- LS = LF – (activity time). For a dummy activity, the duration time = 0; the ES of a dummy activity is equal to the LS, and both are equal to the maximum of the EF of the immediate predecessor(s) of the activity
  - ES (dummy) = LS (dummy) = maximum{immediately preceding EF}
- For a dummy activity, the EF is equal to the LF, and both are equal to the smallest of the LS times of the activity's immediate successors
  - LF (dummy) = EF (dummy) = minimum {immediately succeeding LS}

Using the guidelines above, a PERT chart network presentation for the example is given in Fig. 7.3. It can be seen for path 1 (Start – A – B – E – Fi) that for all noncritical activities, the difference between ES and LS is equal to the slack in that path (17). EF for activity AB is equal to ES + activity time (8 = 5 + 3). The difference between the EF and latest finish LF in noncritical activities in path 1 is the slack in that path (17). The ES for each noncritical activity in Fig. 7.3 is equal to the EF of the preceding activity [5 = ES for activity (A – B) = EF for activity (Start – A)]. LS for all noncritical activities in path 1 is equal to LF

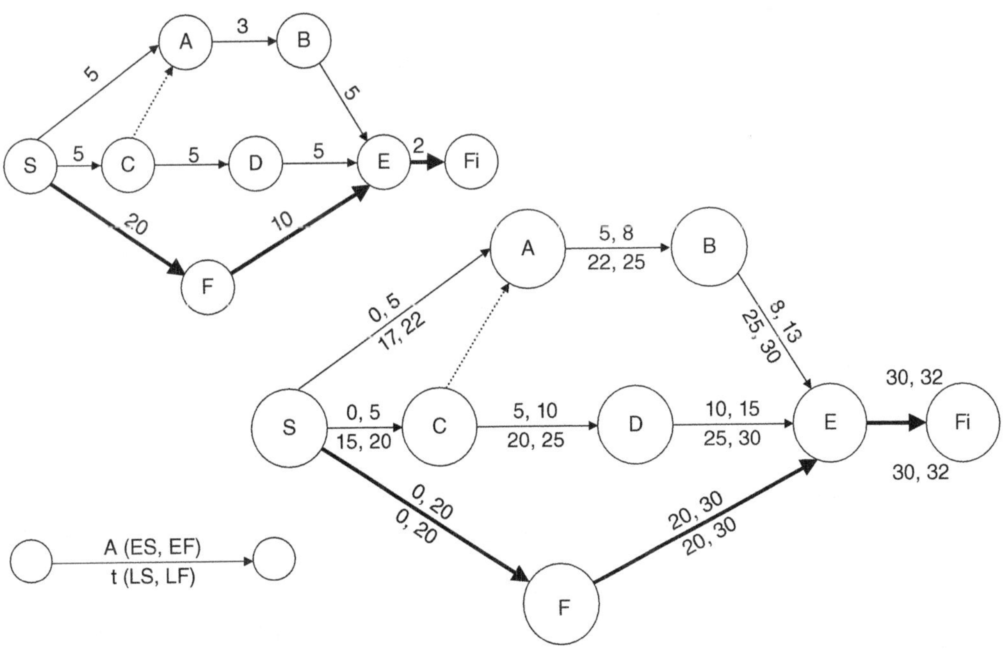

FIGURE 7.3 Example PERT network chart with slack extension.

of the preceding activity [22 = LS of (A – B) = LF of (Start – A)]. For all activities in the critical path = ES = LS and EF = LF. For the dummy activity, the formula for slack extension is $\frac{d(5,5)}{0(22,22)}$.

7.4 Methods and Techniques for Reducing Project Duration and Cost

Several planning techniques could be used to shorten the project cost and duration: (1) resource leveling and allocation; (2) crashing or carefully adding resources to shorten appropriate critical activity(ies); or (3) fast tracking or taking risks in altering the dependency of certain critical activities. Examples of fast tracking could be software training that can be separated and completed before a specialized software activity or getting in contact with suppliers early enough to get a preliminary cost estimate and ESI feedback before formal quotes are issued in a request for proposals (RFP).

A new critical path might emerge from the use of one or more of these techniques. Increased risk will have to be considered and managed appropriately. Finally, probabilistic estimates can be calculated for expected project time, or a cash reserve can be given to the PM to make appropriate decisions.

7.4.1 Resource Leveling and Allocation

These methods can be used to identify the project staff needs versus time and using slack time to move noncritical activities within the slack period to level out staffing needs. It should be noted that the overall project duration is not affected; only cost benefits are accrued by reducing the project staff idle time and leveling the expenditure of funds to support the project. This technique could have less advantage in a high-technology project in view of the following conditions:

- The skill sets of team members are not interchangeable. A skill-set matrix can be used to identify those team members who can work on several activities.
- Usually, the PM should decide on core team members who are permanently assigned to the team for the project duration. The PM may schedule other part-time team members as needs arise.

The resource leveling for project planning using PERT charts starts with identifying the critical activities and listing the manpower needs on a regular time period (weekly) basis. The forward-pass and backward-pass version of the staffing needs should also be laid out. An iterative process should be completed to level out the manpower demand, using the maximum range of demand (highest minus lowest staff needs) as the guideline. An example of leveling is shown below, with crashing information for each activity to be explained in the next section.

7.4.2 PERT Example 2

For the project below, the activities, nodes, manpower, and costs are shown in Table 7.5. The table shows activities, their durations and precedence rules, staff required for each activity; each activity cost per week and the cost of crashing an activity per week. In this example, the activities are shown in letterform, and there are no indicators for the nodes. The slack times are calculated from the different activities based on CPM. For critical activities, the slack time is equal to zero. The slack times for each path are also shown.

Activity	Preceding Activity	Activity Time	Staff Required	Std. $Cost/ Week	Crash $Cost/ Week	Slack Time (w)
A	Start	4	3	1000	2000	1
B	Start	5	2	500	800	0
C	Start	3	1	1000	3000	7
D	A, B	4	2	800	1000	0
E	B	3	4	700	3000	1
F	C	1	2	300	900	7
H	D, E	2	1	1000	3000	0
End	H, F					

TABLE 7.5 Example 2 PERT Chart Data

A PERT chart is constructed in Fig. 7.4 showing slack extension, with forward and backward pass calculations for start and finish of each activity. A dummy activity resulting from the dependencies of Table 7.5 is also shown in a dotted line. The critical path is shown in heavy lines, and it includes the dummy activity.

All of the paths in the PERT are shown below, including the slack time for each path:

Path 1 = ADH; slack = 1 (activity A only)
Path 2 = BEH; slack = 1 (activity E only)
Path 3 = BDH; slack = 0 (critical path)
Path 4 = CF; slack = 7

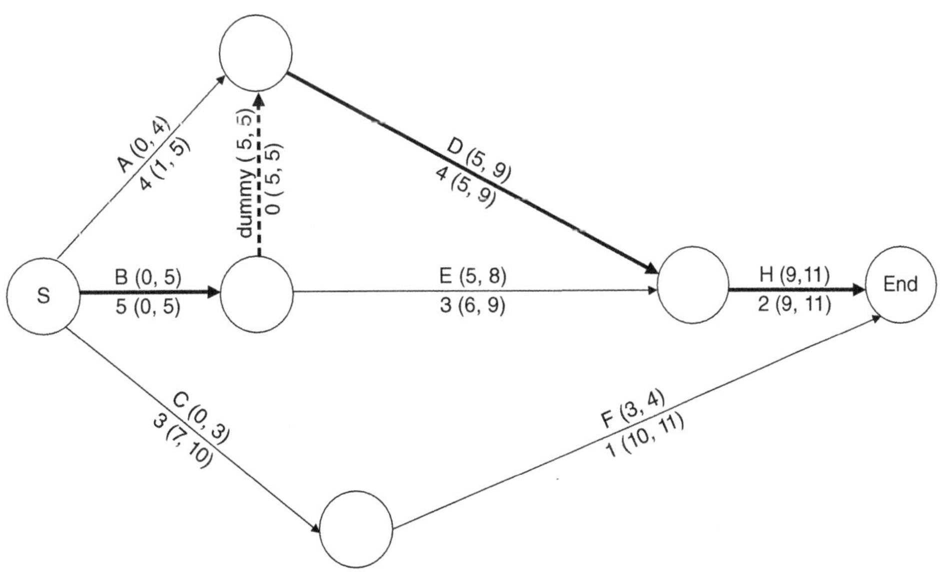

FIGURE 7.4 Example 2 PERT chart.

Week	**1**	**2**	**3**	**4**	**5**	**6**	**7**	**8**	**9**	**10**	**11**	
A	3	3	3	3								
B	**2**	**2**	**2**	**2**	**2**							
C	1	1	1									
D						**2**	**2**	**2**	**2**			
E						4	4	4				
F				2	→							
H										**1**	**1**	Total
Total	6	6	6	7	2	6	6	6	2	1	1	49

Table 7.6 Forward-Pass Staffing Profile PERT Example 2

For the PERT example 2 project duration, the staffing profile in forward pass is shown in Table 7.6, beginning at the earliest time ES for each activity (forward pass). The critical activity staffing needs are shown in bold letters. The staffing profile for backward pass is shown in Table 7.7, beginning at the LS. The critical activity staffing needs are shown in bold letters.

The arrows in Tables 7.6 and 7.7 point to the largest range in staff demand within a two-week span. There is an opportunity for leveling by moving the activity F one week to the right (later) in Table 7.6. Alternatively, moving activity F one week to the left (earlier) also increases the leveling of staff requirements, which can also be done in Table 7.7. These techniques can be used iteratively to find the optimum level staffing needs. The PM use of these techniques can be tempered by the following:

- Forward-pass technique to have all activity front-loaded results in completing many tasks as early as possible, reducing uncertainties earlier in the project. In addition, the staff associated with these activities can leave the project team after completion of activities, thus reducing payroll. This is balanced by the fact the project expenditure profile is also front loaded, making management hesitant to spend the resources early in the project in case it will be cancelled.

Week	**1**	**2**	**3**	**4**	**5**	**6**	**7**	**8**	**9**	**10**	**11**	
A		3	3	3	3							
B	**2**	**2**	**2**	**2**	**2**							
C								1	1	1		
D						**2**	**2**	**2**	**2**			
E							4	4	4			
F										←	2	
H										**1**	**1**	Total
Total	2	5	5	5	5	2	6	7	7	2	3	49

Table 7.7 Backward-Pass Staffing Profile PERT Example 2

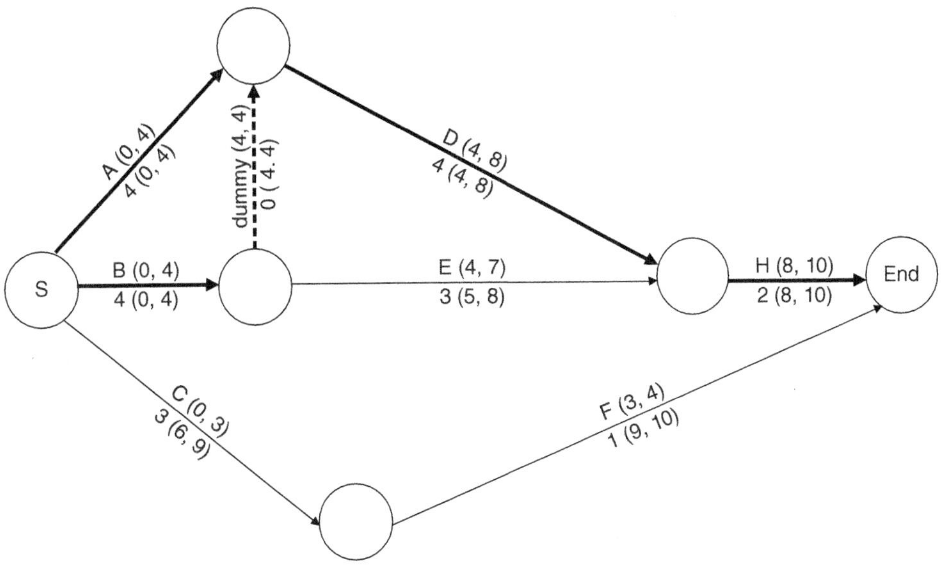

FIGURE 7.5 Example 2 PERT chart crashing results.

- The opposite is true for the backward-pass method. Late activities increase complexity and chances of multiple problems occurring at the same time near the project completion date. The project expenditure profile is also backward loaded.
- The management would prefer a level expenditure of resources in payroll and other project expenses for stability of operations and impact on the bottom line.

The PM can also attempt to reduce total project time by crashing one or more critical path activities by adding more resources to that activity. Crashing should be used iteratively to account in case another critical path emerges. In addition, a cost benefit and risk analysis should be performed to make sure there are no adverse consequences to the crashing decision. In this case, crashing activity B by one week reduces the project duration by one week at the lowest cost. Activity B has the lowest crashing cost. Two (2) critical paths emerge after crashing B by one week: ADH and BDH as shown in Fig. 7.5, Example 2 PERT Chart Crashing Results.

7.4.3 Estimating Expected Project Completion Time

The total project duration time can be estimated from the critical path by the addition of all critical activities times. In order to account for the uncertainties in estimating these times, two probability based methods can be used:

1. Determine three different estimates for each activity time: realistic (most likely), optimistic, and pessimistic. These would represent three probability levels based on linear approximations of the normal curve. The realistic estimate is 4/6 or 66.7 percent of the total area under the curve, corresponding to ± (one) $\sigma_{activity}$. Optimistic and pessimistic linear approximations are 1/6 or 16.7 percent of the area under the normal curve, corresponding to outside of the ± (one) $\sigma_{activity}$.

All three estimates will account for the total area under the linear approximation of the normal curve as follows:

$$\text{EAT} = (\text{pessimistic} + 4 * \text{realistic} + \text{optimistic})/6.$$

The expected activity time (EAT) formula calculates an activity time based on the PM's estimate of the extremes of activity completion variance around the original time estimate. The total project duration time can be calculated from the sum of all expected critical activity times.

2. Alternatively, the PM could estimate the standard deviation (σ) of critical activity time distribution, based on their own experience. The PM can then determine the confidence interval (CI) of project duration by calculating the project variability ($\sigma_{project}$) from all critical activities' σ using root mean square (RMS) methodology. The CI produces a minimum and a maximum project duration estimate based on the desired confidence. The $\sigma_{project}$ can be calculated as follows:

$$\sigma_{project} = \sqrt{(\sigma_{activity1}{}^2 + \sigma_{activity2}{}^2 + \sigma_{activity3}{}^2 + \ldots)}$$

Once the $\sigma_{project}$ is calculated, then CI may be estimated for a selected confidence percentage. Examples are 68 percent confidence ($\pm\ \sigma_{project}$) and 95 percent confidence ($\pm\ 2 * \sigma_{project}$). The 95 percent CI for the project estimated duration is equivalent to declaring that a project completion will fall within $\pm\ 2 * \sigma_{project}$ CI in 95 percent of the time. The end points represent the most optimistic and pessimistic times consequently.

For example, let's assume a project with a critical path of 10 weeks with four critical activities that the PM has estimated to have standard deviations (σ) of 1, 0.5, 1.25, and 0.5. At 95 percent confidence, the minimum project duration (optimistic time) will be 6.5 weeks and the maximum project duration (pessimistic time) will be 13.5 weeks, as follows:

$\sigma_{project} = \sqrt{(1^2 + 0.5^2 + 1.25^2 + 0.5^2)} = \sqrt{(1 + 0.25 + 1.56 + 0.25)} = \sqrt{3.06} = 1.75$
95 percent CI $= 10 \pm 2 * \sigma_{project} = 10 \pm 3.5 \rightarrow 6.5$ (minimum/optimistic)
and 13.5 (maximum/pessimistic project duration). To double check,
$\text{EAT} = (6.5 + 4 * 10 + 13.5)/6 = 10$ weeks.

The most common confidence intervals used for project duration estimates are 50 and 90 percent, corresponding to $\pm\ 0.685 * \sigma_{project} = 50$ percent CI and $\pm\ 1.645 * \sigma_{project} = 90$ percent CI of the project duration estimate.

7.4.4 Gantt Charts

These are bar charts or linear diagrams for documenting and planning projects, what activities are to be accomplished, and by whom and when. Gantt chart planning includes defining the details for each activity: schedules, events, or milestones and tasks to be completed and ownership of each task. Assumptions can be noted for each activity on the chart and contingency plans prepared in advance. An example of a Gantt chart, together with a computer plotted PERT chart, is shown in Fig. 7.6 for the project timeline in Fig. 7.1 using Microsoft Project©.

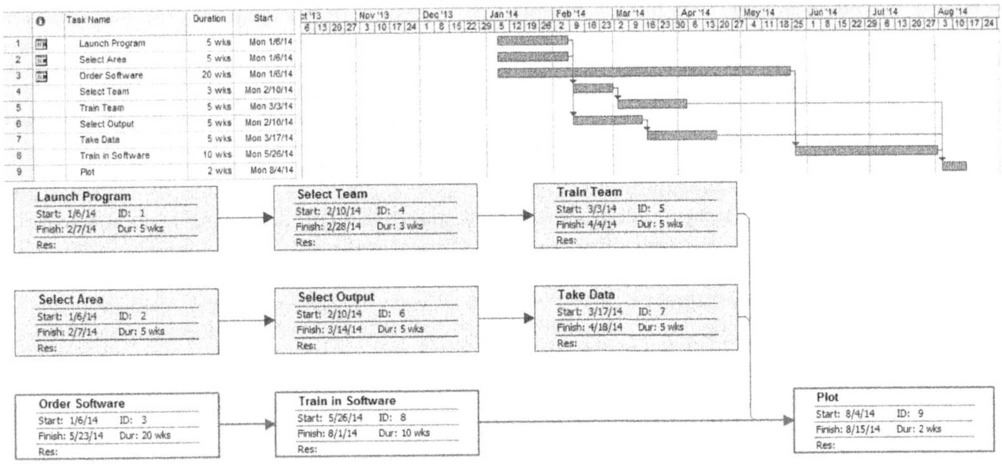

FIGURE 7.6 Gantt chart example using Microsoft Project.(c)

Computers can draw Gantt charts much easier than PERT charts. They show the dependencies of activities by scheduling the start and end times of each activity according to activity precedence. Most project-management software can create PERT and Gantt charts at the same time, though the PERT charts created by software are more difficult to read. They can show the current project progress by filling in the bar chart as activities are completed. They also can provide for additional plans and documentation as shown below.

The major elements for creating a Gantt chart are similar to those of PERT charts:

- Break the project plan into achievable tasks
- Assign responsibilities to each project team member
- Decide how long each task will take and set a realistic completion date
- Document the assumption on which the plan is based and the contingency (risk management) plans to implement if those assumptions are not valid

7.4.5 Plans to Be Completed by the PM Prior to Project Start

Many of these plans were discussed beforehand. Most PMs will integrate these plans into a master plan for the project. The following is a summary of these plans:

1. Work Breakdown Structure (What): List all the tasks and activities to be completed. This is a structured task list from the project definitions, including deliverables and ownership.
2. Schedule Plan (When): This is used to schedule and link tasks using their dependencies. A time-scale chart such as a Gantt or bar diagrams could be used to show precedence. All tasks are shown with a start and end date. Dependencies would include predecessor and successor tasks for each activity and what tasks or activities are influenced by the outcome of each task.

3. Resource Plan (Who): This plan includes the project team members, whether they will be assigned full or part time, when they will be assigned, and when to leave the project and what are their skill sets.
4. Risk Management Plan (What If): This plan includes risk analysis, evaluation of risks, and generated contingencies plans.
5. Plan Optimization (How): Once the preceding plans are completed, the optimization of the project plans can be initiated. Resources and tasks are identified as well as over/under commitment. The PM will analyze the schedule, resource, and deliverables' conflict, revise as necessary, and schedule milestones at appropriate times.

7.5 The Causes of Engineering Project Execution Problems and How to Mitigate Project Delays

Many of the problems in project execution and methods to mitigate them by using proven tools and techniques have been discussed in earlier chapters. These issues are usually grouped into two separate timelines. Those early in the project cycle are focused on the customer/marketing/engineering interface and project planning efficiency. Those in later stages of the project cycle are focused on design/manufacturing interface, the supply chain, product costs, parts procurements, and product delivery.

7.5.1 Engineering Project Design Phase Delay Factors

These issues are summarized in Fig. 7.7, which is a Pareto diagram of early period engineering project delays in the project definition, planning, and early design phases. The rankings of these factors were taken from a survey conducted by the author. Though the ranking of the Pareto diagram might differ in other surveys and in different industries, it is typical of high-technology projects. The PM can use some of the tools and techniques discussed in the early chapters of this book to anticipate and mitigate these project delay factors. The list of factors affecting the project design phase and how to mitigate them are as follows:

- Identification of Market/Customer Needs: This is the topic discussed by the VoC in Chaps. 3 and 4. Using structured tools such as QFD can greatly enhance the process of correctly analyzing the competition and changes in customer needs. Price performance curves can provide additional input into the development of

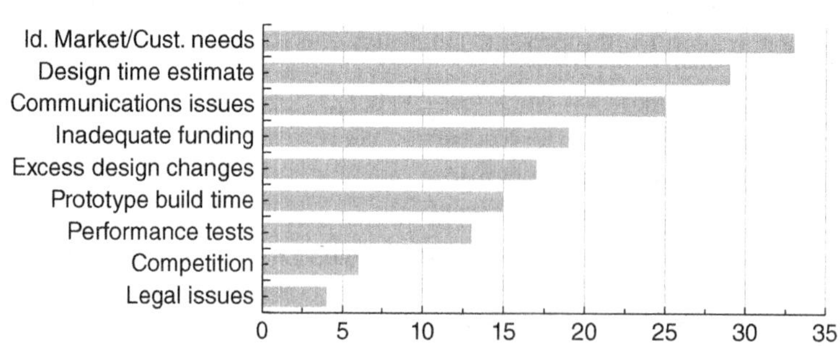

FIGURE 7.7 Engineering project design phase delay factors.

the new product specifications. These tools could mitigate the uncertainty of new product performance and reduce design iterations due to poor specifications.

- Design Time Estimates: The amount of technology adoption in the new product should be critically examined. It is best to apportion new technology into a generation of products, with each successive product incorporating a certain level of the new technology. This would lead to the speed up of development time as well as satisfying customer expectations with a stream of innovative products. Apple Computer effectively employs this strategy. A reasonable technology adoption strategy would mitigate the problem of design time estimates.
- Engineering project communications issues will be examined later in this chapter and in detail in Chap. 10. Good communications is an excellent opportunity for the engineers to voice their concerns and to uncover potential problems in the project.
- One of the important mitigating factors in addressing inadequate project funding is to subdivide the project tasks in great detail and clearly identify the project activities to be performed and the associated costs. Extensive ROI analysis should be performed to ensure management backing of the project. This should be completed in the project planning stage and reviewed by management at the completion of each development phase.
- Excess Design Changes: This issue was examined earlier in the book as to the causes of ECOs. Better project communications, applying performance specification that best match current customer needs, a well-researched business plan, early supplier involvement (ESI) and DFM inputs, and a good design review process should mitigate the issue of excess design changes.
- Prototype Build Time: New additive technologies for prototype build such a DDM (direct digital manufacturing), SL (stereo lithography), and 3D printing could facilitate prototype turnaround time. Good integrated DFM standards within the company and ESI would also reduce the amount of uncertainties in the prototype build. The selection of qualified local suppliers for prototype and production builds could also mitigate this problem.
- Performance test concerns can be reduced with the better specifications definitions, which in turn are derived from better understanding of VoC.
- Competition issues are dependent on the industry sector and product price performance curves in that sector. A good VoC analysis would mitigate the competition issues.
- Legal issues including intellectual property were discussed in Chap. 3.

7.5.2 Engineering Project Manufacturing Phase Delay Factors

These issues are summarized in Fig. 7.8, which is a Pareto diagram of late period engineering project delays, during the project manufacturing and pilot-build customer-deliveries period. The list of factors affecting the project design phase and how to mitigate them are as follows:

- Supplier delays could be based on unplanned schedule changes, other customer demands for parts, poor communications, or poor adherence to the supplier DFM guidelines. These issues could be mitigated by a good supplier selection and well-established communications with the supply chain.

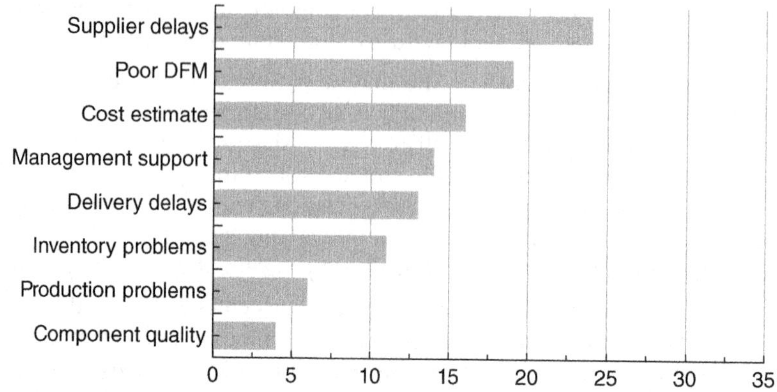

FIGURE 7.8 Engineering project manufacturing phase delay factors.

- Poor DFM could be the result of not actively pursuing ESI and not focusing on a DFM strategy and guidelines for the company
- Product cost estimate issues could be mitigated with the use of cost-based systems such as activity-based costing (ABC); these issues will be discussed in Chap. 11
- Management support could be greatly improved with better communications and justification of project and proposal return on investment (ROI).
- Product delivery delays, inventory and production problems, and component quality are operational issues in manufacturing that can be mitigated with good tactical planning in coordination with the operations department.

7.6 Techniques for Monitoring Project Expense Progress and Estimating Project Completion Profile

The dilemma for the PM is how to find an accurate assessment of the project progress and how to identify potential problems before they have a significant impact on the project schedule. The PM has several tools at his/her disposal to accomplish the project tracking goals. These goals are grouped in several categories for tracking purposes through different information and communication streams.

1. Project Task Expense and Completion Tracking Systems: These systems are used extensively in DTC projects where the customer can demand an accounting for all activities that are being paid for. The data can be collated from the project team members' charges to specific tasks and projects and presented in summary form to the PM and the customer. These systems are sometimes called earned value (EV), which accumulates value as project activities and tasks are completed. Details of task management systems are as follows:
   - Use the project plan of subdividing the project into clustered tasks and activities as the metric for tracking the project by reporting the date of completion for each task. This is the most common method for project monitoring.

The variances of the project schedule and cost can be identified and used by the PM to take immediate corrective action to keep the project on track for cost and schedule.

- Data from the EV can be generated by time allocated by each project team engineer against project expense categories. These expenses can be rolled up and used as the basis for task accomplishments and variances in cost and schedules. The period of time reporting activities varies by product and industry, but it is usually as fine as hourly reporting.
- Intermediate data from the EV can be used as a *snapshot* to predict project completion. Actual variances in cost and schedule as opposed to the original project plan are extracted then extrapolated forward as estimates to project completion.

The EV system is an excellent tool for macro-managing a project. The reporting of schedule information can have shortcomings in terms of gathering up-to-date information and accurately alerting the PM about potential problems:

- Information is only available when a task is completed in the EV system. Problems that can occur inside a task will not be reported until that task is completed or the team assigned to that task has exhausted all resources associated with that task.
- The bandwidth of the task durations can result in not having timely information about progress in completing different tasks.

2. Project Communication Techniques: The PM can use a variety of communication methods to update the project team and the management of the project progress. Formal and informal meetings and reports, as well as regular update meetings, can supplement the project progress information gleaned from formal systems such as EV. These techniques can encompass the following:
   - Regular written reports, authored by the project team members and summarized by sub-team leaders or PMs, can be used effectively in reporting project progress. Project written reports can be used for investigating post release problems, lessons learned, and post completion project forensics. They can be collected on a regular basis such as weekly or monthly.
   - Several techniques can be used to help keep the report short and efficient, such as allowing only a certain number of pages, forcing the authors to use a particular format, or color code the report with a tag for priority of the information and potential for negative impact (red for alerts on critical issues or effects, orange for moderate potential problems, and green for no problems).
   - Unfortunately, engineers and other team members are usually reluctant to write down details or speculate on potential problems, lest they would be looked on as naysayers. This limits the use of written reports as a potential early indication of problems in the project. The collation of data from the various team members could be an arduous and time-consuming activity and therefore not a timely process. Team meetings might be a safer environment for engineers to voice their concerns, without documenting them in writing and possibly being held accountable for unnecessary project delays and false alarms.
   - Team meetings in formal and informal settings. Formal meetings are used to make large project impacts such as financial and management decisions on whether to proceed to the next phase or support additional resources for the

project. Informal project meetings are typically used for quick project status updates, information exchange, and potential project problem alerts and solving sessions. Other distinctions and good use of these communications are as follows:

- Formal meetings such as project milestones, phase reviews, and release meetings have specific agendas that sometimes follow a certain set of rules and requirements. Supporting documentation, if any, should be distributed before the meeting. The PM or team scheduling the meeting should be encouraged to discuss any concerns of the attendees prior to the meeting. This would facilitate and clarify any problems the attendees might have.
- Formal meetings allow for reflection on the topics to be discussed and preparation for questions and discussions before the meeting. Project decisions can be made quickly in, or shortly after, the meetings, as elements of that decision could be discussed prior to the meeting.
- Informal meetings, sometimes called "stand-up meetings," can be held periodically for a specific topic such as a daily project update, with the attendees standing in the project-information rooms and discussing the project status every day. The meeting times should be limited (to less than 15 minutes) in order for the team members to focus on completing the tasks at hand.
- Informal meetings held on an ad-hoc basis are best suited for problem solving or updates with nonproject team members and outside stakeholders such as the supply chain or marketing. Participants are encouraged to be brief and conscious of time with methods such as allowing the attendees to only stand or setting a hard time-line for the meeting.

7.6.1 Earned Value Management System

This is a guideline to help the PM supervise technical projects and to evaluate the impact of current project progress or lack of. It can be used as a yardstick to measure the project progress to date against the original project plan and its use of resources. As each activity is completed, its EV is added to the project total to date. In this manner, the completion of an activity is turned into a dollar value that can be compared with the original planned dollar value.

The EV from completing an activity or a cluster of activities at a point in time can thus be compared with the project planned value. Variances due to project cost or schedule can be calculated, both in dollar terms. These variances are thus projected forward to calculate the expected project completion schedule (in dollars) and cost. Though different terminologies have been used to identify these variances, two major abbreviations are used to denote planned, earned, and actual variance elements.

Variance is defined as schedule or cost deviation from plan. Two types of variances are used:

1. Schedule variance (SV) is a comparison between planned and actual work performed to date, converted into a dollar value. It can be calculated at any time during the project. It is a measure of the planned work versus completed work in terms of tasks accomplished. The SV at any time can be converted to an estimated time at completion (EAC) in dollar values. The calculated variance at

completion (VAC) would give an estimate of the additional resources needed to complete the project with EAC due to variances in the schedule versus the project plan.

2. Cost variance (CV) is a comparison of the deviation of the planned and actual cost of the project to date. It can be calculated at any time during the project. It is a measure of the actual project cost to date versus completed work planned costs. The two variances, SV and CV, are independent. The CV at any time can be converted to a project cost estimate to completion (ETC). The calculated ETC would give an estimate of the additional resources needed to complete the project due to variances in the costs from planned values when actually performing the project tasks to date.

7.6.2 Project Cost Measurement

In the EV system, three basic types of costs are used to track project costs and schedules:

1. Planned value (PV) or budgeted cost of work scheduled (BCWS) is the planned activity cost akin to the amount of work scheduled to be completed at a point in time. Indirectly, it is a measure that is coupled to the planned scheduled cost of completed activities.
2. Earned value (EV) or budgeted cost of work performed (BCWP) is a measure of the dollar value of work actually completed at a point in time. The dollar value is extracted from the planned cost of tasks already completed.
3. Actual cost (AC) or actual cost of work performed (ACWP) is the money spent at a point in time. This amount can be extracted from the accounting systems as the project team members charge their activities to the project on a daily basis.

Figure 7.9 represents an example of all of the elements discussed earlier. It shows that the project is planned according to PV or BCWS in a linear expenditure model of all costs of scheduled project tasks and activities. The total cost rollup for the project based on the project plan is called budgeted at completion (BAC). When a project is delayed, the engineers are still in place assigned to the project, while their output EV or BCWP in terms of tasks accomplished is lagging behind the project plan. As resources are added to counteract the project delay, the AC of the project or ACWP is going higher than the planned cost. The typical result of this scenario is that the project is delayed and total variance at completion (VAC) is high. The EAC refers to estimate at completion, where the PM will estimate the amount of resources needed to finish the project.

7.6.3 Project Variances Extrapolated for Estimates at Completion

The EV system allows the PM to take a snapshot of the project at a particular time, find the schedule and cost variances for that time, and then extrapolate the information forward for determining estimated project schedule and cost-completion estimates. This extrapolation unto the end of the project is linear, based on the current time schedule and cost variances. While this methodology presents a macro level of project control, with some of the shortcomings mentioned earlier, it presents a reasonable measure to control engineering projects. It has to be augmented with additional project-communication

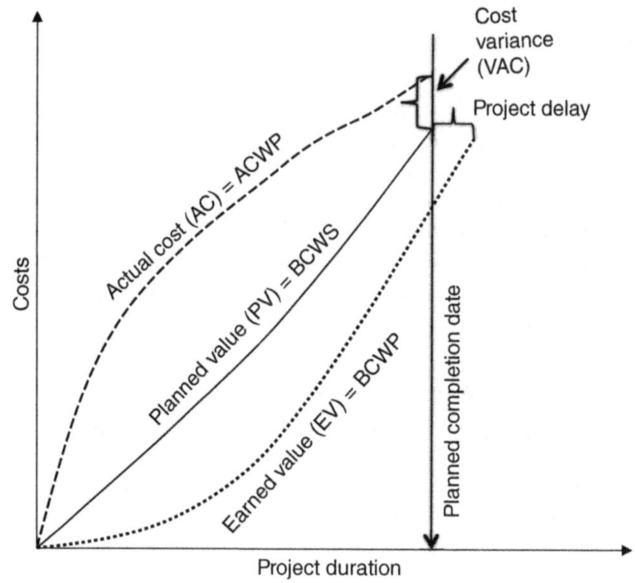

FIGURE 7.9 Earned value system example.

tools in order for the PM to effectively manage the project and keep it on track for schedule and cost. The snapshot data can be used to determine the percentage of completion in cost and of the project tasks at that time.

The extrapolation process from the snapshot time involves the use of two additional indices, which indicate the position of the project at the snapshot point:

1. The schedule variance index (SVI) is a measure of the schedule timeliness at the snapshot point. It is equal to the earned value EV = BCWP divided by the planned value PV or BCWS. A value > 1 indicates a favorable condition. It can be interpreted as the project is ahead of schedule at the snapshot time. The reverse with SVI < 1 is the condition that the project is behind schedule. It is sometimes referred to as the schedule performance index (SPI).
2. The cost variance index (CVI) is a measure of the project cost expended to date versus the planned cost. It is equal to the earned value EV = BCWP divided by the actual cost AC or ACWP. A value > 1 indicates a favorable condition. It can be interpreted that the project's actual expended cost is ahead of the cost budget at the snapshot time. The reverse with CVI < 1 is the condition that the project expended cost is behind the project budget. It is sometimes referred to as the cost performance index (CPI).

Figure 7.10 shows all of the elements discussed earlier. At the snapshot time, the current cost and schedule variances are calculated and then extrapolated as estimates of the project completion in cost and time. EAC is the estimate at completion, which is an indication of the total project cost upon completion based on the original budget at completion (BAC). BAC is calculated from the original project plan. The difference between EAC and BAC is the variance at completion VAC.

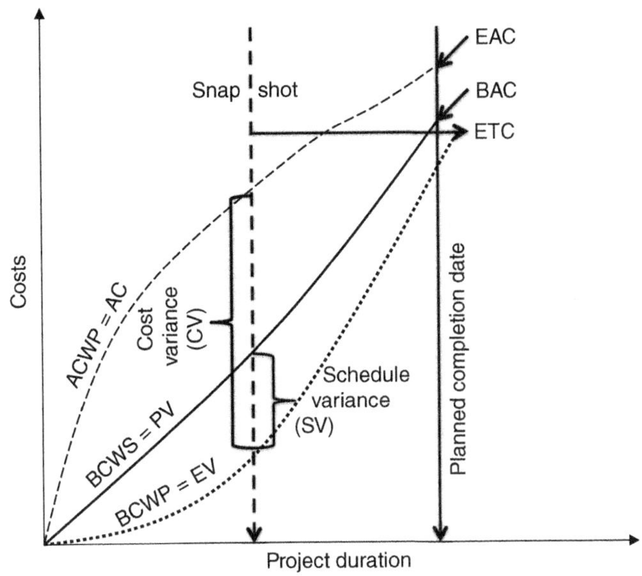

FIGURE 7.10 Project variances extrapolated for estimates of project completion.

The estimate to completion (ETC) is a measure of the remaining tasks to be accomplished starting from the snapshot time until project completion. It is equal to the total EAC minus the AC or ACWP. It is an estimate of how much more to spend from the snapshot point forward to project completion based on the work accomplished so far.

Table 7.8 includes all of the formulas discussed earlier. It shows how to calculate the current status of the project at snapshot time, with three elements for the PM to consider: percent of project schedule completed, percent of project budgeted cost completed, and percent of project cost spent.

The EV system can be a powerful tool for an accurate and up-to-date data set to help the PM make informed decisions on the project progress. In some companies, a management reserve of 10 percent additional value of the project cost is held as a cushion at the discretion of the PM. It will probably be used most of the time to counteract project delays. It could also be used to speed up projects that are within planned time and budget constraints.

7.6.4 Earned Value System Example

A PM for a new product submitted a project plan at project launch consisting of 120 tasks, each costing \$10,000 for 18 months. The PM planned that the project costs are linear with time. At the end of the concept phase of the project (after six months and planned 40 tasks to be completed), the PM called a status meeting and found that only 35 tasks were completed at an actual cost of \$400,000.

The PM submitted the following status report to top management:

Project status at end of concept phase (snapshot time):

PV or BCWS = 40 tasks @ \$10,000 each = \$400,000
EV or BCWP = 35 tasks @ \$10,000 each = \$350,000
AC or ACWP = \$400,000

Control Function	Term	Definition	Formula(s)
Variances (Favorable is Positive; Unfavorable is Negative)			
	Cost Variance (CV)	Earned – Actual	CV = EV – AC CV = BCWP – ACWP
	Schedule Variance (SV)	Earned – Planned	SV = EV – PV SV = BCWP – BCWS
	Variance at Completion (VAC)	Project Cost Variance	VAC = BAC – EAC
Performance Indices (Favorable is > 1; Unfavorable is < 1)			
	Cost Variance Index (CVI)	Cost Efficiency	CVI = EV/AC CVI = BCWP/ACWP
	Schedule Variance Index (SVI)	Schedule Efficiency	SVI = EV/PV SVI = BCWP/BCWS
Project Status at Snapshot			
		% Schedule	= (PV/BAC) * 100 = (BCWS/BAC) * 100%
		% Completed	= (EV/BAC) * 100 = (BCWP/BAC) * 100
		% Spent	= (AC/BAC) * 100 = (ACWP/BAC) * 100
Project Completion Estimates			
	EAC Estimate at Completion (Expected Cost) ETC Estimate to Completion (Left to Spend)		= BAC/CVI = EAC – AC

TABLE 7.8 Summary of Earned Value Management Control System Formulas

Variances calculated at snapshot (negative is an unfavorable condition):

$$CV = EV - AC = BCWP/ACWP = \$350{,}000 - S400{,}000 = -\$50{,}000$$
$$SV = EV - PV = \$350{,}000 - \$400{,}000 = -\$50{,}000$$

The project is behind schedule and over budget based on snapshot variances.

Project completion to date of snapshot time:

$$\text{Percent schedule} = (PV/BAC) * 100 = (BCWS/BAC) * 100 = \$400{,}000/(120 * \$10{,}000) * 100 = 33.3\%$$
$$\text{Percent complete} = (EV/BAC) * 100 = (BCWP/BAC) * 100 = \$350{,}000/(120 * \$10{,}000) * 100 = 29.2\%$$
$$\text{Percent spent} = (AC/BAC) * 100 = (ACWP/BAC) * 100 = \$400{,}000/(120 * \$10{,}000) * 100 = 33.3\%$$

Performance indices (<1 is an unfavorable condition)

$$CPI = EV/AC = 350/400 = 0.875$$
$$SVI = EV/PV = 350/400 = 0.875$$

The project is behind schedule and over budget.

Project completion estimate:

$$BAC = 120 * \$10{,}000 = \$1{,}200{,}000$$
$$EAC = BAC * (AC/EV) = BAC/CVI = \$1{,}200{,}000/0.875 = \$1{,}371{,}429.57$$
$$VAC = BAC - EAC = \$1{,}200{,}000 - \$1{,}371{,}429.57 = -\$171{,}429.57$$
$$ETC = EAC - AC = \$1{,}371{,}429.57 - \$400{,}000 = \$971{,}429.57$$

The project is behind schedule and over budget based on completion estimates.

7.7 Successful Project Execution and Lessons Learned

Project management is an integrated task comprising of many elements such as consensus on the new product specifications, the project approval process, project planning and scheduling, estimating project and product cost, risk assessment, product design, quality planning, and product verifications. This collection of activities is usually referred to as the integrated project master plan (IPMP).

Successful engineering projects, even those with accelerated schedules, share a common basis for success. It is the PM's attention to detail and good planning, coupled with good communications and teamwork in the design team. The true underpinning of managing and controlling an engineering project for success are as follows:

1. Project Planning Phase
    - Establish clear and visible project definitions and goals, including the project statement and the appropriate level of technology to be adopted into the new product
    - Partition the project into well identified tasks with phases and milestones, with specific deliverables at each phase
    - Identify team members and their responsibilities. Keep management and team members informed of project progress
    - Force team members and managers to continuously evaluate and re-plan
    - Identify tasks to accomplish all milestones, with dependencies and proper size and durations; document each task, its input and output, and responsible team member(s)
    - Make a realistic estimate of the project's total schedule, cost, the product manufacturing cost, and its quality and reliability goals; set a realistic manufacturing and automation strategy with different scenarios for product sales volumes
    - Identify a project-wide risk assessment and risk-mitigating strategy
    - Use a small number of documents to manage the project; include PERT and Gantt or Excel, depending on the project size
2. Project Execution and Control Phase
    - Use formal meetings such as phase review process to synchronize the schedule of project tasks and get the go-ahead from management to proceed to the next phase
    - Use the EV system to measure the project performance versus the schedule and cost. Take corrective action if the variances are over 10%

- Use informal and frequent (daily) project team meetings to get project updates and obtain information about potential problems
- Spot potential problems, through the EV system or informal meetings, and help in taking preventative action in the shortest time possible
- Take early corrective action when a problem is detected; use the risk mitigation plan as soon as problems arise
- Improve team communications and delegation of duties and responsibilities
- Help the team focus on the activities at hand
- Other actions in the project execution and control phases will be discussed in Chap. 8, such as improved phase gate process planning, project and team communications, risk management, team problem-solving, and conflict resolution

References and Bibliography

Asberg, B., G. Blanco, P. Bose, M. Garcia-Lopez, G. Toussaintz, G. Wilfong, and B. Zhu, "Feasibility of design in stereolithography," *Algorithmica*, 19(1):61–83, 1997.

Bazaraa, M., J. Jarvis, and H. Sherali, *Linear Programming and Network Flows*, 4th ed., Wiley, New York, 2009.

Budd, C. I., and C. S. Budd, *A Practical Guide to Earned Value Project Management*, 2nd ed., Management Concepts, Vienna, VA, 2009.

Callahan, K., G. Stetz, and L. Brooks, *Project Management Accounting: Budgeting, Tracking, and Reporting Costs and Profitability*, Wiley, New York, 2007.

Campos, J., and J. Balland, *The Voice of the Customer in Product Development*, 4th ed., Multi-Media Publications Inc., Oshawa, ON, Canada, 2011.

Drucker, P., *Management Challenges of the 21st Century*, Harper Business, New York, 1999.

Gebhardt, A., *Understanding Additive Manufacturing: Rapid Prototyping, Rapid Tooling, Rapid Manufacturing*, Hanser Publications, Cincinnati, OH, 2012.

Gibson, I., D. Rosen, and B. Stucker, *Additive Manufacturing Technologies: Rapid Prototyping to Direct Digital Manufacturing*, Springer, New York, 2009.

Hamilton, R., *Study of Methods for Evaluation of the PERT/Cost Management System*, MITRE Corporation, Bedford, MA, 1964.

Haugan, G., *Effective Work Breakdown Structures*, Management Concepts, Vienna, VA, 2001.

Haugan, G., *The Work Breakdown Structure in Government Contracting*, Management Concepts, Vienna, VA, 2003.

Hopkinson, N., R. Hague, and P. Dickens, *Rapid Manufacturing: An Industrial Revolution for the Digital Age*, Wiley, New York, 2005.

Hugos, M. , *Essentials of Supply Chain Management*, 3rd ed., Wiley, New York, 2011.

Kahn, K., *The PDMA Handbook of New Product Development*, 2nd ed., 173–187, John Wiley & Sons, New York, 2004.

Kalpakjian, S., and S. Schmid, *Manufacturing Engineering and Technology*, 5th ed., Pearson Prentice Hall, Upper Saddle River, NJ, 2006.

Kerzner, H., *Project Management: A Systems Approach to Planning, Scheduling, and Controlling*, 11th ed., Wiley, New York, 2013.

Kerzner, H., *Project Management Metrics, KPIs, and Dashboards: A Guide to Measuring and Monitoring Project Performance*, Wiley, New York, 2011.

Lewis, J., *Project Planning, Scheduling, and Control: The Ultimate Hands-On Guide to Bringing Projects in On Time and On Budget*, 5th ed., Wiley, New York, 2010.

Ma, Y., G. Chen, and G. Thimm, "Paradigm Shift: Unified and Associative Feature-Based Concurrent Engineering and Collaborative Engineering," *Journal of Intelligent Manufacturing*, DOI 10.1007/s10845-008-0128-y, 2008.

Mosey, D., *Early Contractor Involvement in Building Procurement: Contracts, Partnering and Project Management*, Wiley, New York, 2009.

O'Neal, C., "Concurrent Engineering with Early Supplier Involvement: A Cross Functional Challenge," *Journal of Supply Chain Management*, 29(2):2–9, 1993.

Page, T., *Design for Additive Manufacturing: Guidelines for Cost Effective Manufacturing*, LAP LAMBERT Academic Publishing, Saarbrücken, Germany, 2012.

Practice Standard for Work Breakdown Structures, 2nd ed., Project Management Institute, Newton Aquare, PA, 2006.

Pritchard, C., *How to Build a Work Breakdown Structure: The Cornerstone of Project Management (Beginnings)*, ESI International, Arlington, VA, 1999.

Proud, J., *Master Scheduling: A Practical Guide to Competitive Manufacturing (The Oliver Wight Companies)*, 3rd ed., New York: Wiley, 2007.

Rad, P., *Project Estimating and Cost Management (Project Management Essential Library)*, Management Concepts Inc., Vienna, VA, 2001.

Reinertsen, D., and P. Smith, *Developing Products in Half the Time*, 2nd ed., John Wiley and Sons, New York, 1998.

Sheldon, D., *World Class Master Scheduling: Best Practices and Lean Six Sigma Continuous Improvement*, J. Ross Publishing, Fort Lauderdale, FL, 2006.

Sidsin, G., and M. Smith, "Managing Supply Risk with Early Supplier Involvement: A Case Study and Research Propositions," *Journal of Supply Chain Management*, 41(4): 44, 2005.

Smith, P., and G. Merritt, *Proactive Risk Management*, Productivity Press, New York, 2002.

Stratton, R., *The Earned Value Management Maturity Model*, Management Concepts, Vienna, VA, 2006.

Wallace, T., and R. Stahl, *Master Scheduling in the 21st Century*, T. F. Wallace & Company, Cincinnati, OH, 2003.

Discussion Topics

1. Discuss the major differences between project scheduling by using Excel, Gantt, or PERT charts. Which scheduling technique would you use and why? Consider the company management style, the industry it operates in, the project size in engineers, and the amount of tasks and the duration of project.
2. When would you use a dummy activity in PERT charts? How do you handle the earliest/latest start and finishing times for dummy activities? Would you use them in project Gantt or Excel based scheduling?
3. Discuss leveling techniques in reducing project cost or schedule duration. Show three actions and their sequence and the methodology you would perform to reduce project cost or durations.
4. Discuss the engineering project design phase delay factors. How would you rate the top three in importance according to your opinion and experience? If you rate them differently than the chapter notes, please indicate your reasons.

5. Discuss the engineering project manufacturing phase delay factors. How would you rate the top three in importance according to your opinion and experience? If you rate them differently than the chapter notes, please indicate your reasons.
6. There are 15 successful project execution ideas in this chapter. How would you rate the top-five items in importance according to your opinion and experience, please indicate your reasons.

Problems

7.1. For the project below, the activities, nodes, and manpower are shown.

Activity	Start Node	End Node	Activity Time	Staff Required	Slack Time
A	S	1	5	2	
B	S	2	4	4	
C	S	End	6	1	
D	1	2	1	3	
E	2	End	2	2	

a. Draw PERT chart with earliest and latest start and finish times. Fill in the slack time of all activities and show the critical path.
b. For the project duration, show the staffing profile in people, beginning at the earliest time (forward pass).
c. For the project duration, show the staffing profile in people, beginning at the latest time (backward pass).
d. If crashing any one activity by one week is equal to $1000, which one activity is best to crash and for how many weeks to shorten total project time? Show the results in the manpower diagram, and all critical paths after you crash one activity. Show new critical path(s) if any and new PERT chart after crashing.

7.2. For the project below, the activities, nodes, and manpower and base PERT chart are shown.

Activity	Start Node	End Node	Activity Time	Staff Required	Slack Time
A	S	1	3	2	
B	S	2	4	4	
C	S	3	3	1	
D	1	End	5	3	
E	3	2	2	1	
F	2	End	1	3	

a. Draw a PERT chart with earliest and latest start and finish times. Fill in the slack time of all activities and show the critical path.
b. For the project duration, show the staffing profile in people, beginning at the earliest time (forward pass).
c. For the project duration, show the staffing profile in people, beginning at the latest time (backward pass).
d. Suggest moving only one activity to best level staffing requirements (from either forward or backward pass).
e. Which one activity is best to crash to shorten the project time? Show the results and the new critical path(s) if you crash the one activity.

7.3. A project manager contracted a supplier to provide 1000 units at an estimated manufacturing cost of $12 each at the rate of 100 units per week for the total duration of the supplier contract of 10 weeks. After four weeks, the supplier delivered 380 units and was very happy to discover that the units only cost $10/unit. Calculate:

Current project status after four weeks:

a. PV or CWS
b. EV or BCWP
c. ACor ACWP
d. CV
e. SV

Project at completion:

a. BAC
b. EAC
c. ETC
d. VAC

7.4. A construction company successfully bid on a contract to repave 20 miles of city road @ $30,000/mile with eight weeks to complete. Two (2) weeks into the project, the company spent $100,000 and completed four miles. Calculate the following terms, making sure that the sign of each term is correct, and indicate whether the company is ahead or behind schedule and budget.

Planned versus snapshot (at two weeks) activities:

BAC
PV or BCWS
EV or BCWP
AC or ACWP
CV
Company is ahead ________or behind ______ budget
SV
Company is ahead ________or behind ______ schedule
CVI
SVI
Project Completion Estimate Activities:
EAC
ETC
VAC

CHAPTER 8

Engineering Project Phases, Control, Communications, Leadership, and Risk Assessment

The need to develop new products at an accelerated rate and shortened lifecycle in many high-technology industries has led to increased focus on innovative design processes managed by well-organized engineering management techniques. This shorter lifecycle is caused by the speed of technology developments and competitive factors. New products are replacing existing products with more capability and performance at a lower cost and higher quality, while expanding the market by satisfying more customers.

There is little time for design quality defects or the resulting engineering changes to fix them in new technology products. Design quality problems can do great harm to a company. They force a delay in introducing the new product as the demand for older products evaporates. A major engineering design quality problem for some of these companies could result in a loss of incorporating a new technology into the current product cycle. The company might have to delay new product introduction until the next technology cycle is available.

This chapter is focused on the implementation of the new product development plan discussed in Chap. 7, using project-planning tools such as Gantt and PERT charts. It includes methodologies such as dividing the development project into phases, communications, leadership and control of each phase, and a plan for risk assessment and management. The methodologies discussed in this chapter can reduce defects in design and manufacturing and protect the company at the critical transition time between products in a product-family lifecycle plan. The elements of engineering project phase communications and control are as follows:

8.1 *The Phase Gate Review Process.* In this section, the product design phase gate review process is outlined, with explanations of the different types of phases and how to transition from one phase to another.

8.2 *Types of Phase Gate Review Processes.* In this section, different types of phase review processes are discussed, with advantages outlined for each type.

8.3 *Implementing a Phase Gate Process.* In this section, the role of the PM in managing the information flow through formal and informal meetings, the elements of the product release information, and responsibilities of each department are outlined.

8.4 *Engineering Project Risk Analysis.* In this section, a plan for estimating project risk and managing that risk during the various development phases is outlined. Traditional quality tools that were the precursor of risk management will also be explored.

8.5 *Managing Engineering Project Teams.* In this section, the characteristics of a successful team are discussed, including how to satisfy the different needs of team members and the different stages of the team. The elements of coaching team members toward self-improvement are discussed, including team member career positions, motivations and expectations, understanding, and nurturing team member skills, which includes technical, adaptive, and transferable skills.

8.6 *Resolving Engineering Team Conflict and Managing a Successful Engineering Team.* In this section, the root causes of team and interpersonal conflict are explored. They include successful team decision-making, how to handle difficult team members and poor performance issues, and methods for developing a successful conflict resolution process. In addition, the attributes of a successful PM and how to develop the skills necessary for leading and motivating teams are discussed.

8.1 The Phase Gate Review Process

The project planning tools discussed in Chap. 7, such as Gantt and PERT charts, are used by the project manager (PM) to plan the project by subdividing it into activities that are defined and monitored as well as determining the precedence of each activity. The use of phases in engineering project management is a good method to organize the work needed for new product creation into smaller more manageable segments, as well as keeping the total enterprise engaged in new product creation. Specifically, project phase subdivision can help in the following:

1. It allows the PM to manage a smaller portion of the project at a time, with visible goals achieved and measured at each phase.
2. A release or milestone meeting usually marks the end of each phase, where all relevant stakeholders agree that the goals of that phase were met. This helps synchronize all of the project activities at one point in time, before the start of the next phase.
3. At the end of each stage, upper management can review the project assumptions, goals, and performance as well as competitive factors. It is a good point to reexamine the original premise of the new product and make the decision of whether to continue or cancel the project.
4. In a multi-project enterprise, the management can commit to the resources for each phase in terms of staffing, material, and equipment. The management can continue to analyze its commitments as phases are completed in different projects. If the resources for the enterprise become limited, then projects can be canceled based on their progress toward full completion and their expected return on investment (ROI) at the current phase.

Goals for Each Phase	Metric Used
Project schedule	Statement of work plan (SOW)
Project productivity	Resource requirements plan
Quality	Six sigma level
Risk assessment	Additional resources to keep to schedule

TABLE 8.1 Goals for Project Phases and Metrics Used

The goals for each phase of the project are to maintain the project schedule as planned, at the level of resources required to complete all activities, and at the quality determined at project inception. In addition, the risk assessment level should be maintained and reexamined as required. Table 8.1 is a summary of the goals for each phase of the project, and the metrics used to measure progress of each goal.

While it is easy to measure the project schedule as opposed to the original project plan using the statement of work (SOW), it is more difficult to ascertain the project progress in terms of productivity and quality. A check of the project expenditures can attest to whether or not the project is on budget, using earned value (EV) techniques. The quality level might be difficult to assess, considering that in intermediate project phases, the final test of the specification of the product cannot be completed. The quality, cost, and risk assessment plans should be modified at the completion of each phase and a new plan be amended for the remaining phases.

The design process phases in most high-technology industry sectors are similar. With increased collaboration from the project stakeholders, each phase can be optimized. A summary of attributes and metrics of success for each design phase are as follows:

1. Project Concept Phase

   This is the period of market research, product functionality definition, development process methodology, performance measures, and supplier selection. The initial cost, sales, and profit targets for the new product are set during this phase, and a small team of engineers and marketers are working on the product idea with constant improvement as a goal. This period is completed when product feasibility is demonstrated and approval is received from the management for commitment to the final concept(s) and to proceed to development. At that time, resources are identified and committed in terms of personnel and equipment, a ROI analysis for the new product is approved, a design schedule is agreed upon, and deliverables outlined for the design completion.

   An important element of this phase is the concept selection process. Many ideas are floated to achieve the product, both in look and feel, as well as achieve product functionality. These ideas should be solicited from everyone in the company, including overseas sites, customers, and the supply chain. A maximum of product design ideas are encouraged in this phase in order to explore the best design selection for implementing the new product. Said one design manager, "No idea is a bad idea at this stage." These ideas are distilled down to one or two concepts that are then pursued into development.

One of the techniques for determining the optimal design in the concept stage is the design funnel, as developed by Professor Pugh in the book *Total Design*. According to the theory of concept design selection, there are three primary components:

I. Concept Generation. This when there are brainstorming sessions to determine the possible concepts that can meet the design criteria. The concepts are then reduced through a process of controlled convergence, depending how they best meet the new product design criteria.

II. Criteria Generation. The criteria for the new product are generated, either from the preceding product in the family or through initial specifications agreed upon early in the concept design stage.

III. Evaluation and Selection of Final Concept(s). This component is similar to criteria ratings, where the concepts generated in Part I are tested against the new product criteria in Part II. The attributes of each design concept are tested against each design criteria. They can be rated either as a numerical value or a ranking of does not meet, satisfactorily meets, or exceeds the design criteria (–/S/+). The sum of the attribute rating can thus determine the best concept among the alternative designs analyzed.

The design funnel process, as shown in Fig. 8.1, is an iterative process where new concept ideas are generated and then reduced through convergence using a criteria-rating methodology. The design funnel process continues as new concept generation of ideas is undertaken and then reduced again through convergence. Each step in the design funnel is carefully analyzed, until a final design concept is selected.

The difference between the design funnel and voice of the customer (VoC) processes, such as quality function deployment (QFD) mentioned in Chap. 3, is that the design funnel can work in revolutionary as well as evolutionary products, while QFD is dependent on customers already familiar with an earlier version of the product. QFD is a guide on how to set the specification for the new product based on customer inputs

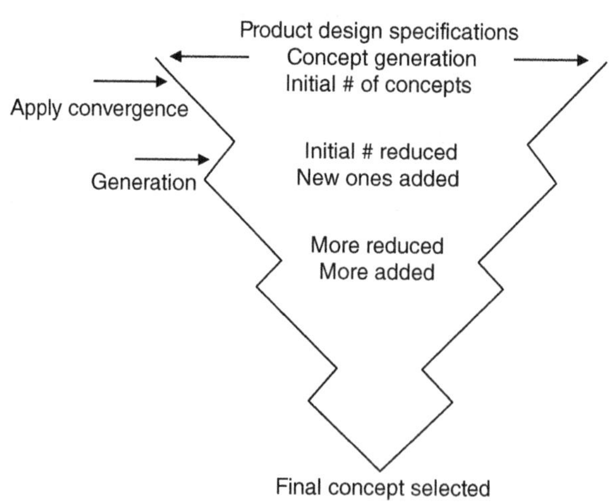

FIGURE 8.1 The design funnel.

through their experience in the use of the older product. The design funnel encourages the evaluation of actual design concept ideas for new products as well as using the older products as a base for evaluating the new product concepts.

Free flow of communications is important at the concept phase, between the core concept team and the other stakeholders of the design process: marketing, various engineering expertise, manufacturing, and the supply chain. While there are no detailed technical drawings of the product, 3D or printer- or CAD-generated outlines can be used to model the look and system layout of the product. These outlines can be the subject of early supplier involvement (ESI) and design for manufacturing (DFM) discussions with suppliers and manufacturing. Local and country inputs can be solicited in order to provide for global rollout of a single product. In addition, the concepts that did not make it to development should be carefully documented so that they could be reevaluated in later generation products.

2. Product Development Phase. This phase is usually divided into two parts:

   a. Early Development Phase. During this phase, activities are initiated to complete the detailed design model on CAD, so that engineering analysis of various disciplines can be performed for mechanical strength, vibration, thermal, and drop/transportation performance of the new product design. Prototypes are made with rapid prototyping or soft tooling in order to evaluate the fit and performance of the product and validate the design concepts. Depending on the desired level of testing and tolerances, machined parts, 3D printed/stereo-lithography prototypes are made to simulate final plastic, cast, or forged parts. Details of the final design can be omitted or substituted in this phase to accelerate the testing of the prototype. An example would be to substitute screw attachments or gluing for snap-fit parts. Most OEMs and suppliers do not use the drawings to actually make the parts. They rely mostly on 3D models and electronic data transfers to manage the manufacturing of parts. Paper drawings are mostly used for annotating features for incoming inspection. Communications at this stage are with detailed initial CAD models of the product, being transferred back and forth from the design team to the design stakeholders. Design experts analyze the design for thermal and mechanical stress points and make suggestions for securing the subassemblies or cables so that the design can pass the drop and transportation tests. Suppliers provide ESI and DFM feedback, such as asking the design team to move some of the features around so that plastic molds would be easier to tool.

      At the same time, the initial bill of materials (BOM) is being loaded into the material procurement system, and purchasing tries to find suppliers for these materials and perform part qualification tests for new parts. Some part suppliers have augmented their services with online assistance. These suppliers have greatly improved their service by adding engineering staffs with ready expertise in parts' attributes such as lifecycle stages, applications, suggested lower cost replacements, and the likelihood of obsolescence.

      The use of engineering communications and fast access of analysis and supplier expertise has reduced the number of engineering changes at this phase. This is coupled with the increased capability of product data management (PDM) systems, which allow for warehousing of documentation and historical progression of the product design. Since most of the changes are made on the engineering CAD model, with no hardware being built, smaller delays result from engineering changes in this phase.

b. Late Development Phase. During this phase, sometimes called the "pilot phase," hard tooling is committed, and parts are made in quantities that are significant enough to test out the manufacturing process capability and readiness. There might be several different pilot runs, with products being made for life testing, local and global regulatory agencies, and preferred customers for field trials. Hard tooling may be started before design reviews are completed to meet tight schedules. Product launch and rollout plans are made for global products at different plants in different countries simultaneously.

   Communications in this phase are critical and involve complete product CAD models between the different manufacturing suppliers such as toolmakers, production shops, and the design team. This is helpful to lessen the risks of early hard tooling commitment. When a problem is detected at this phase, fast, secure, and instant communication is needed to solve the problem because of materials and production schedule commitments. These might involve a second round of expert analysis and redesign.

   Figure 8.2 is a summary of the design process phases, their attributes, and resulting decisions and outcomes of each phase. The connected circles represent the stakeholders involved and how they should interact during these phases.

8.1.1 Attributes and Metrics of Success for Each Design Phase

In most large companies, the concern has shifted recently to fast new product introductions in order to implement a controlled stream of new products. Each successive product incorporates additional technology adoption to keep customers satisfied and leapfrog the competition. Large companies have taken on efficient new product creation cycles as a competitive weapon and part of their strategy to satisfy their customers and maintain market share. Measuring the design efficiency of each phase has become an important part of enhancing new product creation. Table 8.2 lists some of the metrics that can be used to evaluate the efficiency of each phase.

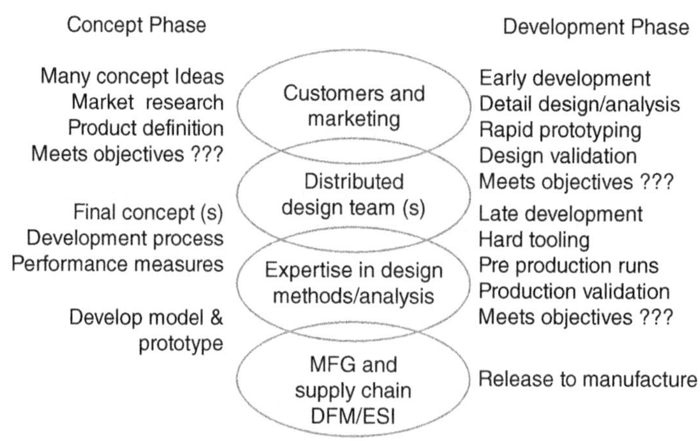

Figure 8.2 Total product realization process phases and characteristics.

Phase	Necessary Attributes for Success	Success Metrics
Concept	Fast access to technology resources Communication with customers/marketplace Feedback evaluating design alternatives Select and negotiate with suppliers Reduce concept ideas to one or two ideas	High success switching to development Number of inputs from MFG, design, sales, and service Maximize number of design alternatives Leverage of new technology in product Concept phase duration
Output	Prototype functionality demonstration	
Early Development	Design analysis and simulation Standardized design methodologies Less formal design reviews Rapid prototyping/soft tooling Minimum design changes Early supplier involvement/DFM Component qualifications	On-time design reviews Metrics (DFM, DFX, cost models) Frequent informal design reviews Turnaround time for prototypes Fast implementation of changes Communications with suppliers Component engineering process
Output	Final concept validation	
Late Development	Detail design models Quick resolution of design/MFG problems Hard-tooling commitments Environmental/life testing plan and analysis Material procurements	Interfacing models to other systems Communication process for resolution Communications with global suppliers Communication with test providers
Output	Product launch plans: Material, manufacturing, and distribution	

Table 8.2 Attributes and Success Metrics for Design Phases

The metrics discussed in Table 8.2 can be measured and successful results adopted by companies to encourage faster new product creation. In this manner, a company can maintain its market leadership in a particular product family. Company management can use these metrics as a basis for rewarding successful engineering PMs in terms of stock options and other benefits, if they achieve or exceed certain metrics in new product creation.

8.1.2 New Product Creation for the Global Economy

The increased focus of technology companies to develop single products for the global market necessitates a transition of the project design teams and the enterprise to a lean organization that is focused globally, as shown in Table 8.3.

The lean organization for developing products in the global marketplace should be as flat as possible, with minimum layers of management. Consequently, the span of control of each level would be much wider, with multiple reports to different parts of the global organization. This would increase communications and feedback from different parts of the enterprise. Preferably, the number of management levels from the design engineer in a development team to the CEO should not exceed five.

Attributes	**Local**	**Global**
Organizational layers	Many	Few
Span of control	Narrow	Wide, intermixed
Skill set	Specialized	Interdisciplinary
Design methodology	Propriety	Standards and roadmaps
Customer base	Local, narrow	Global
Market segments	Few and large	Many and niche
Regulatory issues	Single country specific	Multi-country/region specific
Supply chain	Local, small size	Global, large size

TABLE 8.3 Lean Organizations, Transition from Local to Global Market

The skill sets of the design team members should be multidisciplinary, with decreased emphasis on specialization. Non-core disciplines should be sought as part time resources from other parts of the enterprise or contracted out as needed. Propriety designs should be avoided whenever possible, unless they are part of the core competency. The use of industry standards, technology roadmaps, and standard parts and modules such as power suppliers, cables, and connectors should be encouraged as much as possible.

The customer base should be looked at as a global market, with as many multi-country regulations as possible adopted in the development project. The regulations would include environmental as well as health and safety issues. Product labeling, reference materials, manuals, and packaging should consider local languages and customs. The supply chain should be chosen along strategic global suppliers that can support manufacturing in multiple locations.

The global focus in new product development also necessitates changes to the phase gate process, shown in Table 8.4. The major difference is the global perspective in providing additional design concepts, in-depth engineering analysis, and using online tools for parts selection and acquisition. In addition, the use of software CAD models of the product to perform design analysis and evaluation is of great help in securing input from global stakeholders.

8.1.3 Phase Gate Design Reviews

Formal design reviews (milestone meetings) should be scheduled regularly at the end of each phase. All team members, project stakeholders, disciplines, and design and manufacturing processes should be presented. The major outcome of the design review milestone marking the end of each phase is to solicit agreement from the project stakeholders that all phase goals and requirements are achieved and approval from management is obtained to proceed to the next phase.

There might be a need for interim design reviews, synchronized with the completion of critical activities, to keep the project team and management updated with the project

Phase	Traditional Engineering	Global Engineering
Concept	Design for country specific market	Design for global market
	Limit the number of concepts to explore	Explore maximum number of concepts
	Concepts made from paper sketches	Concepts made in 3D CAD/printing
	Inputs from in-house and customers	Input from global company resources
	Suppliers selected after final concept	ESI on concepts before bid is awarded
	Suppliers bid on detail design	Suppliers bid with cost model or use similar products
	Unused concepts discarded	Unused concepts documented for future
Early Develop.	Design analysis and simulation	In-depth and system analysis
	Drawings are used to make prototypes	Prototypes are made from 3D printing
	Limit and control ECOs	ECOs implemented on CAD model
	Purchasing selects part suppliers	Online suppliers provide more information
	Models routed and discussed serially with experts, suppliers, other teams	Models routed synchronously, discussed Online
Late Develop.	Hard-tooling commitment after design review	Tooling started earlier with communication
	Quick problem resolution process	Secure communications/ remote resolution
	Local material acquisition and production plans	Global Acquisition and manufacturing plans
	Local Regulatory, environmental/ life testing plans	Testing for global requirements
	Local Material procurements	Communications with global suppliers

TABLE 8.4 Changes to the Global Product Development Plans

progress. Many projects use daily stand-up meetings for the PM and core members of the team to review the status and resolve any short-term problems.

Formal design reviews take a long time to prepare and place undue stress and effort on the PM and the team members. There should be a balance between the need to exchange information and the time and effort it takes to properly conduct and report on project meetings. The PM should keep review meetings far apart at the start of the project, replaced by more frequent reviews at the end, as shown in Fig. 8.3.

The PM should document all discrepancies and decisions made at the phase review meetings. Resolution of all problems should be required and signed off by the relevant parties. The next critical activities in the project schedule should not be launched until all problems are satisfactorily resolved in the review meetings. All meeting results should be shared with the team members and the management.

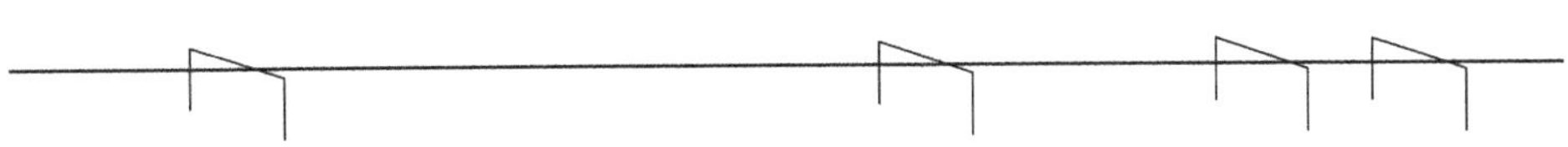

FIGURE 8.3 Phase review meetings time line.

8.1.4 Design Review Preparation

In order to have a successful design review meeting, all relevant information to be discussed should be available to the attendees at least one week prior to the meetings. The PM should confer with the management and the project stakeholders prior to the meeting in order to solicit inputs and clarify any concerns. A successful phase review or release meeting should be a mere formality since all of the disciplines' inputs were solicited and potential issues and conflicts resolved before the meeting. The manufacturing, quality, and service plans for the project should also be discussed by the appropriate representative of each of these departments. A checklist of the design review items encompassing the different stakeholders concerns could be as follows:

1. Development Engineering
   - Activities completed to date based on the original project plan and adherence to the planned schedule
   - Update of project funds expended versus planned budget and cost variances of completed versus planned activities
   - Product cost estimated to date and variances to original cost goals
   - Project design and manufacturing quality plan update and variance to original plan, if any
   - Reliability estimates update and variance to original plan, if any
   - Risk management update and changes in risk assessment if any
   - Up-to-date list of all new product part numbers, current revision drawings, engineering change orders, and contracted design and manufacturing services
   - Up-to-date CAD models, showing the mock-up (MO) of the product, all assemblies and parts, as well as results of stress and thermal analysis
   - Review of original design specifications and tolerances:
     - Performance test results if available, and how they meet the original specifications
     - Worst-case and statistical-tolerance analysis performed for mechanical parts
   - Manufacturing and suppliers' capabilities and feedback in terms of ESI and DFM, and how they were included in the product design
   - Incoming inspection and test process plans
2. Manufacturing Review
   - DFM review and DFM efficiency ratings based on the selected design rating system
   - Critical parameters for ME design:
     - Tolerance analysis, flatness specs, and surface finish specifications
     - Fastening or joining features; encouragement of snap-on assembly
     - Shafts, bearings, pilot holes, and pins for parts alignment
   - Assembly design requirements:
     - Minimum part count (material, movement, or assembly) achieved
     - Minimal or no adjustments for the product in manufacturing
     - Radii and chamfers aid for assembly

- The choice of a base part to locate all components in the assembly of the product
- Minimum handling and fixturing problems due to the design efficiency of the product
- Use of insertion and alignment features in parts to facilitate joining in assembly
- Use of compliant and self-alignment features for ease of assembly
- Use of part feature to facilitate mold making and location of parting lines in the mold

- Electrical test requirements:
  - Access coverage to test points and connectors
  - Built-in test indicators for diagnosis
  - Compatibility with automatic test systems

3. Service Review
   - Number of field service products and major assembly units required for field support
   - Manuals, service, and repair guides availability
   - Minimum service kit tools and parts required
   - Easy access and replacement of modules and parts in the field
   - Mistake proof fasteners, connectors, and ease of reassembly in the field
   - Build in tests or indicators for maintenance in the field
4. Quality Plan
   - Quality metrics such as Six Sigma, *Cp*, *Cpk*, defects per unit (DPU), and first-time yield (FTY) identified for major parts and assemblies
   - Process and product changes well documented
   - Risk analysis, failure mode and effect analysis (FMEA) and fault-tree analysis (FTA) performed
   - Critical parts and processes identified and methods to maintain quality for them reviewed
   - Suppliers and components reviewed and certified
   - Strife and lifecycle testing under controlled conditions and specifications
   - Robustness in temp, humidity, transportation, and electro-magnetic interference (EMI) testing

8.2 Types of Phase Gate Review Processes

Dividing the engineering project into smaller timeline phases could be accomplished through several steps. There is a review process at the end of each phase to be followed by the support documentation for planning the next phase, as shown in Fig. 8.4.

Another name given to this step-by-step approach is "waterfall design," as shown in Fig. 8.5. In this example, each step of this project is to take one year for a total of five years.

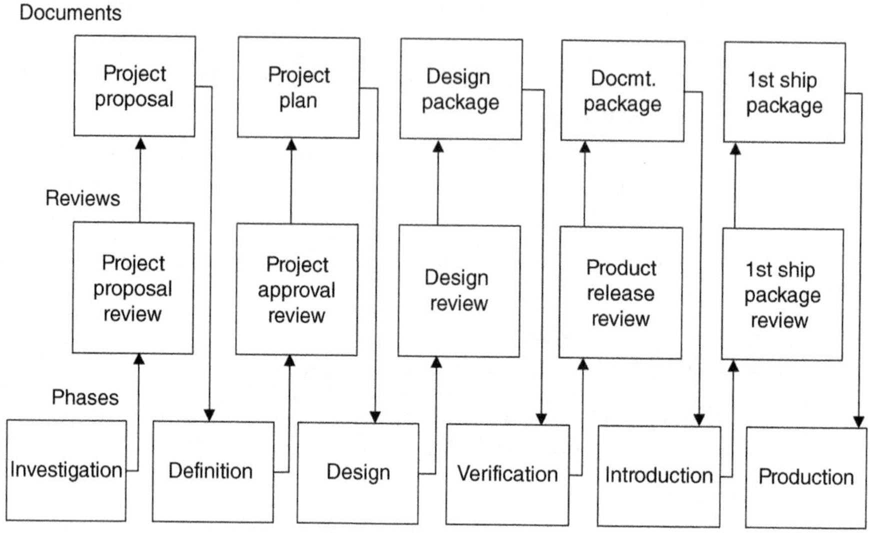

FIGURE 8.4 Sequential phase review process.

Some companies might opt for a more detailed phase review process, with more emphasis at the beginning of the product creation and approval. More details can be performed at the concept stage of the product creation, including in-depth marketing research and competitive analysis, perhaps using QFD. In addition, the adoption of the proper level of technology and the maximum use of standard parts, components, and industry-wide standards can result in a much shorter development phase of the project. These latter parts of the project can then be incorporated into a single or few development phases that can be managed without their associated and time-consuming formal design reviews. The development part of the project can be planned with all required activities and managed through informal meetings and design reviews, as shown in Fig. 8.6.

Another approach to the waterfall design project example is to attempt to perform the project in multiple steps at the same time. Figure 8.7 shows the division of the large

Project Plan

						Total
Waterfall design	Year 1	Year 2	Year 3	Year 4	Year 5	5 Years
Requirements	■					
Design		■				
Build			■			
Test				■		
Support					■	

FIGURE 8.5 Waterfall design project plan.

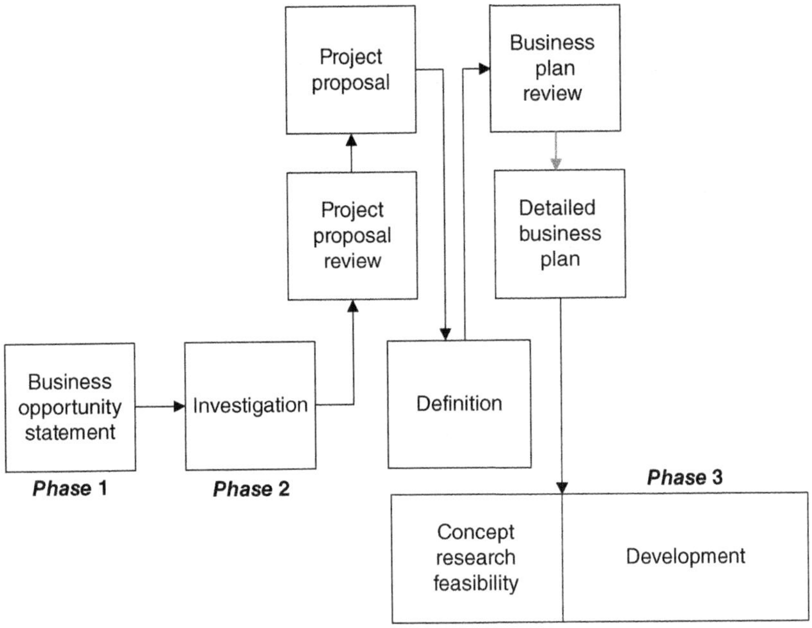

FIGURE 8.6 Front-loaded phase review process.

waterfall project plan in Fig. 8.5 into a cascade of three multiple projects. As soon as the basic product design is complete, the next step of the large project (communications) is launched and proceeds to completion. The next phase of the large project plan (system design) is launched after the communications project design is completed. In this manner,

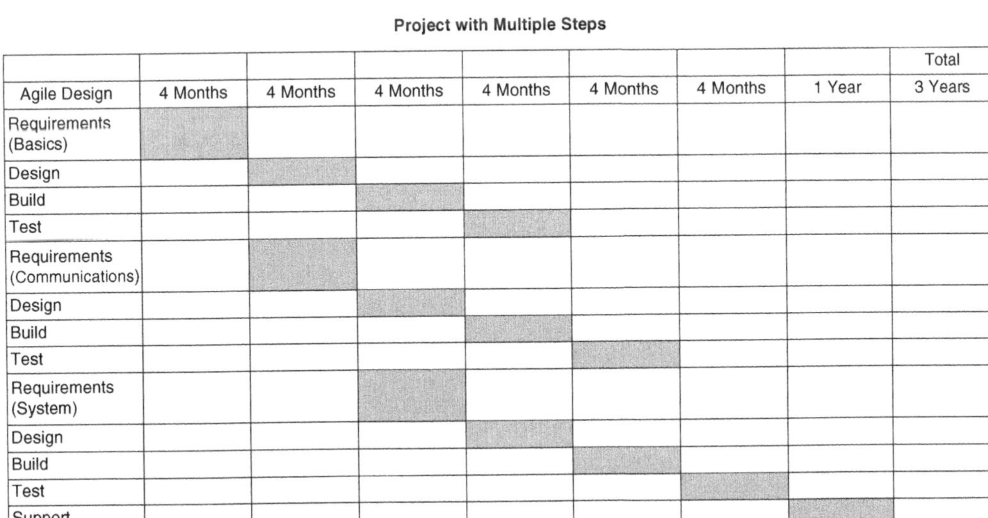

FIGURE 8.7 Waterfall design project with multiple steps.

a large project can be completed by cascading the design portion into separate smaller projects. These smaller projects can start as soon as the design of the preceding smaller projects is completed. So that the total waterfall project-plan time shown in Fig. 8.5 is reduced from five to three years.

8.2.1 Complex Product Phase Review Process

While the front-loaded phase review process shown in Fig. 8.6 can be used for smaller projects, complex projects necessitate more formal phases and subsequent reviews. Expected large sales volumes also need to have the tooling and manufacturing capabilities verified before full production is turned on. An example of a phase review process for a complex product such as a networking device or precise instrument is shown in Fig. 8.8.

The complex product phase review process includes the traditional early phases of the concept, approval, and design (early development), but has additional phases focused on a progression of building prototypes and pilot runs to satisfy different needs for internal testing and verification as well as early shipments to corporate customers and marketing sites. These special builds are used for the following:

1. Mock-up phase build is for a few nonfunctioning units to ensure that design is verified for thermal and environmental testing, such as elevated temperature, humidity, EMI, vibrations, and transportation simulations.
2. Alpha phase build, which could be done concurrently with the MO build, is for a few functioning units consisting of modules and PCBs for hardware and software integration and verification of product performance to design specifications.
3. Beta phase build consists of units to be shipped to selected customers or regulators for verification that the product meets the intended customer needs. Feedback from customers is solicited and evaluated for possible incorporation into the design through revisions to the product.

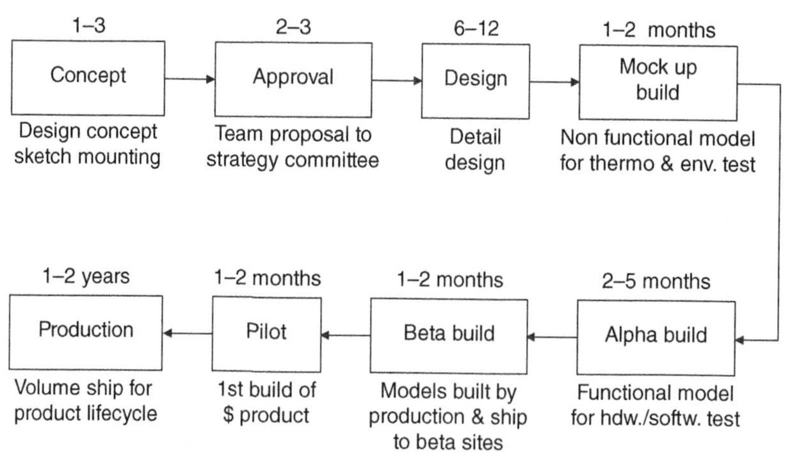

FIGURE 8.8 Complex product phase review process.

4. Pilot run phase. The volume of this run is dependent on the production volume. For high-volume consumer products, a run of 100 to 300 could be made. Pilot units are made in order to test the production tooling and methods. These products are eventually sold to paying customers.
5. Production volume is initiated after a pilot run is completed successfully and could be made anywhere based on the tooling that was tested in the pilot phase.

8.3 Implementing a Phase Gate Process

The PM is responsible for implementing the new product plan through managing the phase gate process and keeping all stakeholders engaged and involved. The PM makes all of the decisions to keep the project on time, within planned budget and specifications, at the planned product cost and quality.

A variety of documents and formal and informal update and review meetings are used in conjunction with the different phases to mark significant milestones and/or the transition into the next phase, then finally into full product release to manufacturing. Formal review meetings are also used to update the project status and reexamine the original assumptions of the project plan. A typical example of project documentation is the product data sheet, generated at the beginning of the project and updated at each phase review meeting given in Table 8.5. It is a summary of all pertinent information about the product, including a tabulation of project cost to date (actual and estimated), completion of relevant project team tasks, and a consensus of all stakeholders (marketing, manufacturing, quality, and service), and their estimates of product quality, cost, future sales, and profit. The project returns are shown in several different estimates, based on return factor, breakeven period, and ROI, including the impact of the project on existing product sales.

The project data sheet is updated at each phase review meeting, including the estimates of the project and product costs, quality, and sales volume. The retrospective column is the final actual product cost, sales, profit, and quality data and is usually updated at the obsolescence of the product or at least a year after release to manufacturing.

The responsibility of each functional area is to input directly into the project review meetings. The type of information that is provided is shown in Table 8.6. In addition, the PM should include relevant reviews at critical times during the project. The concept review should be performed during the concept phase, while the DFM and service reviews could be performed either at the early development (prototype or alpha phase reviews) or updated at the late development (beta or production reviews).

8.3.1 Changing Traditional Design Communications

The advent of Internet communications and the supply chain has provided an opportunity to increase design efficiency. Distributed resources can be easily accessed to collaborate and augment the design effort. These include country-specific requirements for global products, outside design, and analysis expertise and design and manufacturing service providers, ESI/DFM feedback. More on project communications will be discussed in Chap. 10.

It is important that the distributed design team, with associates in other locations, as well as the design and manufacturing service providers, collaborate in real time and

Product Name: Project Manager: Project Staff: R & D			Product No:			
		Date:				
			Project Team Members Manufacturing: Marketing: Quality: Field Service			
Milestone	Concept	Proto	Alpha	Beta	Production	Retrospective
Date (Estimated)						
Date (Actual)						
Development K$ Invested						
Estimated K$ To Go						
Total Fixed Costs						
Nonrecurring Costs						
Lifecycle Costs						
Total Development Costs						
Project Charges						
Tooling Model Shop Material and Labor						
Design Layout						
Specifications Approved						
Material ID (cost, availability)						
Material Forms (cast, molded)						
Operational Loads						
Design Margins						
DFM Review						
Human Factors/Ergonomics						
Inspection and Test Plans						
Reliability/Maintainability						
Failure Rate (%/year)						
Warranty Cost (%/$Ship)						
Estimated Production Cost						
Estimated Selling Price						
Estimated Operating Profit						
Mature Sales (#/year)						
Mature Revenue (K$/year)						
Impact on Other Products						
Net Profit Expected (year)						
Return Factor						
Breakeven Period						
(Profit/Cost; 5 years)						
Project ROI						

TABLE 8.5 Product Data Sheet

Design	Marketing	Manufacturing	Quality
Meeting Customer Specifications	Customer Specifications	Parts Sourcing	Environment Specifications
Disposition of Unused Inventory	Material Pricing	Suppliers Identification	Life/Strife Tests
System Compatibility	Product Support	Supplier Qualification	Quality Approvals
Internal Specifications	Training Customers	Production Plan	Component Review
Life Test Plans (Strife)	Training Service Personnel	Capacity Analysis	Reliability Estimate
Environmental Test	Field Support Plan	Assembly and Test Plans	Environmental Test
Qualification Test	Repair Kits	Production Tooling	Regulatory Agency Approvals
Design Margin Analysis	Product Manuals	Shipping Package	Incoming Inspection

Concept Review	DFM Review	Service and Quality Review
Evaluate Alternative Designs	Minimize Parts Count	Modules Replacement and Stock
Use of Standard Manufacturing Processes	Minimize Assembly Level and Direction	Built-in Test/Maintenance
Use of Existing/Standard Parts	Lowest Cost Fastening	Dis/Assemble with Common Tools
Existing Design Technology	Minimize Handling and Fixturing	Use of Test Points and Connects
Leverage Current Designs	Part Alignment Features	Easy Access for Parts and Test Points
Modular Design	Self-Align and Compliance Features	Tolerances Analysis Completed
Self-Diagnosis	Limit Flexible Parts	Critical Specifications Identified
No Adjustments	Avoid Cabling	CAD Drawings Annotation/GDT
Non Critical Tool Review	Basc Part Locates all Components	Critical Parts Identified and Reviewed

Table 8.6 Phase Review Product Information Responsibility

at the same time. Collaboration should be real time and not asynchronous, able to reach anyone, anywhere in the world efficiently. A problem should be quickly resolved by bringing the team virtually together before it becomes a *showstopper*. As one design supplier CEO said, "One shared mind, one shared moment . . . minds being there when you need them."

Communication within the project teams is also important. These communications should be encouraged to be informal as well as independent of phase review meetings, which are formal meetings and presentations to management, requiring extensive preparations. The informal meetings can be made up of several types:

1. Stand-up Meetings. These meetings are short in duration (15 minutes), scheduled frequently, such as on a daily basis, to discuss the status of the project. They do not have an agenda and usually do not produce status reports, unless major problems are discovered. They are usually held around the project charts, with the PM and the core team members discussing the daily status or any concerns by any members of the team on issues that might have an impact on the project goals.
2. Working Update Meetings. These are regular working meetings scheduled between different working groups in the project team and outside stakeholders. These meetings might occur weekly or bi-monthly, depending on the communications needs. Examples would be meetings within the team subgroups of different disciplines in the project team to assess the interface between the hardware and software portions of the project, or meetings with the distributed part of the team and the supply-chain representatives. An agenda might be necessary as a means of follow-up on open issues from the last meeting.
3. Project Team Status Commutations Meetings. These are interim project status meetings used as fill-in between the formal phase review meetings, with an example given in Fig. 8.9. They are held monthly or even more frequently and require all team members to present the current status of their activities, using a common format. Team members can report progress toward achieving interim goals, accomplishments worthy of note to others, decisions made, and new concerns raised. Any slip in the interim goals should be clearly outlined and that slip's possible impact on the overall project schedule indicated. Any previous concern item should remain on the report until it is resolved. The PM can then roll up the team member reports into a common project status report to upper management. To facilitate management reaction to the reports, they could be color-coded as green, yellow, or red, depending on the urgency of action needed to meet concerns raised by the team members.

Project__________ Manager__________		Date__________
Previous Month Goals:		**Current Month Goals:**
Complete layout rev. #1	✓	Release prototype
Complete material list	✓	Purchase preproduction Material
Schedule design review	→	Schedule design review
Accomplishments		**Decisions**
MFG efficiency score >50%		Build prototype outside
Previous Concerns		**New Concerns**
		Memory IC's obsolete /'17

Figure 8.9 Progress status communications report.

8.3.2 Supplier Control and Communications Needs

The requirements for developing new products in high-technology industries have resulted in faster product development and shorter product lifecycles. Many leading technology companies have created a virtual enterprise, aligning themselves with design and manufacturing outsourcing partners to carry out services that can be performed more efficiently outside the boundaries of the organization. These partnerships enabled a company to focus on its core competencies, its own product brand, its customers, and its particular competency in design or manufacturing.

These global outsourcing companies are providing for cost-effective and timely services for high-technology companies. In manufacturing, they provide multidisciplinary manufacturing, test and support services, including product assembly and test, packaging technology such as sheet-metal and plastic injection molding, software configurations, and support service such as repair depot and warranty exchanges. They also offer lower cost, higher flexibility, and excellent quality, eliminating the need to spend money on capital equipment for internal capacity. This new outsourcing model allows all links in the supply chain to focus on their own core competencies while still reducing overall cycle times.

In design outsourcing, the supply chain offers the flexibility of a single or multiple competencies, including specialized engineering analysis and design validation, testing, and conformance to design standards for multiple countries or codes. Suppliers can offer their own supply chain of strategic alliances in tooling and manufacturing services worldwide. Most of these outsourcing companies offer design feedback in terms of DFM and ESI. These design service providers have reduced the need for companies to purchase or maintain expensive engineering/design competencies, some of which are used infrequently in project design cycles.

The advent of Internet communications has provided for more efficient and cost-effective control and communications with the supply chain. The PM and the supply chain can instantly access critical information about schedule changes, material acquisition problems, engineering design changes, and design feedback in terms of ESI/DFM. Problems can be raised and steps toward resolution can start immediately. Table 8.7 is a listing of the enhancements that can be made in controlling and communicating with the supply chain.

8.3.3 Phase Review Process Communications Needs

The communications revolution introduced by the Internet is a great opportunity for the project manager to leverage the success of the project through using multiple channels of communications within the distributed project team and the project stakeholders. Table 8.8 is a communications summary for the different phases of design that are optimal to use in various stages of the project phases. There are various technologies for communications through the phone system and the Internet. These include teleconferencing, videoconferencing, email, Web-based meetings, electronic bulletin boards and notebooks, white boards, and collaborative geometric model modifications. These could be combined to allow the stakeholders in the design process to collaborate.

It is apparent that collaborative and geometric model modification is the primary tool suited for all engineering design stages, including the concept phase, interfacing with the virtual design team, experts, and the supply chain. It is the most effective means for quick engineering problem resolutions after the virtual meeting.

Traditional	Internet Enhanced
Suppliers placing personnel in OEM factories	Suppliers communicate through the Internet
Suppliers/OEM meet face to face on a regular basis	Meetings as needed via the Internet
Small number of formal design reviews	Many mini reviews, with smaller groups
Suppliers/OEM meet regularly for quality and cost data	Quality and cost data discussed remotely
Large companies do not engage small design suppliers	Design suppliers use Internet leveraged partners
Engineering changes signed off and distributed in days	Engineering changes dissemination online
Operational data flows sequenctially through supply chain	Operational data shared immediately
Engineering data (DFM/ESI) shared serially by phone/fax/email	Decisions and data shared through Internet
Problems in the supply chain resolved serially	Immediate problem-resolution process

TABLE 8.7 Enhanced Controls and Communication with the Supply Chain

Participants can change the model on-line, evaluate the results, reach a satisfactory compromise, and then go back to their individual systems to permanently record the results.

8.4 Project Risk Assessment and Management

In any project, the PM is faced with a certain amount of opportunity and risk in achieving the project goals, on time, within budget, and delivery of a new product at the original specifications, cost, and quality. Opportunity is the potential for improving technical, cost, and/or schedule performance within a project. According to Defense Acquisition University, "Risk is the measure of the potential inability to achieve project objectives, within defined cost, schedule and technical constraints."

Risk assessment involves understating the potential problems in the project, their probability of occurrence, and their impact on a project if they do occur. It involves the identification of all problems that can have a negative consequence with a probabilistic outcome. The problems that are certain to occur need to be characterized, quantified, and dealt with, but not as part of risk assessment or management.

Risk management is the identification, assessment, prioritization, and handling of risks to develop and deliver, on schedule and within cost, a system that meets the customer's requirements. It is a proactive process to identify and quantify risks as well have a ready plan to deal with problems before or when they occur. It is the opposite of crisis management, which occurs when the PM reacts to problems as they are discovered. Problem solving in crisis management can be hasty, under pressure, with mistakes made, without prior planning and forethought. The comparison between risk versus crisis management is shown in Table 8.9.

Communications	Application/Drawbacks	Best Use Method
Teleconference	Telephone line dialing to a predetermined number Lack of prompting to speak Lack of ability to specify an area in a drawing	Adjunct to other Internet systems
Video Conference	Telephone or Internet dialing Can see the other parties Details of drawings sketchy (even if projected) Show physical part or production area	Supplier negotiations and first introductions
Email Web-Based Groups Web Bulletin Boards	Universal in reach Provides a written record Limited interactive capability Asynchronous (serial) in nature Large text files, graphics and CAD models can be attached	Noncritical communications Prompt for net meetings
White Boards and Electronic Notebooks	Allow for sketches and notes to be recorded Allow for multimedia and archiving Capture meetings action items and resolution Asynchronous (serial) in nature	Concept design capture Decision capture
Web Conferencing	Host and manage on-line conferencing Security concerns are addressed One user is the host of the session Different access levels Graphics/data and CAD model viewed/manipulated Ideas recorded/exchanged	Scheduling and negotiations
Collaborative and Geometric Model Modification	Users allowed to share and record data/graphics Interact full CAD models from different systems Ability to identify particular geometry locations Security concerns are addressed Different level of permissions allowed Data can be text, spreadsheet, and CAD models Interfaces to enterprise systems PDM/ERP Neophytes (CAD) users can manipulate data Session recorded and archived	All design stages Engineering problem resolution Manufacturing quality and Inspection

Table 8.8 Communications Summary for Design Phases

Identifying risks early in the development stages allows for less costly mitigation strategies. Typical steps associated with risk management are identification, qualification, analysis, risk-response planning, and risk-detection monitoring and control. In U.S. government contracts, projects are required to identify two risks for every million dollars in the program. Risk identification is the most important step in this process, which is the trigger point for the risk-analysis process.

Risk Management	Crisis Management
Proactive (planned)	Reactive (when problems occur)
Preventative	Corrective
Pay now (quantified)	Pay later (unplanned, probably costing more)
Manager in control	Manager hopes for the best

TABLE 8.9 Risk Versus Crisis Management

8.4.1 Steps in Risk Assessment and Management

At typical example of project risk is the unplanned or unforeseen changes in the competition, projects schedule, scope and resources, product specifications, or project and product quality. The project manager should put in place a plan to assess these risks and manage them through the project execution. A typical step-by-step approach would be:

1. Risk Identification. Identify all possible risks associated with the project. Assign a characteristic and probability (low, medium, or high) of occurrence for each identified risk.
2. Risk Qualification. Qualify severity impact of each risk on the project, if it does occur, in terms of schedule, cost, and quality. Identify any secondary or intangible impact as well. The impact, sometimes called the effect, can be ranked as either qualitatively (low, medium, or high) or quantitatively on a scale of 1 to 5 or 1 to 10, depending on the impact on the project ROI.
3. Risk Analysis. Prioritize each risk as to the probability of occurrence and severity of impact. Select the highest levels of the combined values (or above a certain threshold) as critical risks requiring further attention.
4. Risk Response Planning. This is a risk-handling technique to examine if each critical risk can be mitigated. Are there any steps the project manager can take in terms of changing attributes of the project goals and the scope of its activities to eliminate the risk? If a critical risk cannot be avoided or eliminated, can it be accepted, contained, or transferred?
5. Risk Detection Monitoring and Control. How and when to know if risk materializes (detection); what is the mechanism by which a PM can identify a risk to the project (monitoring or risk triggers). What are the contingency plans (controls) to make in case the risk materializes?

8.4.2 Risk Identification and Qualification

Risks can be categorized based on their possible causes and the probability of occurrence. There are certain risks that cannot be quantified and therefore are called unknown risks. Other risks are predictable and can be accounted for based on their probability estimate or based on the experiences of the PM and the team.

The use of Gantt and PERT charts shown in Chap. 7 are important elements to risk planning and should be analyzed for areas of vulnerability. The critical path is a great starting point because the probability of occurrence as well as the consequences are increased with these events. Also, any choke points or segments where resources are used at maximum capacity are good places to perform further risk analysis. Six Sigma provides for target levels of quality by setting goals for the number of defects in design or manufacturing. This is accomplished through reducing variations by detailing instances that lead to product defects. Relating the number of defects is one way to quantify the consequences of a particular risk during risk analysis. The possible causes of risk can be summarized by the following:

1. Technical. Usually these risks are easier to quantify in terms of probability of occurrence and impact on the project. They include the use of new technology, adopting unrealistic goals or specifications, and the use of nonstandard or unqualified parts in the product.
2. Functional. These risks include those caused by interfunctional mismatch between the different stakeholders of the project. For example, marketing-based risks could include a poor estimate of sales volume and manufacturing; quality risks are associated with poor DFM and their impact on product cost; supply chain risk could be the unavailability of certain parts.
3. Project Management. These risks include inexperience of the PM or critical team members, poor use of resources, and poor or incomplete planning process for the project.
4. Organizational. These risks include poor communications within the project team and the management. Incompatible and poorly prioritized goals of the project including schedule, cost, and scope. Financial uncertainty in terms of poor ROI analysis, competition with other projects, and poor estimate of product sales and benefits as compared with project development costs.
5. External Factors. These include unforeseen changes in organization goals, industry standards, regulatory and safety concerns, standards and requirements, and labor and material availability as well a supply chain issues. In addition, customer tastes, prevailing style, and competitive factors can also have a significant impact on project risk.

These causes can be used to extract a list of possible project risk elements that can be analyzed in order to prioritize risk handling and analysis.

8.4.3 Project Risk Analysis

The risk analysis process can begin once the sources of risk have been identified and the impact of each risk quantified. There are many techniques for project risk analysis. One of the most common is to use the risk factor (Rf), sometimes called "risk priority number" (RPN). It is an evaluation of a risk's severity (impact or effect) or consequence factor (Cf) and probability factor (Pf) of occurrence to determine its seriousness. It is used to compare and prioritize risks and to ensure that the most significant risks get the most attention and resources. The PM and the team should evaluate risks together, and the RPN should be updated at the advent of major milestone meetings and reviews. Once major progress in the risk mitigation process has been accomplished, risk analysis and priorities should be recalculated.

The formula for the Rf is as follows:

$$Rf_i = Cf_i * Pf_i$$

where, Rf_i = risk of factor i; Cf_i = consequence (severity, impact, or effect) of $risk_i$; and Pf_i = probability of Rf_i.

The Rf value is used to prioritize the risks in a project, from the highest to the lowest value. This risk number technique prioritizes responses to a potential problem originated from methodologies of design quality analysis. Potential problem analysis in quality failures can be performed using one or both of two methods: FTA and FMEA. Both methods are similar, with FTA using a top-down method of determining the general methods in which the product could fail and graphically show the possible causes of failures down to the component level, while FMEA uses a bottom-up approach of tabulating all components and their potential defects and the impact of not addressing these problems.

FMEA provides for a formal mechanism to resolve the potential problems by (1) identifying all possible methods a component can fail, (2) ranking these methods with an RPN, (3) suggesting a corrective action, and (4) calculating the resulting improvement in the RPN. The FMEA methodology begins with identifying each component, listing the potential failure modes, the potential causes and the effects of each failure, the initial control conditions, and the conditions contributing to the RPN. These conditions include the probability that the failure takes place (occurrence), the damage resulting from the failure (severity), and the probability of detecting the failure in-house (detection). This analysis suggests a recommended action to eliminate the failure condition by assigning a responsible person or department and recalculating the RPN. FTA is a relational diagram showing possible sources of defects and their interrelationships. It uses logic-type symbols of AND and OR gates to show intermediate and final defect symptoms.

Project risk factor (Rf) analysis is similar to FMEA, which is already well-established in most industries and used for managing potential quality problems. The difference is that the Rf uses only two of the three elements of the RPN (occurrence and severity). The third factor (detection) is treated separately in risk analysis, as part of risk monitoring and control, and is used as a trigger to initiate the risk response plan. In addition, FMEA has a strict set of definitions for the values of severity or impact of each problem, while the project-risk analysis is more qualitative in nature.

Risk analysis is used in many different forms throughout industry. Some follow the format of FMEA closely, while other formats are more simplified. Risk analysis is also called the "risk register", where the occurrence and the severity are plotted against each other, either in graphic or matrix form. Figure 8.10 is an example of a risk register analysis sheet, showing the progression of risk analysis from identification/qualification to analysis to response planning. The risk register shows the Rf analysis before and after risk response planning as well as estimated cost.

The risk register analysis could take the form of simplified division of importance in inverse alphabetical form (ABC) risk classification plot of probability versus impact of risk, with the plot divided into three quadrants: A for low risk, B for medium risk, and C for high risk. Figure 8.11 is an example of an ABC risk-classification plot.

Another form of the risk register is to use a matrix instead of a plot of the risk probability versus severity, using a value of the matrix of 1 to 5 to denote the level of occurrence probability or severity of impact. Color codes of green, yellow, or red instead of ABC are used to illustrate the classification of the project risk analysis process.

Project________________ Model Affected __________ Page ___ of ___
Primary Responsibility____ Schedule Start Date _______ Risk Analysis Date ____
Outside Suppliers_________ Project Manager _________ Approved By ________

				Current Status							Resulting status				
Risk Number And Date	Risk Category	Risk Cause	Risk Effects	Occurrence	Severity	RPN	Risk Impact Cost	Risk Owner	Risk Response Type	Project Phase Acceptance and Date	Occurrence	Severity	RPN	Cost	Response Plan Start Date

FIGURE 8.10 Risk register analysis sheet (blank).

The FTA technique used for project-risk analysis is a top-down approach and has the advantage of showing the interdependence, if any, of the risks being analyzed upon each other. FTA can be used to calculate the expected value of the risk cost with varying scenarios. Each decision branch in the FTA analysis is identified, numbered, and given a probability and payoff cost. Figure 8.12 is an example of when a PM is contemplating making the product in manufacturing using different machines and processes. A probability can be given to each choice, followed by a second probability for the next choice. The total of all different paths probabilities is equal to 1. The expected value is the sum of multiplication of each path probability versus the desired outcome (lowest cost, maximum profit).

$$\text{EV (expected value)} = \sum \text{probability}_i * \text{outcome}_i$$

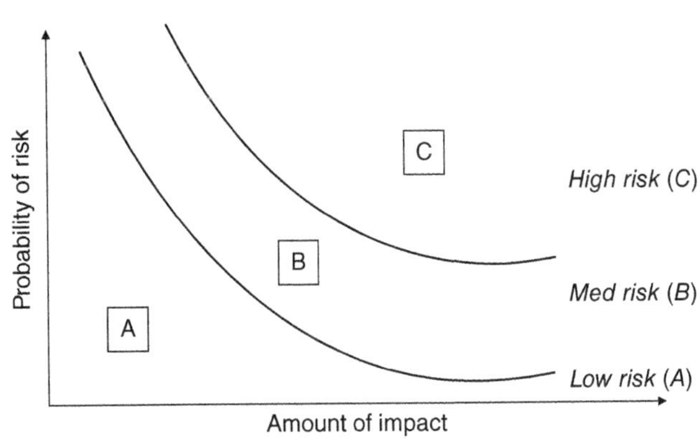

FIGURE 8.11 ABC risk classification plot.

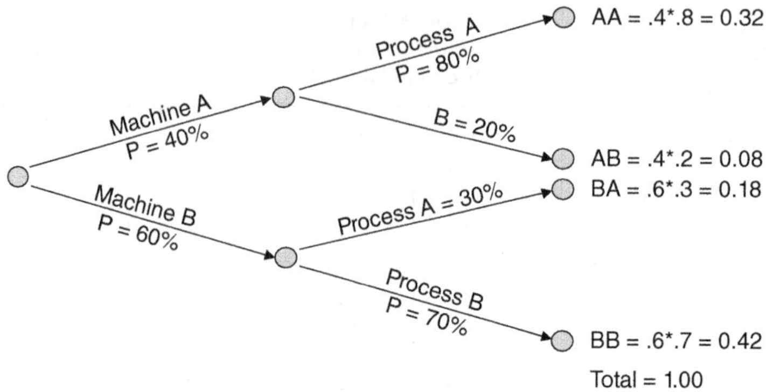

FIGURE 8.12 Fault-tree analysis for risk assessment.

8.4.4 Risk Handling Techniques

Once the project risks have been identified and prioritized, the PM can begin to implement the plans for handling these risks through a variety of techniques. Each risk can be analyzed, and a particular risk-handling strategy can be applied to counter it. The following is a list of options available to the PM to handle project risks:

1. Risk Response Plans. These are strategies that can be used throughout the project phases to respond to specific risks, who owns them, how they will be dealt with, and the resulting reduction in the risk factors:
   - Avoidance is a plan to completely eliminate specific risks by removing the risk causes (moving the schedule, not using new technology, etc.). The PM can develop an alternate plan in the project to avoid these risks.
   - Mitigation is reducing the Rf by either reducing the probability of occurrence or lessening severity or impact if it occurs or both. This can be accomplished by proactively implementing or altering project tasks to reduce the probability and/or consequence of the risk.
   - Acceptance or assumption of a risk is to acknowledge the risk and assume its cost. Risks can be accepted actively by developing careful contingency plans or passively by doing nothing and accepting the consequences (lower profit, delayed shipments).
   - Transference of a risk to a third party that is more suited to handling the risk. The supply chain is the obvious target of transferred risk.
2. Contingency Plans. These are used when a risk cannot be avoided or transferred. They are predefined plans to be taken if risk occurs. Examples would be to purchase additional capacity or switch to higher-cost materials or components. It is a *backup* plan to be implemented if the risk is realized or cannot be sufficiently mitigated.
3. Project Reserve is a buffer resource in the plan to mitigate schedule or cost risk. It is usually in the order of 10 percent on top of the project budget that can be used to purchase additional capacity or resources, if needed. It does not address a specific risk, but it can be used by the PM to resolve unknown or unplanned risks.

4. Residual and Secondary Risks. Residual risks are those not avoided, mitigated, or transferred but are generally acknowledged without a handling or response plan. Secondary risks are those occurring as a result of planned risk response and are unintended consequences of risk management and contingency plans.

8.4.5 Risk Monitoring and Control

An important part of risk management is the monitoring and successful control of risk. The detection of risk occurrence and the need for risk triggers is important for the PM to realize in the event contingency plans have to be enacted to counter the risks. This effort is comprised of three elements:

1. EV Analysis, which was discussed in Chap. 7, is how much work was actually completed versus planned or budgeted. The PM can use the EV tools as early warning systems to indicate the occurrence of risk. In a rapid development process, EV is important because of its ability to track schedule and cost risks. These trends can lead to early indicators of a project's risks. A risk trigger could be reaching 10% variance in EV schedule or cost.
2. Risk Audits can determine how well the risk planning and management is working. Some companies have assigned evaluators or risk managers (outside of the PM and team members) to guide the team through the risk identification, analysis, and response planning. The risk manager's job is to ensure that the project managers have successfully planned and executed risk management.
3. Corrective Action and Documentation: The PM might be forced into unplanned risk response, sometimes called "workaround." The PM should document the action taken as result of a workaround, either planned (anticipated risk) or unplanned (accepted risk).

8.5 Managing Engineering Project Teams

The attributes of successful teams include a good degree of team mission focus, a feeling of commitment to the project goals, management support, and self control over the project progress. Underlying these success factors is a team composition of mixed skill levels and maturity and well-defined roles and responsibilities. The following is a discussion on these attributes:

1. Team Commitment

   The most important element in making teams successful is the clarity of the project mission, task definitions, and the precise outline of project goals and expectations. Team member tasks should be clearly defined: each task should have an objective, time duration, utility, and functionality within the project, inputs and outputs, and a definition of the complexity of the task as well as expected results.

2. Team Autonomy

   The team members should feel a certain level of independence, and members should be allowed to express their creativity and unique contributions. The team should be organizationally independent but supported by management. In a matrix team organization, members should have allegiance to the team but maintain functional ties with their home areas. Team member input should be included in the planning and execution stage of the project:

- Input to the project planning phase such as task definition and duration
- Input to contingency planning such as risk assessment and mitigation plans
- Participation in formal and informal meetings during project execution
- Have access to all relevant information about the project progress
- Team members should be allowed to express their concerns, in project meetings and updates, and have their concerns addressed by the PM. A central repository of project plans and progress toward completion should be available to all team members. Sometimes it is called the project contract. It should contain the following information for the project and be available to all:
    - Project statement including major project goals
    - New product performance criteria
    - The SOW, including summary of all task definitions and deliverables
    - Major project phases and milestones
    - Support requirements including outside resources, personnel, and the supply chain
    - Risk assessment and mitigation

3. Team Membership, Capability, and Skill Sets

    The engineering team members should have a wide array of depth and breadth of resources in different disciplines and with varying levels of experiences. An ideal team should not be composed of all competitive top performers, as it could lead to conflict and disarray within the team. Using a baseball analogy, the team should be comprised of rookies and veterans. The rookies will add enthusiasm to the team and will appreciate learning from others. The veterans should enjoy mentoring the younger engineers and teaching them the tricks of the trade. All team members should have excellent contacts and relationships, linking them to the rest of the organization.

    The team capabilities include problem solving and conflict resolution processes. Problem solving skills should include a thoughtful reasoning process in formulating a solution to the problem, with no jumping to conclusions, panic, or gut reaction. It should be accomplished in a collaborative environment. The goal should be to find a collective solution, rather than finding the person to blame for the problem. A conflict resolution process will be described later in this chapter.

    The results of good team cohesion can be seen clearly in good team characteristics, where a self-reinforcing process can help the team develop further and collectively complete the tasks at hand.

4. Meetings and Relationships

    Successful team meetings are usually relaxed and informal. Team members are supportive of each other. The atmosphere is low-key and enjoyable. Relationships are cordial and respectful. Meetings are short and to the point, with a prepared agenda, and should begin with a brief interpersonal update on the team members' families and friends.

5. Atmosphere and Discussion

   Successful team meetings should be participative with all team members invited. They should foster open dialogue and facilitate new ideas. There should be no rancor or ridicule of any idea. Instead, team members should show mutual respect and constructive criticism. If a conflict arises in the meeting, it should be resolved quickly and efficiently.

6. Decision Making

   The team should develop well-defined procedures and objectives when making a decision. Unintended or adverse consequences of the decision should be explored thoroughly after making the decision. Each decision should be reviewed often and altered if the underlying assumptions were proven inaccurate. The decision analysis process should be consensual, with input from all members of the team. Team members may disagree with the decision, but they should support the team once the decision is made.

8.5.1 Team Development Stages

Once a team is formed to work on a new project, a team development process begins with several distinct stages. All new teams go through these stages, and it is a good opportunity for team members to examine how to satisfy their own needs versus the management goals as the project development continues through its phases and milestones.

Figure 8.13 shows the progression of a newly formed team and the influence of management in the decision-making process versus that of the team members. The influence of management should diminish with time as the team is maturing, until the management relinquishes all of the influence over the project decision making to the team. The PM should become the coach or the cheerleader for the team, advocating for its decisions up to the management. The stages that newly formed teams go through are as follows:

1. Team Start-up

   This is the starting position of the team. Members are selected and given their assignments. In this stage, the team is undergoing a procedural transformation. The mission for the team and each member is given, roles and responsibilities are identified, and the team is directed by the PM. Discussions within the team are more formal and contractual and one on one with the PM.

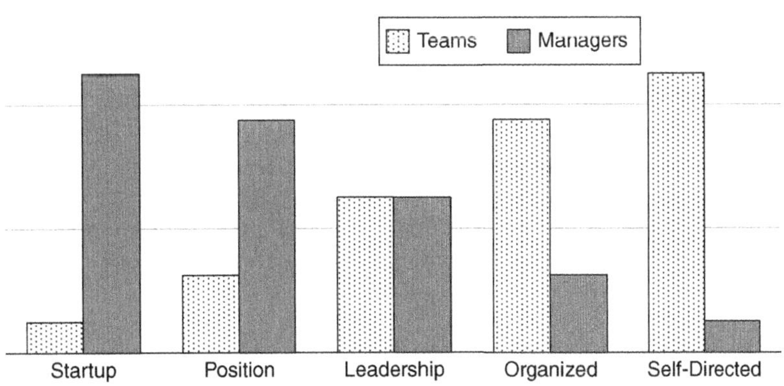

FIGURE 8.13 Team development stages.

2. Team Position

   In this phase, the team members are allowed to test and position their relationships. Roles and responsibilities of team members are clarified, with some assignments redefined. These roles are also reinforced, making sure that tasks assigned to each team member are a natural fit to their talents and interests.

3. Team Leadership

   In this phase, natural leaders emerge from the team members. Some members take on the helm due to their multidiscipline competency or technical skills. Other team members are elevated because of their superior organizational and communication skills. These team leaders could be selected on a rotating basis, in synchrony with the project design progress. Early in the project, there is a need for technical competence leadership. In later project stages, there is more need for organizational and communication skills. These leaders are not appointed for the duration of the project, but they can be rotated as the project lifecycle progresses.

4. Team Organization and Maturity

   In this phase, the team reaches a level of maturity that allows for team processes of problem solving and conflict resolution to kick-start automatically. The team has developed successful operating procedures and techniques that are used innately, without being conscious of them, much like tying your shoes. The team operates smoothly and takes on additional responsibilities without much fanfare. Team member concerns are addressed as quickly as possible. When new problems arise or the team becomes aware of new issues, the team is already organized to tackle these problems quickly and efficiently.

   At this stage, team influence expands beyond immediate team issues. Through self-motivation, team members take on additional responsibilities and communications with stakeholders outside the team. They create business relationships with other parts of the organization or the supply chain. They can take on decisions on behalf of the team in their areas of expertise.

5. Self-Directed Teams

   This is the ultimate in team autonomy. A team can become self-directed after successfully reaching the organization and maturity phase. The self-directed team members collaborate to make project lifecycle decisions as opposed to single or multiphase tasks' perspective. Leadership is rotating from within the team, depending on the project development stage.

8.5.2 Team Leadership and Interactions with Team Members

A PM can motivate team members through seeking team input, consulting with team members and asking them to participate in the decision-making process. Consequently, the decision-making process can be delegated to leading team members. This incremental delegation of power and authority from the PM to the team members follows directly from the evolution of the team development process depicted in Fig. 8.13. The four steps in this process follow stages similar to those in the team development process:

1. Command Management Style

   This is the initial process during the team start-up phase in Fig. 8.13. The PM makes all the decisions and then informs team members. They have a choice to respond, but only after the fact, which often leads to disgruntled and unhappy employees. This is an efficient management style, but lacking in team inputs and can lead to wrong decisions and employee alienation.

2. Consultation Management Style

   This style can be used during the position stage of the team development. The PM makes decisions after team member inputs. There is increased team member involvement, but resentment could build if only certain members of the team are consulted. Some employees will continue to feel not involved in decisions that have an impact on their team.

3. Participation Management Style

   This style is used in the leadership phase when the PM seeks team member input through a formalized process. The process might include regular meetings with the team to discuss current problems and ways to solve them, and the formation of sub-teams to tackle specific problems and solutions. The PM ultimately makes all of the decisions, but team inputs are strongly solicited. Team members feel more satisfied that their opinions are being respected and used in the decision-making process. This style can be used during the leadership phase of team development, allowing natural team leaders to develop within the ranks to solve specific problems.

4. Delegation Management Style

   This style can be used during the organized stage of the team development. Problem solving and decision making becomes part of the team members' responsibilities. The PM switches their role to becoming a leadership coach and to assist the newly recognized team members in making sound decisions. This style could be a precursor to self-directed teams. Team members enjoy learning the ropes while having the PM around for advice and encouragement. It is a good stepping stone to joining the management ranks and learning the tools of the trade in a safe environment. Care must be taken to encourage these developments while rewarding team members with salary increases as their responsibilities increase.

8.5.3 Engineering Career Stages

While these management styles previously mentioned are appropriate for different stages of team development, the individual team members are also in various stages of their own careers. Not all team members aspire to become PMs. Companies recognized the need for dual career development for engineers: one for a technical or engineering track and another for a management track. This dual track has been formalized in many companies with the recognition of highly talented and motivated senior engineers who are in charge of or manage certain technical disciplines, as opposed to managing technical teams.

Table 8.10 shows the progression of an engineer's career from entering the field as an apprentice (neophyte) into reaching high technical maturity and achievement, sometimes referred to as a "consultant" or "technical fellow." It also marks the increased confidence in the engineer's technical abilities and judgment. Using a recognition system, as shown in Table 8.10, allows for several additional benefits for career minded engineers as well as a good career progression system, as described below:

1. The apprentice (neophyte) is usually a recent engineering college graduate who joined the company and needs introduction to the company's systems and procedures. Though he or she might take a formal human resources course,

Stage	Activity	Role	Criteria
Apprentice/Neophyte	Directed by PM	Dependent	Follows
	Routine tasks	Learning new skills	Instruction
Contributor	Self directed	Independent	Confident
	Significant tasks	Problem solver	Responsible
Mentor	Helping others	Competent	Credible
	+ own work	Task manager	Selfless
Consultant/Fellow	Critical tasks	Product/process	Technology
	Discipline influence	Champion	Leadership

Table 8.10 Technical Career Stages

that person needs a confidant to show them the ropes. Apprentices are usually teamed up with mentors while getting their feet wet in the company. This would relieve the PM from daily interaction with beginning engineers and can be used as a great teamwork relationship and motivational tool for both apprentice and mentor. This stage allows for a beginner to develop confidence, technical experience and knowledge, problem-solving and decision-making skills under the tutelage of an experienced team member. This stage should be brief (three to six months), with the engineer expected to follow direct guidelines in performing the job. Beginners are tasked with routine assignments, such as data analysis or supply-chain communications. They are expected to learn quickly and begin to contribute directly to the team.

2. The contributor stage is the dominant position of the technical staff in the company. It begins after the initial apprentice stage for a recent graduate or an introduction phase for a transferring experienced engineer into the company. The engineer at this stage is expected to take on any assignment or solve any technical problem in the project. He or she is expected to provide a plan for the means to complete the assignment, problem solution or technical decision, obtain the team or management approval for the plan, and implement the plan. This results from the engineer's self-confidence in their ability to contribute significantly to the team and the project. These engineers have the necessary technical and organizational skills to complete all tasks with no errors and within schedule. The team and project manager have total confidence in the contributor's ability to perform his or her tasks.
3. The Mentor has the same responsibilities as the contributor, with the addition of mentoring new team members and apprentice engineers. In most companies, this position is not an official title, but acts as recognition of the engineer's contribution. Mentoring should not be a significant portion of the mentor's time, amounting to a maximum of 15 to 25 percent. Too much time spent in mentoring can have a negative impact on the mentor's own work and can lead to concerns about the project schedule. The mentor's contribution should be recognized in the project planning phase by not fully loading the mentor's time with project activities.

4. The Consultant or Fellow is a special track for technical leaders who do not wish to pursue management careers. They have the task of technical leadership in certain areas of expertise or core competencies for the company. They can set the company strategy for new product families and technology adoption. They can be tasked to manage new technology investigations or evaluations. They represent the company at technical conferences and standards-setting organizations.

Normally consultants or fellows do not participate in project management activities for new product creation. They might attend design reviews or act as technical consultant to the PM or design team. Their numbers should be few, less than 5 percent of the total technical staff.

8.5.4 Team Motivation and Compensation Policies

The career stages mentioned in Table 8.10 have to be augmented by sound compensation policies. Not all of the career stages are recognized officially as titles bestowed on team members. The PM has to use available tools to motivate the team, including the financial rewards of the positions. Companies practice a wide variety of compensation policies to administer the pay practices for the technical staff:

1. Compensation policies should reflect the engineer's career stage, with the engineer being paid at a salary commensurate with their career position. Many companies use fewer titled levels in salary administration to make it simpler to manage. In most cases, only two positions are used, such as member and senior member of the technical staff. The beginner/neophyte position is considered a trial period (it is a transition into the technical staff). The consultant or fellow salary is administered as part of the management salary curves. The mentor position might not be recognized as an official salaried position, but is used to instill teamwork and leadership skills among the team members.
2. Salary curves are developed for each titled position. Salary curves are developed from national salary surveys and administered as salary curve versus the number of years' service in the profession or in the company, as shown in Fig. 8.14. They are usually administered evenly across the global corporation to encourage the engineering staff (i.e., design and manufacturing) to respect each other, eliminate pay jealousy, and allow for easy transfer of all engineers across the company.

   It can be seen in Fig. 8.14, the salary curve for a particular technical position allows for flexibility in administering the technical staff compensation:

   a. A new salary curve is initiated each year, and it usually follows the previous year's salary curve with additional increases due to inflation or cost-of-living increases. It might also change due to economic factors in global industry growth or due to competitive factors in certain industries.
   b. The curve is steeper for the early years of experience to allow for faster salary increases for beginners/neophytes in that technical position.
   c. The salary curve flattens out sometime after reaching maturity in the position. Yearly salary increases for experienced engineers begin to follow the cost of living increases in the economy, as recorded by the new salary curve for the following year. The period of curve flatness might begin as early as 10 years of experience. When an engineer reaches that level, and

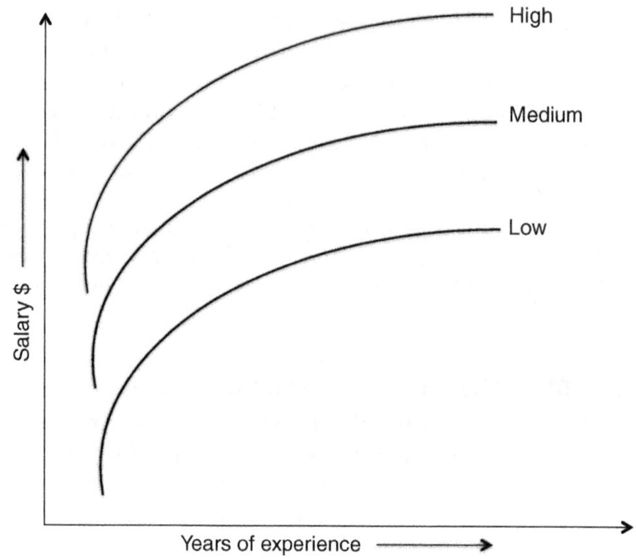

FIGURE 8.14 Typical salary curve for a technical position.

yearly salary increases are steady, other motivational strategies could be applied such as asking the engineer to become a mentor for new members of the team or tasking the engineer with more responsibilities.

d. The salary curve also includes a minimum and maximum range of salaries for that position (versus years of experience). The ranking of engineers within that position determines their actual salary within the range.

e. The salary curves allow for administering wages for a variety of technical training or education of the engineers by increasing the number of years of experience. For example, a master-degree engineer can have two years of experience added to his or her service in order to compensate for the higher degree than a bachelor's only engineer.

3. Other means of compensation or motivation for engineers could take the form of rewards and benefits for the individual as well as the team. Several means can be used for that purpose, outside of salary increases, including:

   a. Timely retreats (get-togethers) outside of the work place for project teams could be organized to build team spirit, motivation, and communication. It could be centered on physical activities (skiing in the winter, hiking, and water sports in the summer), with meetings and discussions early or late in the day to allow the members to enjoy the activities during the day.

   b. Stock options for targeted individual team members deemed to be contributing far and above completing required tasks. Stock options have an additional goal of retaining valuable employees, since they can only be exercised fully into the future (four to five years).

   c. Additional rewards for above average performance. These might occur during non scheduled critical activities or at the end of the project and after release to manufacturing. These include encouraging project team members to travel to trade or technical conferences to meet customers or fellow engineers working in the same industry and to observe the latest technological trends.

d. Special recognition awards such as competitions for advanced training and education. These are usually off work site and in university settings. An example could be to encourage successful team member(s) to enroll in a master's program in engineering or business school. The designated engineer might draw full salary while attending school.

e. Sabbatical ideas such as spending time pursuing personal interests or service activities.

It can be seen that there are many choices for the PM to motivate and reward the design team members, outside of the obvious monetary rewards. An integrated approach of salary compensation strategy and motivational techniques will result in satisfied and productive project teams.

8.5.5 Understanding and Nurturing Team Member Skills

The PM has the opportunity to recognize each team member's strengths in terms of adaptive skills, while they can focus the team member on improving transferable skills.

Adaptive skills are innate skills developed through a living environment. These skills include self-caring, adapting to change, social, and communication skills. They are the positive traits a team member brings to the job. They may include flexibility, reliability, or patience. The PM can help the team members by identifying which adaptive skills are the most value to the team and make each team member conscious of the skills they are lacking. The yearly performance evaluation is an excellent opportunity for the PM to give feedback to team members on adaptive skills or lack thereof. Quick reminders of the importance of these skills can also be noted by the PM to team members on a daily basis for positive reinforcement.

Transferable skills are acquired during the team member's professional career that the company would compensate for, such as analyzing data and gaining knowledge in technology, systems, equipment, or procedures. The transferable skills necessary for success in the design project include knowledge of the company's procedures, products, and systems along with industry regulations and certifications. In addition, transferable skills include knowledge of data analysis, design, test, and assembly as well as equipment, tools, and the software systems needed to run them. The yearly performance evaluation is an excellent opportunity for the PM to give feedback to team members on transferable skills needed to increase team member development and contribution to the project.

Table 8.11 is partial summary of some of the adaptive and transferable skills. Some of the skills are grouped together into subheadings. Reviewing the strength and weakness of the team members in relation to their contributions to the team is an excellent opportunity to seek improvements. This review can be undertaken with the help of the PM and team members. Lack of some of these skills could be detrimental to the team member's professional advancement. Discussing it with the PM and other team members is an excellent opportunity to see a reflection of one's self unto others. The lack of some of these skills can also lead to poor communications, judgment, and decision-making along with personal problems and conflicts in the team.

Adaptive	Transferable
Personal Skills	*Communications Skills*
Diplomatic	Report Writer
Charismatic	Presenter
Optimistic	Enthused Speaker
Good Listener	Follows Through
Conversational	Motivator
Competitive	Answers Clearly
Expressive	Consensus Builder
Team/Leadership Skills	
Takes Direction	Collaborator
Responsible	Analyst
Intuitive	Evaluator
Resourceful	Reasoned Decision Maker
Self-Motivated	Delegator
Risk Taker	Goal Setter
Work Style Skills	
Organized	Planner
Dedicated	Anticipator
Reliable	Early Problem Identifier
Flexible	Innovator
Objective	Computer Literate
Realistic	Error Free
Accurate	Satisfies Customers
Punctual	Safety Conscious

Table 8.11 Summary of Adaptive and Transferable Skills

8.6 Resolving Engineering Team Conflict and Managing a Successful Engineering Team

In a team process, where there are many members of varying needs, styles, personalities, skill sets, experience, motivation, and cultural backgrounds; the probability of conflict arising in the team is high. It is the responsibility of the manager and the project team to identify strategies of conflict prevention, resolution and settlement, and finally post-conflict team reconciliation.

Understanding the sources of conflict in a project team requires a thorough knowledge of the reasons behind the conflict. The reasons could be substantive; that is, there is an identifiable or measureable cause for having a conflict of opinion; or cultural, based on differing backgrounds and methods of communication. The most difficult conflict to resolve could be emotional, based on team members' adaptive skills, personality, and temperament. Once these sources are identified, a mitigating process could be implemented to reduce potential conflict. Recognizing and understanding these reasons is a first step toward conflict resolution.

8.6.1 Understanding the Sources of Conflict and How to Mitigate Them

Substantive issues for conflict are usually the result of poorly identified project goals and responsibilities, either within the core project team or with the supporting organizations. Emotional issues leading to conflict could be the results of cultural misunderstanding and satisfying the individual needs versus the team needs. The following is a listing of the reasons behind team conflict:

1. Substantive Issues. These conflicts might stem from poorly defined roles or responsibilities during the project planning stage. They could also arise from poor definitions of organizational boundaries and responsibilities:
   - Poorly defined boundaries of the organization might results in conflicting loyalties to the project or the organization. This problem becomes more acute in the matrix management projects. In addition, this could result in competition between team members reflecting the functional competition for future influence, turf, and status within the organization. This is especially severe if the other parts of the company (marketing or manufacturing engineers) are paid on a different salary curve than the engineering design project team. The obvious mitigating factors include better understanding of the responsibilities of each function, reviewing the logic of the different pay scales in the company (if there are any), and the emphasis on global factors such as ROI and company profitability to fold the functional concerns into the project goals.
   - Perceived conflicting goals between the project team members and other members of supporting and functional areas. Conflicting goals with manufacturing/operations include the documentation status of the new project and readiness for production. Conflict could also arise from disagreements on project schedule and estimated product cost in material and labor. Conflict with marketing could stem from the need for better specifications, options, and project cost and delivery times for new products. These factors could be mitigated with better design review processes, where the other functional areas have a chance to voice their concerns and resolve any discrepancies in the project schedule and deliverables.
   - Poorly defined roles and objectives within the team. These could be due to poor project planning and execution. The problem solving and decision making processes should be well-defined, with no need to assign blame, only to resolve the problem and make a reasonable decision.
2. Cultural Issues. Conflicts based on cultural and ethnic issues might result from bias toward, and misunderstanding of, different ways of communicating and behavior in certain cultures. These issues might not be recognizable by one of the conflict parties, and there is a need for self-awareness of cultural and personality sensitivities. The PM and the team could mitigate these potential conflicts by conducting sensitivity and cultural training for the team.

   For example, in certain cultures, nodding of the head might mean that, "I understand what you said," while the other party might think nodding of the head means, "I agree with your position." This could cause future problems since no agreement was made. In some cultures, the word "no" can usually be followed by a counter idea or alternative solution. One party might answer an opinion or a question with a sentence beginning with "no," meaning, "But consider my point," or as in, "No, here's a different alternative or my idea." In some cultures, the word "no" might have several connotations or nuances, such as "maybe" or "but." This scenario might result in the original party thinking its view was rejected out of hand, leading to anger or frustration. Repeating an opinion just made by one party and paraphrasing it back to the other party is normal in some cultures. This might cause impatience and annoyance, leading to conflict.

3. Emotional Issues. These conflicts arise from incompatible styles derived from adaptive skills of personal, teamwork/leadership, and work styles mentioned in Table 8.11. They might also involve a personal need for recognition and/or independence. These potential causes of conflict can be mitigated through understanding of the team members' individual personal goals versus team goals and creating a strategy to address these incompatibilities. Understanding each team member's adaptive skills and putting them to use for the best combination in assigning responsibility during the project planning phase might reduce potential conflicts in the future.

8.6.2 Conflict Resolution Strategies

An important aspect of conflict resolution is to understand the reasons underlying the conflict. For substantive issues, the key to conflict resolution is to investigate the basic reason and try to delineate the three factors mentioned in understanding the three sources of conflicts of substantive issues: organizational boundaries, team goals versus other functions' goals, and poorly defined roles and responsibilities within the team. For cultural issues, the official policy of the company should be that no bias would be tolerated. The company should conduct sensitivity and cultural training for all employees.

Conflicts based on emotional issues are more difficult to resolve. Most of these conflicts stem from adaptive skills centering on personal versus team goals. Resolving interpersonal conflict can take the form of negotiations based on discussions in Chap. 12 (conflict modes are discussed in Fig. 12.4).

The process of negotiation to resolve conflict can take several forms, depending on how the parties engaged in conflict view their perspective positions. Five scenarios and strategies can take shape in resolving conflicts. They include avoidance, competition, accommodation (or cooperation), compromise, and collaboration:

1. Avoidance. One or both parties involved in the conflict will avoid the conflict. This is an undesirable outcome. One or both of the parties have low expectation in terms of personal goals or relationship importance within the team. The technical competence that a party can use to influence the problem resolution is lost. This strategy rarely solves conflict and leads to resentments.

2. Competition. In this strategy, one party attempts to force a favorable decision at the expense of the other party. Usually there is no compromise, and everyone digs in their heels and tries to overpower the other party with arguments and intimidation. A competitive team member highly invests in achieving personal goals at the expense of the relationships with others in the team. Competition is a win/lose situation that often results in antagonism within the team.

3. Accommodation or Cooperation. In this strategy, one or both parties place a greater emphasis on team relationship over personal goals. They would give up an important *want* so the other party could win in order to reach a resolution. The danger of this strategy is that the proper negotiation process is aborted in favor of maintaining team relationships. This could possibly result in the wrong decision being made for the sake of reducing tension in the team.

4. Compromise. This strategy is similar to accommodation. Each party gives up a little, with a resulting resolution that can be the wrong one, due to compromising by all parties to balance team and personal goals.
5. Collaboration. This would be the perfect blend of simultaneously achieving personal goals while maintaining good team relationships. In this strategy, original goals could be reexamined, and the parties renegotiate the goals to achieve common ground and balanced decision. This strategy creates a win/win situation where everyone will be satisfied with the decision made.

8.6.3 Conflict Resolution Methodology and Settlement

Once a conflict is qualified as one of the three types identified earlier (substantive, cultural, or emotional), a resolution process can begin. Is the conflict based on misunderstanding of the functional boundary, confusion about project goals, or the responsibilities of the project team members? Is it caused by culturally based miscommunications or terminology? Or is it based on the conflict between the individual members' personal goals and team relationships?

A reconciliation process can begin by reconsidering the other party's position, restating the problem clearly, and concisely brainstorming single or multiple solutions and then using criteria rating and decision analysis techniques covered earlier in Chap. 6 to make a reasoned and acceptable decision. It is important that the decision has the consensus of all parties involved in the conflict. In summary, the conflict-resolution methodology can be accomplished in the following steps:

1. Identify the sources of conflict discussed above (substantive, cultural, or emotional) and set the stage for further steps:
   - Make sure all parties understand the sources of the conflict, whether it is lack of clarity in substantive issues, lack of communications or poor terminology in cultural issues, or a conflict of personal versus team goals in emotional issues
   - Remove passion and allow for time out before continuing the problem definition step
   - Practice active listening and be respectful and not dismissive during the discussions
   - Visualize the other party's concerns and arguments and rephrase their position
   - Restate the problem clearly and concisely
2. Generating alternative solutions:
   - Brainstorm possible alternatives once the problem causes are identified
   - Participate with all parties in the brainstorming session
   - Do not advocate your own solution
   - Do not jump to conclusions or dismissive or make derogatory comments while ideas are presented
3. Classifying alternative solutions:
   - List the criteria for the decisions into musts and wants
   - Rank the *wants* into levels of importance (either by numbers 1 through 10 or by satisfactions [+/S/−])
   - Remove alternative solutions that do not meet the musts for the decision

4. Evaluation of alternative solutions:
    - Rank each alternative solution versus each criterion
    - Keep the discussion focused on one alternative solution and one criterion at a time
    - Add the total score for each alternative solution
5. Deciding on a mutually acceptable solution:
    - Select the solution with the highest score
    - Revisit other solutions if the totals are close
    - Think of any adverse consequences if the solution is implemented
    - Make sure all parties agree that the decision process is fair and the solution is justified
6. Settlement and implementation of the solution:
    - Make a thorough plan to implement the chosen solution with tasks identified and committed to completion through a certain schedule
    - All parties should commit to the speedy implementation of the solution
    - Agreement to consider when the problem solution is indeed completed and resolved

While this conflict resolution process is formal and quite lengthy, it can be turned into an informal and smooth resolution process once it is practiced several times. Conflict resolution can become an unconscious process when the project team appreciates the great benefits and personal energy reserve saved when it is practiced correctly.

8.6.4 Managing a Successful Team

A successful PM leads by example. The most important attributes for a manager are the same as those for a successful team member, with greater emphasis on communication, planning, organization, teamwork, leadership, and good decision making skills. Most PMs in technology companies get promoted to their first management assignment, such as project management, based on their perceived technical skills. The challenge is to have the PM become successful and grow further into the management track, based on their management and technical skills.

First-level managers such as PMs have direct access to information about the project and the team they manage. Typical engineering teams in a high-technology project consist of three to 15 engineers. Program or section managers manage larger projects. Program/section managers typically have three PMs reporting to them. Moving up the management chain, there are lab managers up to functional managers that manage up to hundreds of engineers. At the upper levels, decision-making becomes more diffuse, relying on less information than the first-level/project managers. Second-level management is dependent on the project reporting and communication systems that are set up to quickly inform the top management whether their attention is needed when a project is not meeting their goals.

PMs need to be conscious of their roles in mentoring their teams and keeping communications open to the team, other functional areas, and top management. Other skills they need to focus on include matching the team skills with the project needs during the planning stage, organization, and planning project tasks as well as anticipating and managing technical problems/risks and resolving team conflict. The keys to successfully leading project teams are as follows:

1. Communication Skills
    - Become an active listener. Be respectful of all opinions. Make sure to follow up on all concerns. Listen, Listen, Listen.
    - Build relationships with all functions and areas. Take time out to meet managers at other functions and projects, including shared meals and scheduled face-to-face meetings.
    - Document and archive everything:
        - Project specifications and documentation and their evolution since project inception
        - All emails and communications
        - All problems and how they were identified and resolved
        - All decisions and why they were made as well as adverse consequences and history of any changes to the decisions
        - All concerns and by whom and how they were handled
        - Milestones and meeting minutes and summaries
    - Coordinate project meetings (daily stand-up, weekly updates, phase reviews, and milestones).
    - Publish weekly updates of project status to all management and functional departments. Include areas of concerns, project slippage, and successes.
    - Respect the project team efforts by thanking them and making their successes and achievements visible to all.
    - Be prepared for monthly status reports and on the spot updates for upper management.
    - Always answer an email or telephone call within 24 hours whether or not you have the information.
2. Organization and planning project tasks skills:
    - Decide what stage the team development has achieved
    - Decide the type of management style to practice commensurate with the team development stage
    - List and discuss with each team member his or her current career stage
    - Select the project tasks that best meet each team member's skill set
3. Anticipating and managing technical problems:
    - Use all information at your disposal, daily stand-up, weekly project, and formal milestone meetings as well as financial based systems such as project accounting or EV systems reports to obtain a clear view of the project progress.
    - Work on an exhaustive project risk analysis plan and keep the plan up to date.
    - Modify the agenda or meeting minutes to ensure there is room for team members and others to include concerns and successes.

- Make sure to converse with each team member at least once a day to get personal updates.
- When a problem materializes, act quickly and efficiently to make a decision or reach a conclusion. Make an assessment on the impact of the problem on schedule, cost, or project performance. Estimate any adverse consequences due to problem resolution.

4. Anticipating and resolving team conflict:
   - Identify the source of the problem (substantive, cultural, or emotional).
   - Put in motion a plan to mitigate substantive and cultural issues through better definition of functional boundaries, project task definitions, and team member assignments.
   - Identify the adaptive and transferable skills and potential for conflict based on team member personal goals versus the team relationship matrix.
   - Allow team members to resolve conflicts on their own. If the problem escalates, then facilitate a solution to the conflict using the principles of decision analysis.
5. Motivating team members and handling rewards and career concerns:
   - Encourage the team to become cohesive by adopting the attributes of successful teams noted in the early part of this chapter
   - Strive to mentor the team progress toward the more mature stages of team development
   - Concurrently seek to change the management style towards delegation management
   - Allow natural team leaders to emerge and coach them for success, based on the current team development stage and the manager's own leadership style
   - Identify each team member's current career stage and encourage him or her to advance by creating a career plan
   - Use a skills matrix to document each team member's adaptive and transferable skills
   - Work with each team member on a career plan, including filling in the gaps for training on agreed-upon and needed transferable skills
   - Encourage experienced team members to mentor new engineers
   - Explain thoroughly the direct compensation policy and pay curves for the company as well as indirect compensation such as stock options policies
   - Always give credit to the team member who anticipates or resolves a problem, recommends a successful solution, finishes tasks ahead of time, or delivers better than planned project performance

8.7 Conclusion

The design phase is usually broken down into smaller parts, with the completion of each part being recorded either in a formal checkpoint meeting or at the completion of a particular task or milestone. It is important to have each checkpoint or milestone be of some significant and measurable progress in the project, to add to the project team and the management confidence in the progress toward achieving the project goals.

The impact of the Internet and the global reach of the supply chain in design and manufacturing are changing the product development process. Examples of the effect of the Internet on project communications internally within a distributed team as well as outside stakeholders were shown.

Project risk assessment and management were discussed in order to manage potential problems in executing projects and reduce risks in the project. A full plan for identifying, prioritizing, and minimizing risk to the project was outlined in this chapter.

Characteristics of successful teams and PM styles were discussed as well as identifying team members' career stages, needs, and motivation. Sources of team conflict were discussed, and methods of successful conflict resolution and keys to the PM success were presented.

A PM's success is mostly due to the success of the team and confidence in his/her own ability. As one successful PM told the author when asked, "What is the secret of your success?" He answered, "Two things: make sure to publicly acknowledge deserving team members, and if the management does not like my work, they can find someone else!"

References and Bibliography

A Guide to the Project Management Body of Knowledge (PMBOK® Guide), Project Management Institute, Newtown Square, PA, 2008.

Anderson, D., *Design for Manufacturability & Concurrent Engineering; How to Design for Low Cost, Design in High Quality, Design for Lean Manufacture, and Design Quickly for Fast Production*, CIM Press, Cambria, CA, 2010.

Carlson, C., *Effective FMEAs: Achieving Safe, Reliable, and Economical Products and Processes using Failure Mode and Effects Analysis (Quality and Reliability Engineering Series)*, Wiley, New York, 2012.

Clement, J., A. Coldrick, and J. Sari, *Manufacturing Data Structures: Building Foundations for Excellence with Bills of Materials and Process Information*, Wiley, New York, 1995.

Ericson, C., *Fault Tree Analysis Primer*, Create Space Independent Publishing Platform, Charleston, NC, 2011.

"DAU Systems Engineering Fundamentals Guide,"Chap.15 in *Risk Management, Defense Acquisition*, University, Fort Belvoir, VA, 2001. Download at: http://www.dau.mil/pubscats/Pages/sys_eng_fund.aspx.

Garwood, D., *Bills of Material for a Lean Enterprise*, Dogwood Publishing, Incorporated, Marietta, GA, 2004.

Haines, S., *The Product Manager's Desk Reference*, McGraw-Hill, New York, 2008.

Harris, T., and J. Sherblom, *Small Group and Team Communication*, 5th ed., Pearson, Upper Saddle River, NJ, 2010.

Hillson, D., *Effective Opportunity Management for Projects: Exploiting Positive Risk*, Marcel Dekker, New York, 2004.

Humphrey, W., *TSP: Leading a Development Team (The SEI Series in Software Engineering)*, Addison-Wesley Professional, Upper Saddle River, NJ, 2005.

Humphrey, W., and W. Thomas, *Reflections on Management: How to Manage Your Software Projects, Your Teams, Your Boss, and Yourself*, Addison-Wesley Professional, Upper Saddle River, NJ, 2010.

Langford, W., *Logistics: Principles and Applications*, McGraw-Hill, New York, 1995.

Lessard, C., *Project Management for Engineering Design (Synthesis Lectures on Engineering)*, Morgan and Claypool Publishers, San Rafael, CA, 2007.

Lipson, H., and M. Kurman, *Fabricated: The New World of 3D Printing*, Wiley, New York, 2013.

Monk, E., and B. Wagner, *Concepts in Enterprise Resource Planning*, Course Technology Cengage Learning, Independence, KY, 2007.

Poli, C., *Design for Manufacturing: A Structured Approach*, Butterworth-Heinemann, Woburn, MA, 2001.

Raines, S., *Conflict Management for Managers: Resolving Workplace, Client, and Policy Disputes*, Jossey-Bass, San Francisco, CA, 2013.

Ramroth, W., *Risk Management for Design Professionals*, Kaplan Publishing, New York, 2007.

Rausand, M., and A. Hoylan, *System Reliability Theories, Models, Statistical Methods, and Applications*, 2nd ed., Wiley, New York, 2004.

Reid, D., and N. Sanders, *Operations Management*, 457–458, John Wiley & Sons, New York, 2002.

Risk Management Guide for DoD Acquisition, 2nd ed., Defense Systems Management College Press, Fort Belvoir, Virginia, 1999.

Robinson, P., *Bills of Material, Design and New Product Introduction (Business Excellence)*, 2nd ed., BPIC, e-book by Amazon digital services, Kennewick, WA, 2012.

Rosen, E., *The Culture of Collaboration*, Red Ape Publishing, San Francisco, CA, 2009.

Rothwell, D., *In Mixed Company: Communicating in Small Groups and Teams*, 7th ed., Cengage Learning, Independence KY, 2009.

Runde, C., and T. Flanagan, *Becoming a Conflict Competent Leader: How You and Your Organization Can Manage Conflict Effectively*, 2nd ed., Jossey-Bass, San Francisco, CA, 2012.

Smith, K., *Teamwork and Project Management*, 3rd ed., McGraw-Hill, New York, 2005.

Stamatis, D., *Failure Mode and Effect Analysis: FMEA From Theory to Execution*, 2nd ed., ASQ Quality Press, Milwaukee, WI, 2003.

Whitcomb, C., and L. Whitcomb, *Effective Interpersonal and Team Communication Skills for Engineers*, Wiley-IEEE Press, New York, 2013.

Discussion Topics

1. List the engineering project phases for a company that you may work in or are familiar with or theorize the list of phases for a product you use. How do you describe the different phases for product development and the reason why you think these phases are appropriate to the product being developed?

2. Discuss the engineering project communications summary used in Table 8.8. List items you would recommend as changes or alterations to the methods listed and their uses for communications between project stakeholders at different phases in the project.

3. Discuss the differences between the quality analysis methods to reduce product defects (FMEA, FTA) and the project risk analysis equivalents.

4. Discuss the three types of sources of conflict and how to mitigate them. Use examples in your personal work or in school.

5. Cultural issues resulting in team conflict have become more frequent as the diversity of teams has increased. The chapter describes several instances of cultural misunderstanding leading to conflict. Can you recall at least one example

of similar cultural misunderstanding that happened to you personally or you noticed at work, school, or otherwise?

6. List the top three project manager personal skills needed to successfully manage project teams and your reasons why you selected these skills?

Problems

8.1. Write a generalized view of a phase review process for a new evolutionary (follow on) high-technology communications product such as a cell phone that has a short lifecycle and limited technology adoption. Please consider the following items in your report:

- List the number of phase(s) required and the communications needs for the project within the distributed project team, the supply chain, outside technical services, and the management.
- Identify the different types of meetings required, their frequency, and the agenda needed for the meetings, if any.
- List the types of reports needed to manage the project and update the project stakeholders and management. Show the format and content for the reports and when they should be issued.

8.2. Risk assessment and control have become an important part of project management, especially in the execution, control, and implementation part of the project. If you can hearken back to the planning stage of the project, would you recommend that risk assessment become part of the project planning methodology?

Please write a summary of your opinion, considering the following:

- What tools discussed in this chapter would you recommend to keep in risk management planning? How would you change them to use in the planning stage, if any?
- What new tools would you use to add risk management to project planning, if any?
- How would you change the PERT or Gantt chart planning to account for project risk? Can you think of using a Six Sigma type of quality prediction to quantify risk?

8.3. You have been asked to consult with a major tobacco company. It informed you that it has a product idea, but it does not want to tell you what it is, only that it is high-tech. Once you arrive at the headquarters in Richmond, the company takes you to tour the facilities, including the automated lines that make the cigarettes from the raw materials of tobacco and paper, then deliver the finished cigarette packs right into the delivery trucks. The lines are all dual input, so if one line is stopped for maintenance or breakdown, the other line instantly takes over. The company proudly mentions that all the design and assembly of the lines are made in-house with an excellent manufacturing automation internal engineering capability.

After some initial conversations, you realize that you guessed right. The company is thinking of an electronic or e-cigarette, which is a large task for a commodity company. It has excellent marketing and manufacturing engineering competencies, but it has have never developed a high-tech product. The company asks you to help in creating a development phase gate process, based on a QFD process it used to develop the new product specifications. What are the phase/gate stages that you would recommend based on the following?

- The company has committed large resources and budgets into the project.
- It has a competent internal and professional engineering staff used for the design of automated lines. However, its engineers are not experienced in developing consumer products.
- The company has a large and active marketing/sales department.
- The company is not averse to subcontracting design or manufacturing, but it has concerns over trade secrets (not divulging the new product).
- It plans extensive consumer testing during the product development.

Write about the different phases/gates that you would propose to the board, including estimates of times to completion of each development phase and methods of project communication within the development team and the board.

8.4. Write a generalized view of the hallmarks of a successful team that was selected to attempt to complete a project for a follow-on product in half the time it normally takes to complete. Show how the PM would motivate and encourage the team to accomplish these difficult tasks on time and within budget and project plans. Please consider the following issues in your report:

- The team is a mix of veterans and rookies with a large ethnic, racial, and cultural diversity and with comprehensive interdisciplinary skills.
- The project manager is an experienced veteran with many successful projects completed on time, within budget, and within planned specifications and product cost and quality.
- The project manager has individual interactions with each team member.
- The progression of the team from project launch to successful release to manufacturing.
- The corresponding style of the project manager as the team develops over time.
- How to foster a good problem solving and conflict resolution processes within the team.

8.5. A multinational company based in Germany makes cell-phone network-communication products. The company updates its products with each new telephony standard upgrade of the global system for mobile communications (GSM). It is undergoing tremendous changes based on stiff competition in the telecom industry such as:

- The company is planning to reduce its costs by subcontracting its manufacturing to Portugal for European sales and China for the rest of the world.
- It wants to actively involve its North American branch in the design and development of new products.
- The company feels that its German engineers need development in working with global distributed teams.

You propose to interview the management team and some development engineers for the next month. At the completion of the interviews, you will prepare a plan to improve teamwork and communications for the new projects in the company. You are provided with two secretaries, an experienced woman who does not speak English, and a young woman who is fluent in German and English (to translate). You note the following:

- Most of the managers and engineers, especially the young ones, speak excellent English.
- Some engineers are concerned about potential loss of jobs due to the inclusion of outside resources.
- Some engineers are resentful of the coming and unforeseen changes and are suspicious of the competency of the outside engineering teams.
- Teamwork and communications are lacking in the organization.
- Most projects are not completed on time, and there are many delays in projects due to low spirit, poor teamwork, and inadequate communications.

Write a recommendation to the company management to improve and increase communications and teamwork in the company and prepare its engineers to work in global distributed development teams. Suggest some ideas for team building and conflict resolution within the development teams.

CHAPTER 9

Project Monitoring and Control Case Study

Project monitoring and control is one of the most important elements in the project life cycle. A project schedule is only as good as the efficient and complete execution of planned tasks. A project plan really only exists to exert control over the entire project. What is the point of control? As a project manager, you need to know where the project is headed and where you are supposed to be at any given point in the project lifecycle. By understanding the underlying project constraints on a daily basis, which include costs incurred to date, meeting project requirements, and the work is completed on time and within scope, the project manager will be far more effective in managing the project.

Many, but not all, companies have various types of project-information systems that are intended to aid the project manager by capturing, tracking, and modeling changes to the project. These systems range from home-brewed systems using Microsoft SharePoint™ designs or commercial off-the-shelf solutions, including Microsoft Project and other commercial solutions such as Primavera.™

More complex tools such as earned-value management systems are often used in support of projects. Each of these tools has pluses and minuses about how they help (or hinder) a project manager in assessing the true status of her or his project and monitoring progress. Tools are only effective when used properly and may not present a true and complete picture of project status. How does the project manager keep his or her finger on the pulse and pace of the project?

As early as 2000, the legacy Goodrich Corporation developed and continued to evolve its own methodology and processes to develop a project monitoring and control process that was applied to projects. Beginning July 2012, Goodrich Corporation was acquired by United Technologies Corporation (UTC). UTC is a diversified company that provides high-technology products and services to the building and aerospace industries.

In mid-2012, two industry leaders, Hamilton Sundstrand and Goodrich Corporation, combined to form UTC Aerospace Systems (UTAS). UTAS is comprised of two segments: Aircraft Systems (AS) and Power, Controls, and Sensing Systems. Aircraft Systems include Actuation Systems, Aerostructures, Air Management Systems, Interiors, Landing Gear, Propeller Systems, and Wheels and Brakes. Power, Controls, and Sensing systems

This chapter was written by David A. Nolte.

(PCSS) include Electric Systems, Engine Components, Engine and Control Systems, Kidde Fire Systems, Intelligence, Surveillance & Reconnaissance (ISR) Systems, Sensors & Integrated Systems, and Space Systems. This combination of two industry leaders has created an organization with key positions on a wide range of aircraft flying today and substantial content on various unmanned airborne vehicles, satellites, and ground and naval vehicles.

The case study presented in this section will focus on the use of project monitoring and control as originally developed by legacy Goodrich Corporation that is currently used at ISR Systems, a business unit within PCSS. While there are many forms and flavors of project monitoring and control that have been developed by a wide range of industries, this case study will refer to the methods and processes developed by Goodrich and currently practiced at United Technologies Aerospace Systems (ISR) Systems, Westford, Massachusetts.

9.1 Key Project Monitoring and Control Processes

At ISR in Westford, several key processes are part of the overarching monitoring and control of a project. These follow, in general, the guidance described by the Project Management Institute's (PMI) *Project Management Book of Knowledge* (*PMBOK*) as ". . . all processes required to track, review, regulate the project against the project plan." For ISR Westford, these include:

- Monitoring and Management of Work Within a Given Project Life Cycle. This is done by first understanding the project baseline. When earned value management systems are applied, the project baseline is the approved project schedule and work breakdown structure. The Integrated Project Team (IPT) and the project manager are constantly comparing the actual work performed against the project baseline. If actual results and baseline show variance that is beyond an acceptable margin, the project manager must make adjustments to bring the project back to the baseline, if possible.
- Monitoring and Management of Project Schedule and the Budget. The IPT and project management need to pay close attention to this area. Realistically, a project schedule presents the best representation of a particular point in time on how to complete the remaining work. Schedules are dynamic and frequently change. A daily stand-up area can provide the manager (and team) with a daily status on tasks due that day or week. This gives the IPT and project manager more time to react to changes.
- Monitoring and Management of Project Scope. Project scope reflects the summation of the work contained in the project and the overarching boundaries of the project. The IPT and project manager need to understand the high-level scope and the business requirements of the project that should be reflected in the internal project plan and the external agreed-upon contract with the customer. It is up to the team to identify potential or actual scope changes from the customer and ensure these out-of-scope attributes are captured as modifications to the contract such that there is compensation for the scope changes. Monitoring and control of scope also means that internal-driven *scope creep* is kept at bay to prevent unnecessary costs to the project.

- Monitoring and Management of Project Risk. This is another critical area. The IPT and project manager should have a well-developed risk register that captures risks, assesses probability, and impact and, most importantly, has countermeasures in place that can be implemented if the risk occurs. The daily stand-up process is an excellent opportunity for the IPT to raise new risks, capture into the risk register, create countermeasures and adjust the integrated master schedule.
- Monitoring and Management of Project Quality. Some projects have quality functions as part of the IPT. It is their responsibility to monitor quality throughout the project lifecycle and update the quality management plan for the project as required.
- Managing and Monitoring Communication. Both Internal and External to the Project. Communication should follow the project communication plan at a minimum, but successful projects usually have in place other mechanisms for communication at all levels, both internal and external. The daily stand-up area is a key element that promotes communication among key stakeholders.

9.2 The Daily Stand-Up Board and Area

At ISR Westford, each project is required to set up and maintain a daily stand-up board for the integrated project team members to meet and utilize for project monitoring and control. The purpose of this area is:

- Provide a focal area for daily monitoring of project tasks and deliverables associated with the project schedule
- Facilitation of the capture and display of status information for the integrated project team (IPT)
- An efficient, focused meeting point for the IPT and any other key staff or subject matter experts (SME) as required

This focal area can be a large area or small (Figs. 9.1 and 9.2). It can support single or multiple projects. The notion of the stand-up area is just that—it is designed such that people stand and has a limit to the number of people it can support. It is the *command and control* area for the IPT that will display critical project information and status.

The stand-up process should be held to a consistent 15 minutes. This is not a problem-solving session but a status and opportunity for the team to capture new issues, risks, and highlight solutions. If problem-solving or coming to a root-cause solution is needed, it should be done immediately after the stand-up as, most likely, the right people are there already and can immediately attack the problem.

9.2.1 Area Design Essentials

The area should be chosen with thought and consideration. It is efficient to have a site-wide standard for the design, location, and setup of these areas. Project managers, leadership, and IPT members should be involved in its design and implementation. Of course, the rule of creativity over capital applies here. There might be temptation to create an electronic wonderland of monitors, computers, and other equipment to support

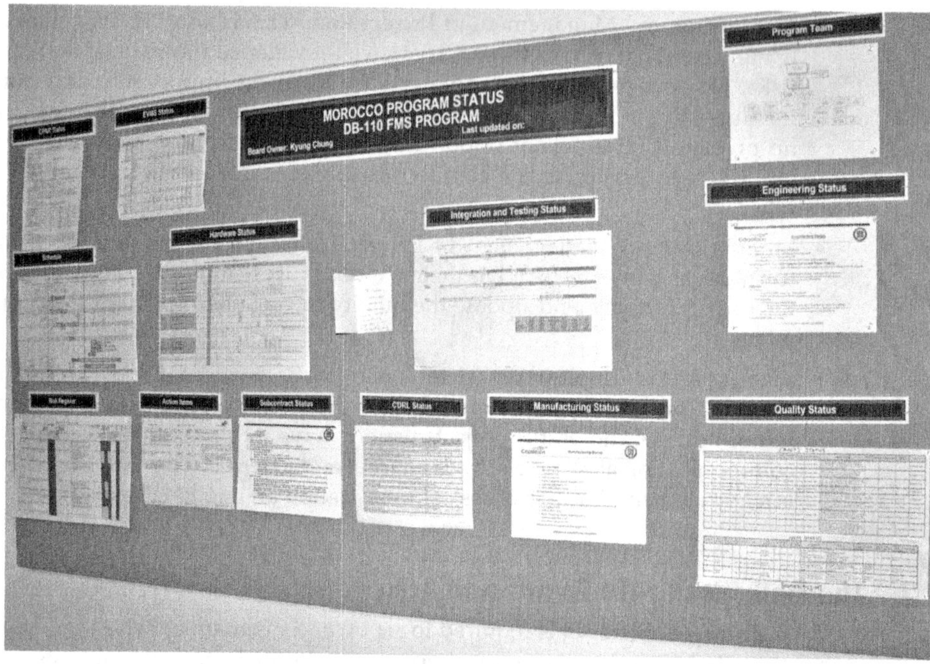

Figure 9.1 Single project management stand-up area and visuals.

Figure 9.2 Multiple projects daily stand-up area.

these areas. You may find that posting printed large visuals that an IPT can update manually will suffice and be very practical to use and update. Also recognize that the standards and processes for using these areas can evolve over time as IPT members will provide ideas to improve efficiency.

- It will need enough area to handle the visuals.
- It should have a level of security if required for display of sensitive information.
- The area should be easily accessible by staff.
- The area should be well-lit so the visuals are easily read.
- Support equipment should include a phone, computer, monitor, and small printer near the stand-up area or, if the stand-up area is in a room, then inside the room.
- A visual timer is useful. The timer can be used by the IPT to limit the stand-up meetings to 15 minutes.

9.2.2 Metrics and Status Elements

What elements of information should be addressed on the visual boards? The team tasked with the standardization of these stand-up areas should *storyboard* out the visual boards. This can be done on a large foam board and even temporarily displayed and used by a team to understand the utility of the board (Fig. 9.3).

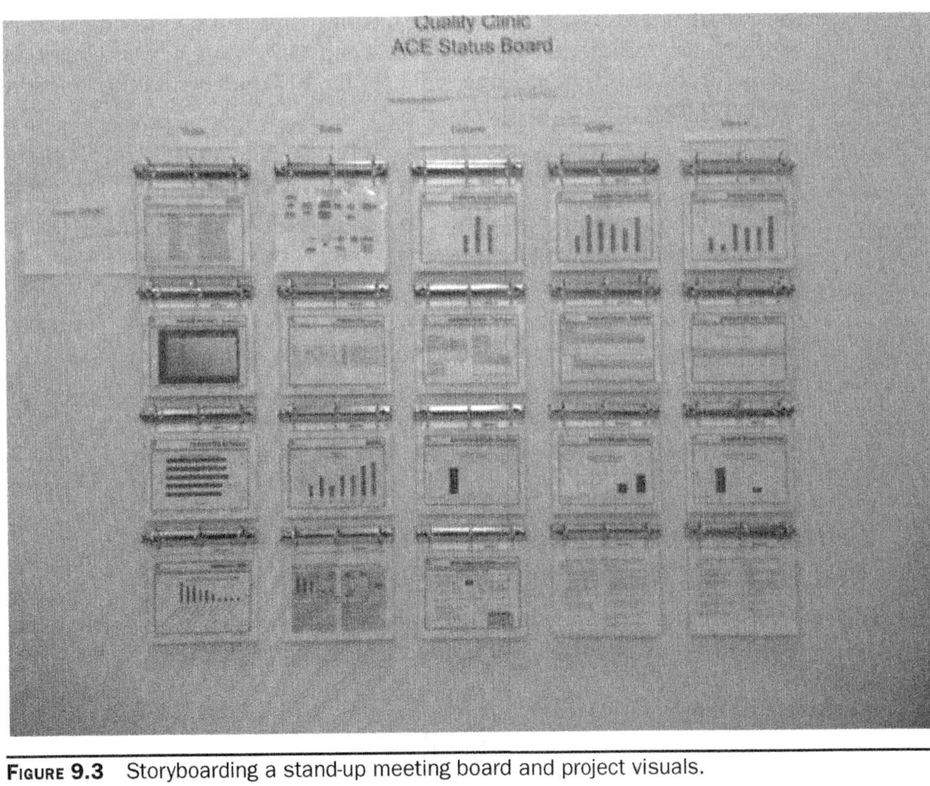

FIGURE 9.3 Storyboarding a stand-up meeting board and project visuals.

9.2.3 Setup and Operation

Once the area has been set aside and created to support the daily stand-up, the IPT should be made aware of the standard work around use of the area:

- 15-minute rule for the length of the stand-up; use of an Andon-style timer is useful
- IPT cost account managers, technical leads, and the project manager should attend
- If the project uses a planner, this is an opportunity for the planner to capture task changes in real-time
- Either the project manager or project engineer leads the stand-up session
- The key questions covered in the daily stand-up are:
  - What task(s) are due today?
  - Is the task(s) done?
  - If not done, when will it be done?
  - Do you need help?

If there are no schedule tasks due that day, the team should use the time to status the project risk register. Are the current active-risk countermeasures being accomplished? There should also be some time reserved for the IPT members to raise new issues or potential problem areas for the team to be aware of.

9.2.4 Lessons Learned

- One of the key lessons learned is that holding to a 15-minute status meeting can be a challenge at first. Use of strong meeting discipline is important. The status area can have a timer for the team to use and displaying *ground rules* or standard work for the stand-up meeting will help the IPT understand the function of the area and their role using it.
- The stand-up meeting is less effective if it lacks the multidiscipline elements of the team. Attendance should be strongly encouraged by all IPT members.
- IPT members should help the meeting leader by self-policing for solving problems at this meeting instead of the status. Ideally, problem solving can start immediately after the status meeting as the right people in part are already there.
- The paper visuals posted should be manually updated by the team daily and at least refreshed/printed out electronically and on hard copy weekly.
- If the IPT has a planner/scheduler, that person can efficiently capture schedule changes during the stand-up meeting directly from the IPT, which eliminates the need for the tracking down people at some other time.

9.3 Other Uses for Stand-Up: Supply Chain, Operations, Red Flag, and Risk Register Reviews

This type of status board and stand-up area can also be used by support services to manage the flow of information within their teams and to provide an efficient area for daily status. Figure 9.4 is an example of how the supply-chain functional group manages key information for its team. It shows key metrics used, including supplier quality and delivery. Operations areas can use roll-around easels or permanent boards in their areas to share daily stand-up board meeting information, as shown in Fig. 9.5.

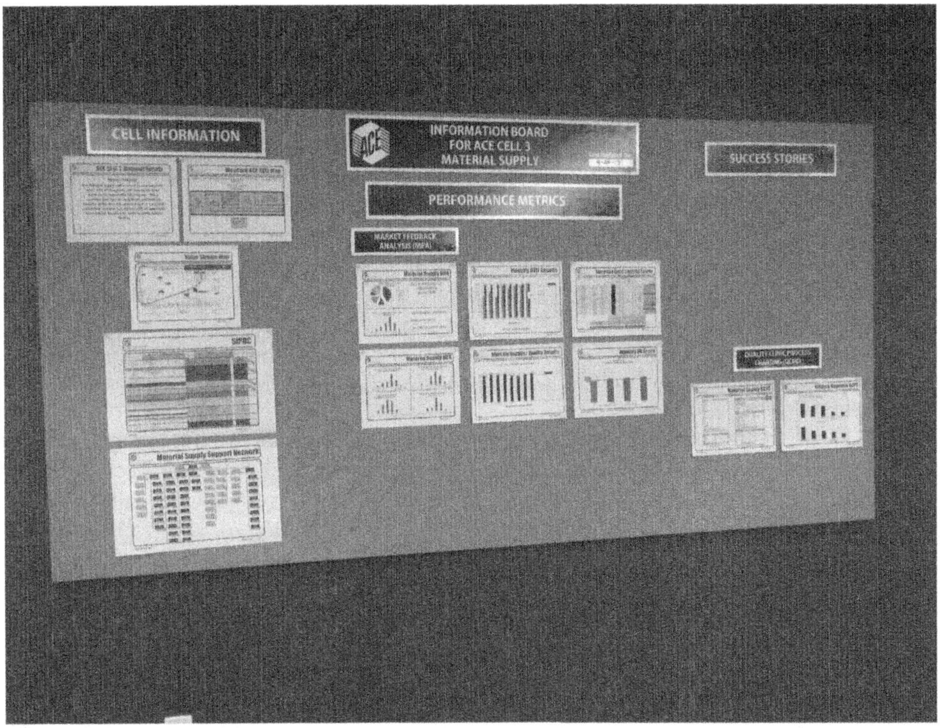

FIGURE 9.4 The supply-chain team monitors key metrics on quality and delivery daily.

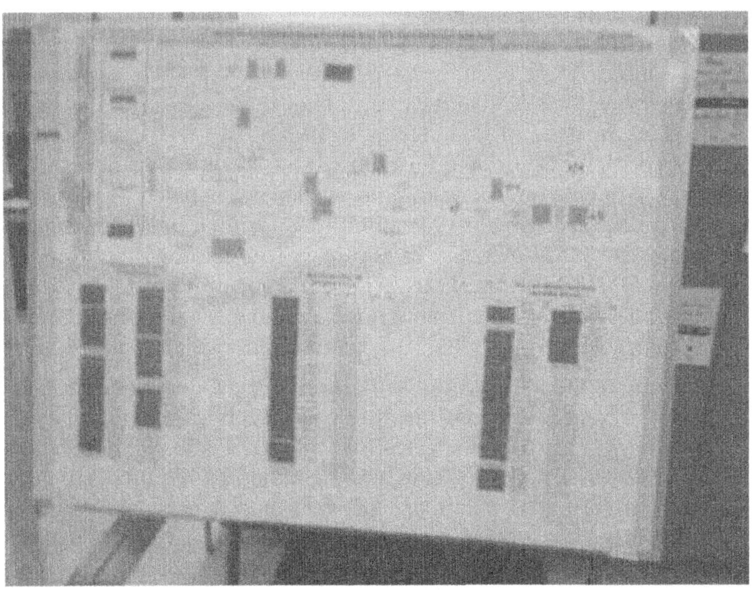

FIGURE 9.5 Operations daily stand-up board.

9.3.1 Red Flag Reviews

One element of project management that also can be reviewed during a daily stand-up at least once per month or even once per week by the IPT is the risk register and associated *red flag* review. The ISR Systems red flag process and risk register used at ISR Westford defines a program-risk register, risk assessment, and review methodology to allow for executive overview of new and current program status. The intention of this process is to provide a management alert of programs that have excessive risk and the potential for a negative estimate at completion. The process uses a cross-functional team, which includes program management, engineering, contracts, operations, and quality and other disciplines as required by the project to ensure a comprehensive review. This assessment was derived and expanded from the legacy Goodrich Enterprise red flag process. A summary of the ISR process for red flag review is shown in Fig. 9.6. There are several key steps to ensure the application of the red flag review:

- Create the Program Assessment Team. The project management office and/or the value stream director assigns an assessment team for each program. The program assessment team is a representative cross-section from the program being assessed (the IPT). It typically consists of a team of five to 10 members from the functional areas of program management, engineering, contracts, operations, and quality or other discipline that is part of the IPT. Each member responds to the entire red flag questionnaire. The worksheet is designed such that it has a tab for each IPT member, and the resulting scores are merged into an overall team score.
- Provide Team Training if Required. Typically, prior to the review, a refresher training session on the process is given to the IPT. Training materials are available from the shared business management system. If a team member is not able to attend the training, he or she can self-train by reviewing the training package.
- Complete the Red Flag Questionnaire. The process is for the assessment questionnaire to be completed in two phases. Each member of the program's assessment team separately completes the red flag questionnaire.

  The purpose of this step is to make sure that each member has thought about each question and answers it based on his or her understanding of the program. The questionnaire has worksheets labeled #1 through #8 for the team members to complete. Questions are answered by selecting one of the answers from the pull-down list in the response column. If the team member can't answer a specific question because it is not applicable for the program or he or she does not have sufficient knowledge to answer it, then it is answered as "Not Applicable" (N/A). Notes can be added to explain reasons for the answers.
- Complete the Joint Response Portion of the Questionnaire. The next step is for the program assessment team to collectively complete the joint team response for the program. It is expected that this will be a collaborative activity with open discussions among the team and facilitated by the program manager. It is up to the team to come up with one answer for each question. If the team is divided on an answer, use techniques such as asking, "Can everyone live with this answer" to reach a consensus. If a question is "Not Applicable" to the program, then answer it as not applicable and provide a comment explaining why it was not

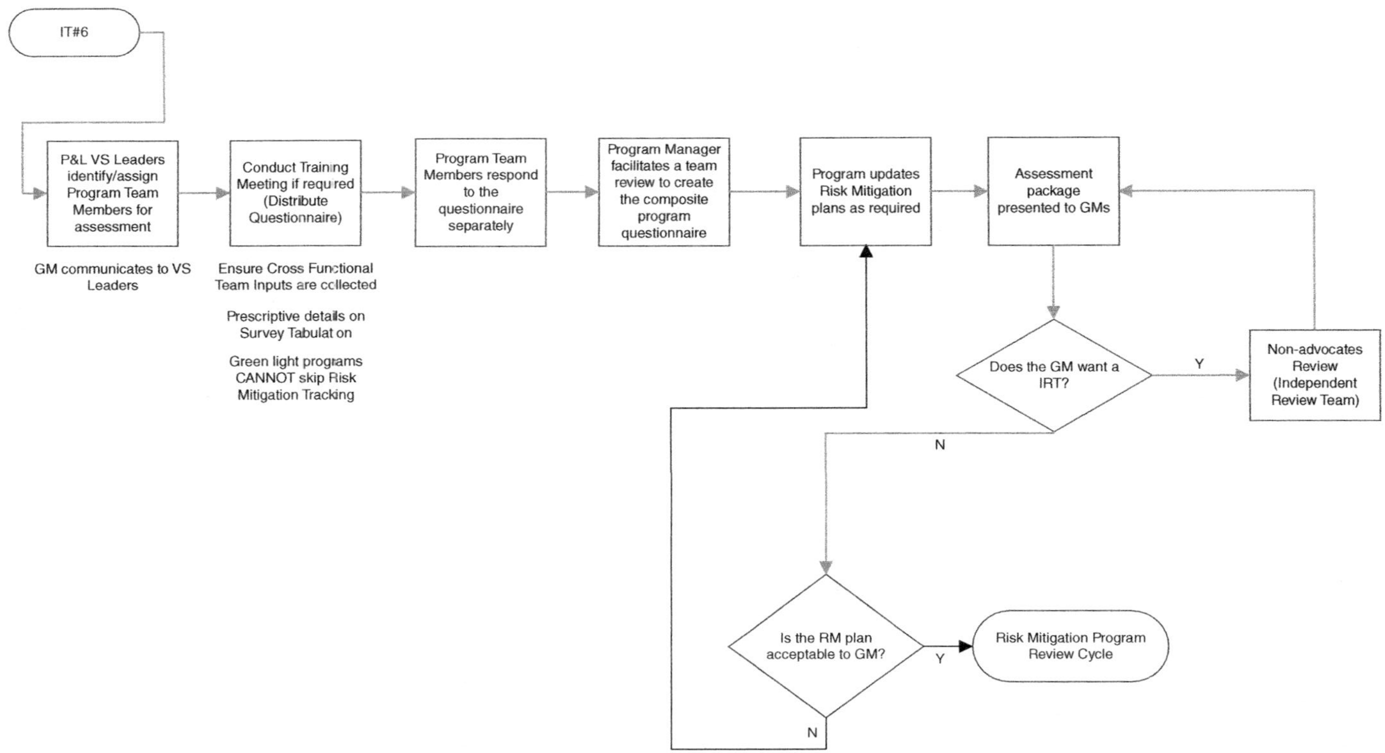

FIGURE 9.6 ISR Systems red flag review process.

applicable. Use the joint team response column on the consolidated worksheet for the team's answers.

- Complete the demographic information that is part of the questionnaire. The team should also complete the demographic information in the questionnaire. This includes:
  - Name of the program
  - Program number
  - Value stream
  - Location
  - Phase of the program
  - Program manager
- Capture Risks from the Red Flag and Add to the Project Risk Register. If the process of completing the assessment questionnaire has identified new risks to the program, those risks should be added to the program's risk register and processed as any other program risks following the risk-management process used by the program, including developing appropriate risk-mitigation plans.
- Summarize the Results. At this step of the process, the value-stream leader will review all of the program assessment packages and prepare a summary for presentation to the General Manager (GM). To accomplish this, it is recommended that each program manager review the program assessment questionnaire (or summary of it), the top program risks, and mitigation plans/status for those risks with the value-stream leader. After review of all of the assessments, the value-stream leader prepares a summary package, addressing all program assessments, for review with the site general manager.
- Plan to address shortcomings as required. The value-stream leader will then schedule and conduct a review of the assessments with the site General Manager. If the General Manager (GM) is not satisfied with the risk assessment/mitigation plans, a follow-up review with the GM can be planned after the shortcomings identified are addressed. For high-risk programs, the GM may choose to have an independent review team (IRT) review of the program. It is the responsibility of the GM to select the IRT.

9.3.2 Basic Elements of the Red Flag Review

The questionnaire is a spreadsheet containing multiple worksheets. There are separate worksheets for each member of the assessment team to complete, which are labeled #1 through #8. The results from each individual assessment worksheet are tallied into the consolidated worksheet. In addition to the questions, the questionnaire contains other useful information as is shown in the partial illustration in Fig. 9.7. Column C is where the responses to the questions are entered. The response is entered from a pull-down list in each cell. The next column identifies what the low-risk response is. The low-risk response is provided in case the question is not clear to the person completing the questionnaire. Columns E through G provide additional guidance to the person completing the questionnaire.

There is also a worksheet that provides a consolidated view of the results. A partial illustration of the consolidated worksheet is shown in Fig. 9.8. Columns C through J are copied from the individual team member worksheets. The consolidated worksheet is where the joint team response to the questionnaire is entered.

	A	B	C	D	E	F	G
1		Questions	Program Manager	Low Risk Response	Guidance	Risk Triggers (Cause)	Impact of Risk (Effect)
2	1	Are LPD tools or equivalent employed in the program?		YES	The intent is to understand the customer requirements. Tools such as Voice of the Customer, Impact matrix, Deliverables Map or equivalent. Has the team worked as a team on the program planning and understand each others needs?	Customer requirements not well understood or defined.	EAC growth late in the program when it's most significant.
3	2	Does the contract have explicit wording that provides contractual relief from late customer CFI or CFE.		YES	Customer is required to provide timely responses to CDRL submittals, TBR and TBD resolution in customer documents, change requests, etc	Customer responses do not support schedule	Late information results in cost and schedule impacts late in the program when it's most significant.
4	3	Is the customer team stable, e.g. technically mature, non-adversarial?		YES	Customer recognizes obligations and addressed them in a timely fashion. Customer communications effective and frequent.	Customer counterparts change frequently or are combative	Non-optimal working relationships impact cost and schedule
5	4	Is the customer a potential competitor for the product we're making?"		NO	Customer and Goodrich compete for same business	Exchange of critical program information is not optimal	Vital information may be withheld impact cost and schedule

FIGURE 9.7 Sample portion of the ISR Systems red flag review.

After the team has completed the consolidated portion of the review, a site-developed rating scale can then be applied to the results, with Table 9.1 as an example. It is important that the IPT captures the red flags, move them into the risk register, and prepare countermeasures and rationale for each.

Name	Big Box
Number	489283
Phase	pre CDR
Value Stream	Space
Location	Danbury
Manager	Joe Smith

<<<<ENTER Program info here

Tab	Participants	Role
1	J Q Program	Program Manager
2	Stuckee	Project Engineer
3	Guru	Technical Lead
4	mr Goodwrench	Manufacturing
5	etc	Operations
6	etc	Parts Engineer
7	etc	Contracts
8	etc	Quality

	Questions	Program Manager	Project Engineer	Technical Lead	Manufacturing	Operations	Parts Engineer	Contracts	Quality	Low Risk Response	Joint Team Response
1	Are LPD tools or equivalent employed in the program?	0	0	0	0	0	0	0	0	YES	
2	Does the contract have explicit wording that provides contractual relief from late customer CFI or CFE.	0	0	0	0	0	0	0	0	YES	
3	Is the customer team stable, e.g. technically mature, non-adversarial?	0	0	0	0	0	0	0	0	YES	
4	Is the customer a potential competitor?	0	0	0	0	0	0	0	0	NO	
5	Are there cultural barriers to effective execution and communication.	0	0	0	0	0	0	0	0	NO	

FIGURE 9.8 Consolidate worksheet example.

Banding Range (%)	Risk Level	Characteristics
0–20	Low	Program has track record of 100% OTIF with no negative EACs. Design/requirements are stable with no outstanding ECs or performance TBDs. SPI and CPI are green both over last six mos. Program plan, risk register, deliverables all nominal. Team is intimately familiar with the customer and product. Subcontractor/supplier risk is low-moderate trending toward low.
20–40	Low-Moderate	Nearly 100% OTIF with no significant negative EACs. Design is nearly stable with key performance parameters demonstrated. SPI and CPI are green both over last month with a long-term trend to green. Program plan, risk register, deliverables all have < 10% variances. Team is familiar with the customer and product. Subcontractor/supplier risk is low-moderate trending toward low.
40–60	Moderate	OTIF is sporadic. Forecasting negative EACs unless risk plan is executed. Design/requirements are not complete with < 10% key performance parameters not demonstrated. SPI and CPI are yellow or red with no green trend forecast. Program plan, risk register, deliverables all have < 10% variances. Subcontractor/supplier risk is moderate with no trend established. Team is somewhat familiar with the customer and product. Subcontractor/supplier performance key to program success.
60–80	Moderate-High	OTIF is red for three months running. Forecasting negative EACs unless risk plan is updated. Design is not complete with < 50% key performance parameters not demonstrated. SPI and CPI are yellow or red with no green trend forecast. Program plan, risk register, deliverables all have < 50% variances. Executive communication has occurred concerning ISR inability to execute. Team is not familiar with the customer and/or product. Subcontractor/supplier risk is moderate-high trending negative. Subcontractor/supplier performance critical to program success.
80–100	High	OTIF is red. Forecasting negative EACs with unfeasible updated risk plan. > 50% Key performance parameters are not demonstrated. SPI and CPI are yellow or red with no green trend forecast. Program plan, risk register, deliverables all have > 50% variances. Multiple executive communications have occurred concerning ISR inability to execute. Team is not familiar with the customer and product. Subcontractor/supplier risk is high with no plan for recovery. Subcontractor/supplier performance critical to program success.

Table 9.1 Example Rating Scale for the Red Flag Review Analysis

9.4 Lessons Learned and Chapter Conclusions

A lack of monitoring and control processes for a given project will make project management and recovery from change difficult to overcome. Some areas to consider:

- Starting a daily stand-up regime may be difficult at first, but the payoff is effective team communication and understanding of the project.
- Team understanding of risks and how the risks will be managed is a critical contributor to overall project success.
- ISR Systems has found that the red flag tool increases overall team understanding and awareness of risk elements and helps the team develop more effective countermeasures.
- There are many flavors of risk registers and risk logs out there. Choose or create one and use it not only to capture the risks but to help review risks, status, and effectiveness of the risk countermeasures. An example is shown in Table 9.2.

Stand-Up Board Exercise

Participant's Tasks:

Based on your knowledge and understanding of the project management process, design a daily stand-up board and stand-up area for your integrated project team that is addresses a medium-complexity technical project valued at $8.7M that will be completed in 14 months after receipt of order. You have eight leads on your integrated project team covering project management, engineering, operations, quality, supply chain, finance, contracts, and lean transformation (continuous improvement). The experience of the team ranges from seasoned veterans of past programs to new hires just assigned to the team. An example of stand-up board is given in Fig. 9.9.

Think about what you, as a project (or program) manager (PM), might want to share with your management, the IPT, and the general population at your site.

1. For each element that you place in your stand-up area and on the board, describe why you have created it.
2. For the stand-up area and visuals, what are you trying to communicate and to whom?
3. How do you expect each stand-up area user to use the area and the information?
4. Could this design change over time?
5. As program manager, what are the three most important elements of the area to you? To your team?
6. Would a virtual meeting space and electronic communication work more effectively than a physical area? Defend your rationale.

Discussion Points for the Stand-Up Board Exercise:

1. For each element that you place in your stand-up area and on the board, describe why you have created it.

   There are many ways to place the board horizontally or vertically. Status elements should include the following items:

	A	B	C	D	E	F	G	H	J	K	L	M	N	O	P	Q	R	S	T	U	V	X	Y
1		1) Risk Identification:				2) Risk Analysis:				3) Risk Response Planning:			Current Date:				Sort by Current Risk Score				Total Current Risk Score:		Risk Forecasting:
2		Sort by Risk Item #			Sort by Risk Score				Total Cost of Effects:	Sort by Risk Owner										Total Cost of Responses:			
3						TOTAL RISK:		15	$25,000								RISK AFTER:		6	$5,000	12.75		
4	Risk Item #	Entry Date	Risk Category	Cause (If this happens...)	Effect (Then this may happen...)	Probability (1 - 5)	Impact (1 - 5)	Risk Score (1 - 25)	Risk Effect Estimated Cost $	Risk Owner	Response Type	Summary of Response to Risk	Project Cross Ref. #	Milestone when Risk Must Be Acceptable	Due Date	Response Status	Estimated Probability After (1 - 5)	Estimated Impact After (1 - 5)	Estimated Risk Score After (1 - 25)	Response Cost $	Current Risk Score		Response Plan Start Date
5	1.0	1-Jan-13	Supply Chain	If optics are not delivered by supplier on time	There is a day-to-day slip in schedule that could affect final delivery	3	5	15	$25,000	John Doe	Avoid	Supply Chain Management will work with supplier to 1)do an inventory buy earlier than planned; 2) Incentivize delivery performance	Para. 2.1	PDR	4/1/2013	◔	2	3	6	$5,000	12.75		1/2/2013
6								0								○			0		0		
7								0								○			0		0		
8								0								○			0		0		
9								0								○			0		0		
10								0								○			0		0		
11								0								○			0		0		

TABLE 9.2 Risk Register Example

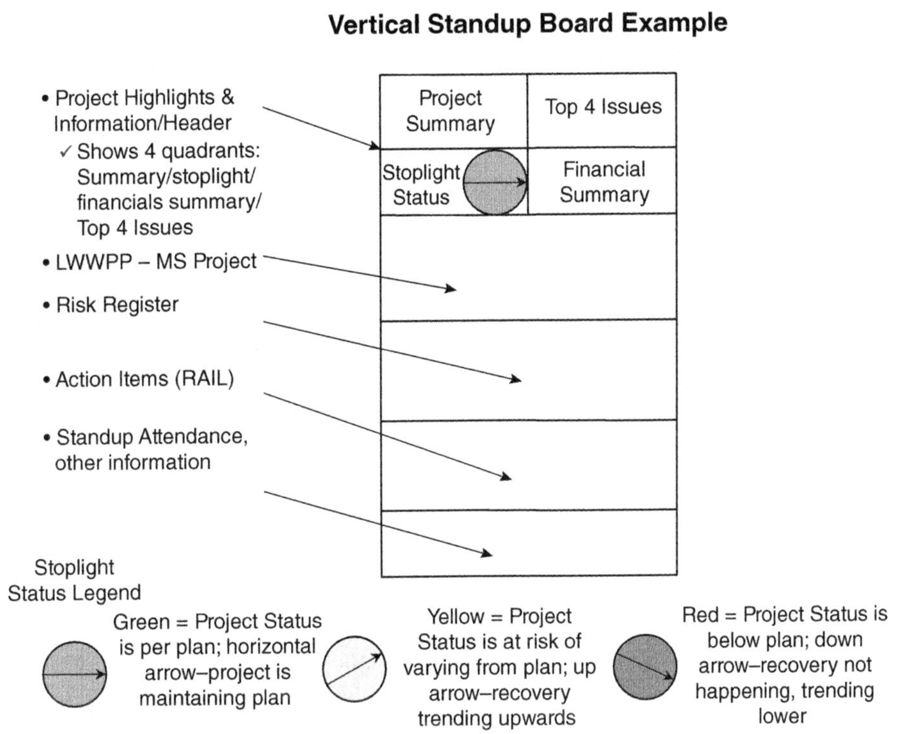

FIGURE 9.9 Stand-up board example.

- Title, board owner, date of last revision.
- Some type of symbol to reflect overall health of the project. You should be able to determine the overall health status of the project within a few seconds of looking at the board. There should be a legend on the board to explain this status or *stoplight* type of status.
- Some type of overall project summary information about the project will help the general viewer understand the scope of the project.
- Schedule: posting a complete Microsoft (MS) project schedule is one way but can be too overwhelming. Either post a rolling-wave portion of such a schedule or create a simple visual timeline showing, for example, key deliverables, key tasks, and key milestones. A detailed schedule should be made available to the IPT but through electronic form.
- Risk Register: same idea. Some risk registers are extensive. Try posting only the high-risk items or show the top-10 or top-five risks (and status) for the program.
- Action Items: these could be done weekly, but items not completed at the end of the week would be rolled over into the next week. Tracking action items as a team is critical to show that work is getting done, risks are being closures, and *pop-up* issues are addressed using real-time problem resolution techniques.

- Attendance can be tracked if the team is not as disciplined as it should be.
- Other Information: having a portion of the board reserved for this can be used to share information to the team that is important for all to see and to understand.
- Management Directed Elements: these might include schedule performance index (SPI), cost performance index (CPI), a list of contractor deliverable line items, contract deliverables against a payment milestone schedule, key suppliers list, etc.

2. What are you trying to communicate and to whom?

   The stand-up area needs to support several levels of communication. For all viewers, it should present an immediate status of the overall health and performance of the project. If the status is not good, the viewer should quickly be able to determine what the drivers of poor status is and what is being done to recover, regardless of whether it's an IPT member or general management.

3. How do you expect each stand-up area user to use the area and the information?

   Everyone should be able to quickly understand the status of the program and any drivers that are impeding progress. At a schedule/task level, each IPT member should understand what tasks are due and when based on a weekly, rolling-wave-type display of project information. Each IPT member should also know which action items he or she owns, so they can work those tasks to closure. Each IPT member should also know that he or she can manually update the information on the board such as action items, new risks, or other information on a daily basis. Management should be able to view the boards and to quickly understand status and corrective actions at a minimum.

4. Could this design change over time?

   Yes. It is a good idea for a site to develop some form of guidance, checklist, or standard work around these daily stand-up areas. For example, having a standard *design template* for the boards makes it easy for all staff to understand the information being presented. That said, as with any tool, if it is not working for the team, the team should make adjustments to their standup process and visuals to make it more effective.

5. As program manager, what are the three most important elements of the area to you? To your team?

   This should be answered by each student—the instructor can comment on their responses in a group setting.

6. Would a virtual meeting space and electronic communication work more effectively than a physical area?

   Defend your rationale. This should be answered by each student—the instructor can comment on their responses in a group setting. There is not *one* way about the application of stand-up areas. In some cases, the only method open to an IPT may be virtual. Even with a virtual stand-up area, there should be at least one physical area maintained by the project manager if necessary. Experience shows that having a live, face-to-face status meeting is a powerful communication and project management tool.

CHAPTER 10

Engineering Project Communications

Communication is intrinsic to all aspects of project management and execution, as managers must coordinate the efforts of individuals as well as entire organizations to achieve a common goal. The right information must be available to the right people at the right time to help them react to change, make sound decisions, and work effectively. Because it is implicit that people will talk, meet, and share ideas as necessary to get their work done, communication is often taken for granted: time is not taken to understand the information needs of stakeholders and contributors; investment is not made in training, technology, and resources to enable those needs; and measures to ensure the system works and people use it as expected are not implemented. Instead, communication is largely ad hoc, left to the devices and preferences of individuals. The eventual communication breakdown contributes to budget overruns, schedule delays, and outright project failures.

For organizations that recognize the importance of communication and invest in planning and support accordingly, it can offer a competitive advantage. The advantage increases with the complexity of the endeavor and the network of resources necessary to accomplish it, allowing the lead organization to leverage the capabilities of disparate groups—increasingly separated by time, distance, infrastructure, and culture—to achieve goals faster and more efficiently than they could individually.

The *Project Management Body of Knowledge* (*PMBOK*) provides a high-level overview of methods to develop a project communications plan and implement related activities of managing stakeholder expectations, distributing information, and reporting performance. But a project manager's communications responsibilities go beyond that limited scope. This chapter builds toward a discussion of those activities, with a specific focus on engineering projects, by first defining the role of the project manager in communications, introducing a communications model, and then using the model to discuss the challenges and opportunities communication poses. These challenges include:

- Collaboration and concurrent engineering.
- Outsourcing and Globalization: There is no monopoly on the laws of physics, and product development is subject to the same factors that influence decisions

This chapter was written by Robert A. Campbell, Jr.

to outsource manufacturing. A global market not only creates new competition risks but also opportunities for collaboration. Organizations must invest in communication to counter the competitive pressures and take advantage of the collaborative opportunities.

- Sharing complex engineering data.

10.1 The Role of the Project Manager

While project managers' duties may include particular communications, such as regularly reporting progress to senior executives, their most important functions are planning and facilitating the communication of others. This entails:

- Identifying the formal and informal communication needs of the stakeholders and participants
- Establishing expectations and procedures for the conduct of communications
- Providing the necessary training and resources
- Ensuring that expectations are met and information needs are satisfied during the life of the project

The first step is to develop a project communication plan that identifies stakeholders and their information requirements, the most efficient and effective means for communicating with them, the individuals responsible for specific communications, and the expectations—format, frequency, distribution—for specific communications. The content and development of the plan is described at the end of this chapter.

According to the *PMBOK*, "Effective communication means that the information is provided in the right format, at the right time, and with the right impact. Efficient communication means providing only the information that is needed." Communication is not without cost—it takes time to prepare and send; time to process and reply; and time and money to implement and support the necessary processes and technologies. Including extraneous information can make it more difficult to ensure that stakeholders were sent correct and complete data—and that the data were received and understood. If overwhelmed with data, stakeholders may miss critical information or choose to ignore the message altogether; one person's information is another person's noise.

The project manager can avoid top-down information overload by adhering to a well-thought-out communication plan. But information overload also can be a bottom-up problem, as individual contributors feel they are demonstrating their productivity by sending unnecessarily detailed status reports and copying anyone who might be remotely interested. This relates to the project manager's next responsibility: establishing expectations and procedures for conduct.

Consistency in form and application is key, especially in formal project communications. If status reports adhere to a standard format, recipients can either quickly find information they require or safely conclude there is no information available. Conversely, if the practice of publishing minutes for a technical working group meeting lapses, stakeholders may think nothing important transpired or that the group's work has been completed.

Formal communication (meeting minutes, status reports, design reviews) is an important but relatively small piece of the volumes of communication necessary for a project's success. Most of these communications take place in the course of and are inseparable from doing the work itself. The exact information needs of those doing the work cannot be anticipated. But the communication tools and processes those workers need to perform the job—and what is needed to support their peers' and the project manager's efforts—must be identified. The project can break down over a failure in either of these facets. The project manager needs to enable the communications and ensure expectations are met throughout the project life cycle.

Enabling communications includes providing resources and technology, establishing procedures for the proper use of these resources, and providing training and support. Email is an often-overlooked example. Because of its ubiquity, it is taken for granted. But this very ubiquity is a sign of its misuse: workers rely on email when a different medium may be more appropriate, subject lines are nondescript, the wrong people are included in sensitive discussions, and chains grow too long, etc. Training on the proper use of email can alleviate this.

Critical communication channels and processes should be tested occasionally to ensure they work end-to-end and are understood by those who need to use them. This includes traditional aspects, such as filing status reports, holding meetings, and publishing minutes. Companies should also test the process of sharing information with an outside partner and of holding remote and off-hour meetings. Sharing CAD data with a partner organization can require approvals on both ends, multiple translations of data, and the passing of large amounts of data over networks and through security layers. If this is critical to the success of the project, it should be tested before it is actually needed. This is true of any critical communication process.

10.2 A Communication Model

While all projects require effective communication, engineering projects have additional challenges in that they require those involved to share and act on technical, confidential data that is often encoded in closed, proprietary formats. A simple model, shown in Fig. 10.1, will be useful in discussing the requirements, responsibilities, and challenges of communication. The model applies equally well to simple and complex

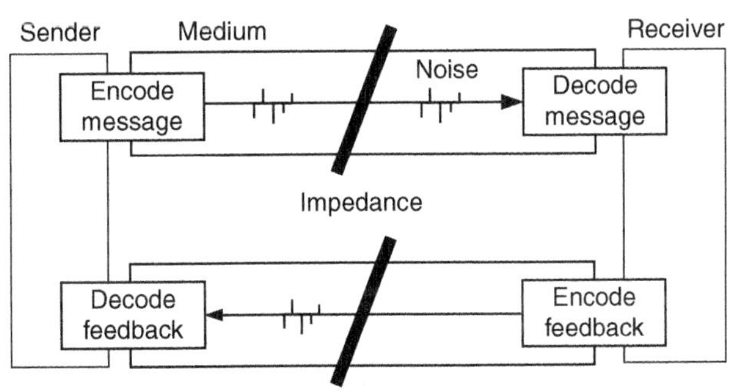

Figure 10.1 A communications model.

interactions, such as two people engaged in a face-to-face meeting or two organizations collaborating on an interface control specification in an electronic format.

The message is the information or query being communicated.

The medium is the method used to convey the message. The medium for a meeting may be in-person discussions, a teleconference, or an online chat. For an interface-control specification, the medium could be a text document, a 2D drawing, or a 3D model.

The sender is responsible for ensuring the message is clear and complete and that it is in a form that can be understood and used or acted on by the receiver.

The receiver is responsible for confirming that the message has been received in its entirety and understood.

A feedback message may simply confirm receipt of the information, or it may restate the information in a way that demonstrates correct understanding. In a conversation, this could mean using active-listening techniques. In the case of CAD model representing the mechanical interface of a system, it could mean showing that the model was properly imported and aligned in the mating assembly.

Encoding and decoding is the process of translating the message to and from a form that is compatible with the medium and that the receiver can understand. However, information can be lost in the process. For example, messages received during a teleconference will be void of physical cues such as the sender's body language and facial expressions. In sharing mechanical CAD data, if the sender and receiver do not use the same CAD program, he or she may receive only the geometric envelope, stripped of the history tree and design intent. Encoding could also include translating the message into a foreign language or clarifying technical language for a nontechnical audience, such as an executive review committee.

10.2.1 Noise

Noise interferes with the reception or understanding of the message. The term noise refers to not only actual sound but also environmental factors; hot or cold temperatures, for example, that can interfere with reception. Smartphones, televisions, and other stimuli and sources of information can compete for the receiver's attention. In remote communications, these distractions can be difficult for the sender to recognize and account for; the receiver may not catch the full message yet may signal that it has been understood.

Emotional and social factors can also affect reception, such as:

- If the parties involved in a meeting are in different time zones, it may be early morning for one group and after normal business hours for the other; energy levels will be at opposite peaks, and the latter party may no longer be mentally vested in work.
- If the parties don't agree on the purpose or the need for a meeting, they may never discuss the actual issue.

Missing context or background information can prevent parties from correctly understanding the message. Often this is because the sender does not clearly explain operating assumptions or incorrectly assumes there is common knowledge or a shared belief. The sender may also leave out elements of the message he or she considers unimportant, such as the reason for an action or decision. Without understanding *why*, it can be difficult for the receiver to accept the message or to understand its importance.

Noise can also interfere with the feedback process:

- A project leader may listen to a team member's grievances without giving a definitive "No." This can be misinterpreted by the team member as agreement.
- Certain receivers, due to cultural or individual bias, may believe saying "No" is impolite or an admission of weakness. Therefore, these individuals may answer "Yes" to confirm understanding or the ability to satisfy a request, regardless of their actual certainty.

10.2.2 Impedance

Impedance resists the free exchange of information between parties. Impedances are often encountered where communications cross organizational boundaries. They may reflect:

- Cultural differences in a bureaucratic or closed organization engaging with a nimble or open one
- Legal issues and safeguards to protect intellectual properties; laws restricting the export of certain information, technology, or goods
- Technological issues—the need to duplicate data in independent business systems
- Security measures such as firewalls restricting access to certain collaboration tools; inability to grant partners access to business systems or other resources
- Environmental issues such as distance or differences in time zones

Impedance tends to increase with the frequency or urgency of communication. Sharing revised CAD data with a vendor may not cause trouble. But if two parties are actively collaborating and exchanging models on a design with a high degree of flux, high levels of impedance will prevent both sides from reacting quickly to the change, perhaps causing schedule delays and increasing costs due to redesign, rework, and scrap. The project manager must understand what communication paths will be critical to the success of the project and minimize impedances along them. This could entail establishing procedures, seeking preapprovals, and implementing systems. These critical pieces of communication infrastructure should be tested in advance with real data to ensure they work.

10.2.3 Choosing the Right Medium

A medium, or the method used to convey a message, has two senses. Both should be considered when deciding the best way to deliver a particular message. The first sense is technological: a meeting might be conducted in-person, over the telephone, or via text-based chat. The primary differentiators between these technology options are bandwidth (the capacity to carry data) and information richness (the capacity to carry detail). The next section explores these concepts.

The second sense is structural: deciding to use a meeting in the first place also represents a choice of medium. If the message is a dramatic shift in the direction of the project due to rapidly changing circumstances, for example, project managers could have communicated this through a broadcast email. That method may be efficient but not effective. A broadcast email is an impersonal, one-way communication medium. For such a major announcement, one would anticipate questions.

And those questions could trickle in by email (asynchronous). In the meantime, productivity and morale could take a hit as recipients try to sort out the meaning themselves. Additionally, the project manager would not be able to observe how the team responds to the announcement. A meeting is a synchronous medium, providing the opportunity for immediate feedback, both verbal and nonverbal, to take the *temperature of the room*. The meeting leader can even adjust and clarify the message in real time to respond to feedback.

The project-communication plan should establish guidelines for the choice of medium for common events and formal communications. For exceptional events, the project manager must match the medium as well as possible to the situation at hand, weighing its effectiveness and efficiency. Individual contributors will need to make countless such decisions while doing their work and supporting the work of others. The communication plan should specify training to help managers and team members make informed communication decisions.

10.2.3.1 Information Richness and Bandwidth

Information richness is the capacity of a medium of expression to carry detail. A transcription of a speech communicates only the words that were spoken. Audio also communicates the inflection of the speaker and the response of the crowd. Video captures the speaker's expression and gestures. Being present, one could see the crowd and the speaker as well as feel the literal and figurative temperature of the room.

Bandwidth is the capacity of a medium of transmission to carry data per unit time. Even a data connection with unlimited bandwidth may be unable to communicate the full richness of an event. Consider an expert consultant participating in a prototype failure analysis via video-conference.

- A multi-camera setup cannot follow the presenter's actions or switch focus from the macro to the micro—from the speaker to a circuit board buried in the frame, for example—as fluidly as a spectator in the room.
- The remote consultant cannot smell the effects of unintended friction; cannot feel for signs of wear.

The data connection cannot transmit the full experience of being there; information is lost. However, everything has an opportunity cost. The consultant's expertise may be invaluable in this situation but is cost-prohibitive to maintain in-house due to his or her specialization. The urgency of project troubleshooting may preclude travel on-site. Organizations must anticipate these eventualities, plan accordingly, and make the best use of available tools. While it's not as ideal as in-person interaction, a multi-camera setup, a high-speed connection, and the planning to support these activities is lacking in most companies.

10.2.3.2 Other Characteristics of Media

In addition to bandwidth and information richness, you can evaluate media using a number of other criteria. Most represent a continuum, such as a status report residing at one end of the formality continuum and an Internet chat residing at the other end. Figure 10.2 presents a number of these characteristics and where certain media might fall on the continuum. When choosing a medium, the sender should consider the effect these characteristics have on reception and interpretation.

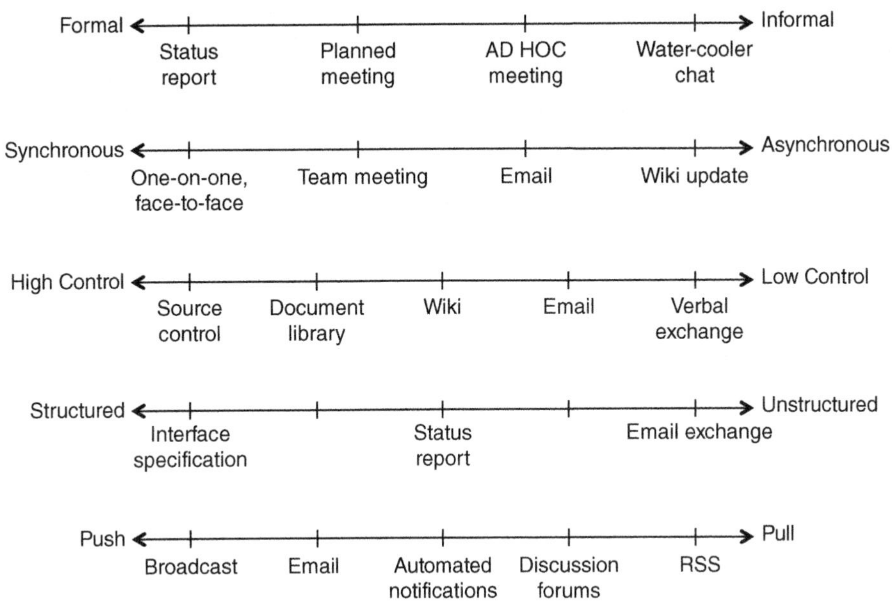

FIGURE 10.2 Characteristics of media.

Formal communications such as status reports are important and typically addressed in the project communications plan. But less formal communication methods, such as daily stand-up meetings, email messages, and impromptu discussions around a project artifact, such as a failed prototype, are critical to complete the project. Formal communications typically become part of the project record and are self-documenting; the status report or the specification is archived in a controlled system. It is generally counterproductive to burden informal communication with too high a documentation standard, but participants should recognize when something is important enough for formal documentation and follow up.

Synchronous communication is real-time, demanding that all parties involved be present (physically or virtually) at the same time. A near instantaneous medium is required to support it.

Asynchronous communication allows different parties to interact at different times and could involve exchanging email or collaborating on a shared document stored in a workspace, making edits and offering comments. Asynchronous may be a desirable attribute for some communication, such as when the parties are separated by 12 time zones. It may also be an undesirable side effect of the communication medium; even a *real-time* email exchange is slightly asynchronous due to the time required to comprehend the message, compose a response, and transmit the reply. There may be intervening messages in that time that change the context or facts.

Asynchronous communication also allows individuals to schedule the sending and processing of communication. Not every communication needs immediate attention; interrupting work to read every email as it arrives typically hurts productivity. Because email is asynchronous, email activities can be batch-processed at selected times during the day.

Push communication is when the sender controls who receives the message. This is often associated with broadcast media such as television; however, the sender can better target his audience via email and other instantaneous technologies. The key is the recipient cannot choose whether to receive the message, though he still must interpret it and could simply ignore it. Push communication is useful for urgent messages, but loses its effectiveness if overused. It is also useful for automation, such as workflow systems that notify recipients that a document revision is awaiting their approval.

Pull communication provides access to information but requires the receiver to take action to obtain the information. A discussion forum, which has many of the properties of email but does not have a distribution list, is an example of pull communications. The first step in being able to access information is knowing that it exists, so a combination of push and pull methods are often used:

- An email is sent to a distribution list, requesting members to respond to a survey. The email also provides a link to a web-based survey.
- A specification document requires input and edits from the team. Rather than sending the specifications to the team as an email attachment, it is placed in a collaborative workspace that allows simultaneous editing with version control. The link is sent only to necessary team members.

A key advantage of pull communication is that a person can choose to participate: a buyer elects to be notified by email when a major revision of the specification is released, ignoring draft revisions; a functional manager subscribes to a daily digest of discussions about the specification and any associated changes. This selective participation enables a person to be more informed about more relevant communications than a typical all-push email strategy. It also raises the signal-to-noise ratio of push communications. As email becomes less used, recipients likely will pay more attention to incoming messages and respond accordingly.

Internet technologies and media such as online discussions boards, social media streams, and rich site summary (RSS) feeds enable pull functionality in the enterprise. Recipients can use social media techniques such as *mentions* to alert them of a potentially relevant discussion and then follow or ignore it as they see fit.

Structured communications promote consistency, which can assist information gathering and automation. If a status report has a consistent format, for example, a person knows where to look for relevant information. Managers can use search and other technologies to make unstructured information more accessible; team members can setup automated daily searches of project databases and discussion forums about relevant terms.

Different degrees of control are required for certain communications to ensure they become part of the project record and that changes are not arbitrary or unexpected. An interface control document (ICD) is a form of communication that assists independent resources to work on related portions of a product or project in parallel. It is important that changes to the ICD be made in a controlled fashion, and that affected parties are provided adequate notice of those changes.

10.2.3.3 The Use and Misuse of Email

Email, which has become so commonplace and important in project communications, is also the source of many communication breakdowns. While this section focuses solely on the use of email in corporate communications, the practices herein can be applied to the use of any medium within a communications plan.

Email seems to have become acceptable for any task:

- Document collaboration: An original document is attached and sent to a distribution list; one or more recipients reply with changes, some of whom change the filename; the document owner integrates changes while another person replies with edits to one of the second-generation files.
- Distribution of documents that should be part of the project record, such as meeting minutes.
- As an extranet, such as sharing CAD models with design partners who cannot access the product data management (PDM) system behind the corporate firewall.
- Workflow and approval systems.
- Involved discussions of design alternatives: this could replace having to gather a group of experts into one place to discuss a problem or design.

This is not to say email should not be used for such tasks; it is just a poor medium for many of them. Email should not be the default choice for sharing or communicating. This use and misuse is symptomatic of two problems:

1. Lack of training on the proper application and convention for using email, including the need for meaningful subject lines and sending messages to limited distribution lists;
2. Weakness in other aspects of an organization or project's communication infrastructure and plan, such as a controlled means for sharing complex data with external partners or a formal workflow approval system.

10.2.3.3.1 Characteristics of Email Email has certain characteristics that are an advantage in some situations and a disadvantage in others. Email is nearly instantaneous. Barring technical difficulties, a message will arrive in its recipient's inbox soon after it has been sent. This does not necessarily reflect its urgency, but individuals tend to treat all email messages the same—they stop what they are doing and check each new message as it arrives. This may in part be due to the potential urgency, but it is more likely out of habit. Individual email clients often allow senders to flag messages as urgent; however, the receiver must use a compatible client and the term *urgent* must be defined prior to use. Additionally, email is a productivity killer, especially for those workers who stop what they are doing to read each new message that enters their inbox. To help alleviate interruptions, it's important to automate how email messages are handled based on the flagging system.

Email is asynchronous. Since most project communications are not urgent, this can allow email to be processed in batches and triaged. Asynchronous communications can be advantageous when participants in the conversation are otherwise occupied, have different schedules, or are in different time zones.

But this can be problematic in a highly interactive and dynamic conversation. Key participants may not be present at critical times, and, even when they are, responses can take a long time to compose and deliver. Urgent, critical information may arrive too late, or its meaning may be misinterpreted due to changing context based on other information that was received in the interim.

Email conversations are unstructured. Each email message is a stand-alone container of information, only connected to other messages in a conversation thread by metadata that is used to deliver it and non-strict criteria such as matching subject lines. This simplicity contributes to its robustness and pervasiveness as the lowest common denominator method of electronic communication. But email conversations do not really exist as structured data objects that appear the same regardless of perspective, i.e., who is reading the conversation, which email client they are using, and when they receive or read the message.

Some email clients attempt to create a structured view of email conversations based on the subject line, the message replied to, and the time the reply was sent or received (depending on whose perspective of the conversation is being viewed). But threaded views are artificial and the result of an algorithm the client applies to its service. Some clients do not support such views, while it's an option you must enable in others—there is no standard method between clients. This could change a recipient's interpretation and understanding of details contained within the message.

Individual email messages are unstructured and inconsistent, especially in how they handle the original text in a reply. Typically, the reply is prepended to the message, which reads awkwardly but is reasonable in a threaded view of a conversation. Individual users may intersperse their replies in the original text. It may read more naturally, but it is not the de facto standard and can be easily missed. There is also no consistent way to flag in-line replies and separate them from the original message.

The length of original, quoted text grows with each reply. When reading an email, this *cruft* can usually be ignored. But when performing a search within an individual email or within an entire inbox, it is noise. The search returns not only the original email, it returns every reply.

10.2.3.3.2 Email and the Project Communication Plan Used the way it is, email chains become important project artifacts. Design alternatives are generated, evaluated, and selected. Documents evolve into their final form. Key decisions and actions from meetings are documented as minutes. Yet, despite these artifacts' importance, they are not:

- Preserved as part of the project record or product definition;
- Centralized to provide access to everyone necessary, nor in such a way that they present a single view, regardless of who views them or how they are viewed;
- Well-controlled.

Does this mean organizations should eliminate the use of email? Of course not. Email is a vital and efficient form of communication. It can be used as the virtual equivalent of, "Come here and look at this a minute." It is useful for urgent communications to raise awareness of a problem, control an issue, or adhere to a critical deadline such as a shipment (as long as the urgency is properly communicated). In general, informal communications that do not need to be part of the project record or design history are good candidates for email.

Email should not be used ad hoc for more critical and specific project communications. According to Prichard, (2004), the project-communication plan should define guidelines and protocols "to ensure that project messages are elevated to visibility and to ensure that project team members have a consistent vision as to what information is appropriately transferred (and, as a result, maintained) by this medium and what information is not." These guidelines should include:

Email Considerations	
Subject Line	Never blank
	Concise but descriptive
Distribution List	Focused list
	Do not include recipients "for information"
	Guidelines for stakeholders
	Guidelines for forwarding to stakeholders and outsiders
Replies & Quoting	Prepended or inline replies
	Indicate if inline replies exist
	Method for differentiating inline reply from original text
	Guidelines for trimming earlier replies
Change of Subject	Create new email with new subject line
	Update distribution list as appropriate
	Quote relevant portion of original chain

TABLE 10.1 Email Considerations

- Details specific to the mechanics of email, such as subject lines, distribution lists, and reply formats. These apply to informal as well as formal communications. Table 10.1 lists a number of considerations. Pritchard provides an example of a systematic subject-line convention for formal project communications and guidelines for handling long email "trails."
- Appropriate language, topics, and uses.
- Explicit rules of what email should be not used for and alternative communications methods. For example, email should not be used to distribute documents for editing and review. The document should be placed in a collaborative online workspace; a link to the workspace can be sent by email, allowing interested parties to subscribe to the workspace, so they can be notified of document changes.
- Document retention and archiving: there should be a protocol for archiving a single copy of the exchange.

Guidelines should be established for distribution lists (the To, CC, and BCC fields). To reduce noise, lists should be focused; do not *copy the world*. The same level of care taken when specifying the original list should be taken when adding people to replies. More significantly, the inclusion and addition of key stakeholders and outsiders should be addressed, even to the extent of restricting who may add certain stakeholders to email messages. Consider an internal discussion about a customer issue, which includes unflattering comments about the customer. After a resolution is reached, someone adds a brief description of the resolution to the last email in the chain and forwards it to the customer.

As with other critical project-communication tools and methods, training on the proper use of email should be included in the communications plan. An important skill is recognizing when to move a conversation off email and to a more suitable medium. Such instances could include:

- Before attaching an editable document, consider posting it in a document workspace and sending a link to it via email. Sending an invitation for individuals to subscribe to the workspace and receive notifications of document updates may be a better solution.
- Once a crisis has been resolved, call a meeting to discuss the root cause and prevention.
- Once discussions change from informal and shallow to in-depth, significant matters, such as discussions of design alternatives, transition to a more structured asynchronous medium such as a wiki, discussion forum, or document workspace or conduct a meeting.

If the transition to a different medium is not made, the email exchange should be treated as though it were a meeting. Prepare minutes summarizing the content of the email discussion, resulting decisions, and actions assigned. This can be more challenging than meeting minutes because it is less structured and does not have the same definitive end, but the effort must be made. The email exchange should also be archived in a central, accessible location to ensure it is preserved, and there is one, official view for all parties to reference.

10.2.4 Using the Communication Model in Planning and Execution

Consider a company, ACME Systems, developing a new, modular platform for performing a related series of steps on semiconductor wafers. ACME will design and manufacture the central material-handling unit (MHU) and several key modules. The design and manufacture of other modules will be outsourced to companies that specialize in the process steps they perform; one of those companies is Beta Systems. The MHU and associated modules will be developed simultaneously.

To ensure the completed modules fit and interact properly with the MHU, ACME has created an ICD to share with its partners and its own internal design resources working on the modules. The ICD includes a mechanical CAD model representing the key features that must be present on both sides of the interface, geometric envelopes, and tolerances.

The ICD is a message from the sender (ACME Systems) to the receiver (Beta Systems). Representing the ICD as a CAD model is a choice of medium; a PDF of a 2D drawing could have been an alternate choice.

If Beta Systems uses a different mechanical CAD program than does ACME systems, it will first need to translate the ICD model. This is impedance, causing delay and introducing lag into the communication process. If it is a one-time operation, the effect will be minimal and can be planned for. But if the design of the interface itself is not stable, the message needs to be sent and translated multiple times at unpredictable intervals, causing significantly more disruption. The impact of impedance worsens with frequency.

Because of the incompatible CAD systems, the ICD is translated to a lowest-common-denominator format that communicates only the geometry. Parametric features, geometric tolerances, and historical links to earlier versions (the design history) are lost. In other words, information is lost. This is noise.

To close the loop, Beta Systems imports the translated ICD model into the model of its module. The company sends screenshots showing the ICD properly inserted and aligned in the module as feedback to confirm it has received and understood the message. It will design to this interface and will later reverse the process, sending a model of its module back to ACME to integrate into its model, through the same sets of impedance and noise.

If the translation step poses such difficulties, why not select a partner that uses the same CAD platform as ACME? While this could be a consideration, the primary goal is to give each module best-of-breed capabilities, which will be a lasting benefit of the product in the market. If the capabilities of Beta far exceed those of alternatives that use the same CAD platform as ACME, Beta would be the logical choice.

The communication model can be used to anticipate and account for these issues in the development and execution of not just the project-communication plan, but of the project as a whole. Critical communication processes should be identified and mapped out, including sources of impedance and noise. This can inform choices in planning, organization, and collaboration so as to minimize their number and mitigate their impact, but they cannot always be avoided. Necessary investments in technology, training, and other resources can be planned for and implemented. And the entire system can be tested from end to end before it is needed in practice and again periodically throughout the project to ensure it continues to function properly.

This is a greatly simplified example. Pulling back slightly reveals more considerations that could apply to this or other communication processes, especially when those communications cross-organizational boundaries. For example:

- Bureaucratic: Has the team put in place non-disclosure agreements (NDAs) and approvals from senior management to share data?
- Political and Cultural: Are functional managers within either organization hesitant to share information they control as a means of maintaining power? Is the organization as a whole secretive?
- Legal: Are there restrictions on sharing data with partners not in the same country, as with defense-related data in the United States?
- Technological: Are file-sharing methods other than email available, especially for sharing large CAD models? Do firewalls, account access, and other security measures hinder communication?
- Change Control: How are changes to artifacts such as an ICD approved and communicated?

Brainstorming such issues should be part of the communication-planning process.

10.3 Distance and Communication

Distance is a recurring theme in project management and communications. Collocation is a technique used to improve project performance, in part due to enhanced communication through physical proximity. Yet outsourcing and collaboration on a global scale make it difficult or impossible to achieve project-wide; the challenge turns to understanding and attaining as much of the benefit of collocation as possible. And with increasing distance comes additional challenges, such as time-zone differences and cultural clash.

Thomas Allen's *Managing the Flow of Technology* reports that the likelihood and frequency of communication drops off dramatically with distance. The publication of the often-referenced book in 1977 predated the widespread use of email and the development of other remote collaboration tools, and the scale of distance in the study was only feet. Today, for better or worse, the most common mode of communication between two people in adjacent cubes could be email. So only the most general trend—that frequency of contact declines with increasing distance—is relevant today.

The lack of opportunity for face-to-face/in-person communication is not the only explanation for this decline. Though separation may preclude personal contact, proximity does not ensure it. Nor is a face-to-face meeting necessarily the most efficient or effective mode of communication.

There are also social, technological, and practical factors associated with being in close proximity that increase the likelihood that one person will reach out to another. Familiarity and social ties that develop from informal interactions (even just passing one another in the hall) and secondary relationships, such as mutual acquaintances, increase the chance of being aware of someone and how one might be able to help the other. It also reduces the resistance to reach out to a stranger.

On the practical and technological side, something as simple as a phone list or a company email address book could reduce the significant burden of establishing contacts. Being inside the same organization eliminates the chance an email will end up in the spam filter.

Team-building and social media (built on an internal or external platform) can be used to reduce the social distance between team members and partners at different locations. If a project website exists, populate it with personnel Web pages using photographs, biographical information, contact information, and role information. An interactive map that shows the location of personnel will give team members an additional sense of each person and assist understanding time-zone differences that may exist.

Special team-building exercises that utilize teleconference, Web meeting, and other technologies should be planned. Pairs of team members in different locations could be assigned to collaborate on early, straightforward tasks. This would accomplish useful work while building relationships and testing communication systems that will be helpful when more complicated collaborations take place.

It is also worth spending some of the travel budget early in the project to bring people from different locations and organizations together to work on planning, specifications, and other key documents. This will not only help meet project goals but could also help build strong relationships among team members early in the process.

10.4 Collaboration and Concurrent Engineering

According to Highsmith (2010), communication is sharing of information; coordination is using that information to align activities and time events. They are key elements of collaboration, but neither, alone, constitutes collaboration. Collaboration is "working together to jointly produce a deliverable . . . or make a decision" (Highsmith). In a serial environment, product development and manufacturing coordinate activities, using the product development (PD) schedule and bill of material (BOM) to drive material for a pilot build. In concurrent engineering, product development and manufacturing collaborate on a design that better meets the customer's needs by taking into account the input of downstream functions. Collaboration is about using information—not just sharing it—in a joint creation.

A project is a high-level collaboration, and many milestones and intermediate work products within the project are best achieved by collaboration, rather than as the efforts of an individual or a well-defined, homogeneous group. The scale of collaboration may range from pair programming in an agile software-development methodology to a virtual organization outsourcing development, manufacturing, fulfillment, and service. Through collaboration, an organization intends to achieve a better outcome than it could on its own. For a given project, better could mean lower production cost, shorter time to market, or higher return on investment.

Collaborations could be based on:

- Expertise: An organization might leverage the expertise of another who specializes and excels in a certain area, rather than rely on or invest in making improvements to a mediocre internal capability. This also permits the lead organization to focus on its own specialization, while pursuing opportunities it may not otherwise be able to.
- Capability: An organization may partner with another company that possesses a capability it lacks or that possesses a distinct advantage in a specific area. This typically relates to an operational capability or a physical resource. Outsourced manufacturing is a familiar example, but the partner organization could have access to unique testing equipment or proximity to a university research lab that is relevant to the effort.
- Capacity: The lead organization may have the expertise and capability to perform the project in its entirety, but not the capacity to complete it in the time required. There may not be time to invest in expanded capacity and meet the project schedule, or it may not be economically viable because it is only needed to meet a peak load in this project.
- Strategy: The advantage gained may be outside the scope of a single project. For example, a North American developer of semiconductor manufacturing equipment may outsource its manufacturing and services to Asia, where most of its customers reside, so it is more responsive to its customers' needs and can shorten the supply chain. But this has a direct effect on each new product development project.

The very point of some forms of collaboration, such as concurrent engineering (CE), is to promote better information sharing to improve outcomes. Other forms require excellent communication to realize the intended benefit of the collaboration. Communication benefits (in the first case) or challenges (in the second) of collaboration vary with the interdependence, simultaneity, and uncertainty of the work.

Concurrent engineering is an example of collaborating along the value chain; sharing information between functions that usually perform their work sequentially. In addition to sharing information, some of the work by downstream functions that normally occurs in later steps now occurs in parallel with the work of upstream functions—using incomplete and fluid information. But the downstream functions are still typically consumers of information, mitigating the communication challenges: data translation, if necessary, occurs only in one direction; change primarily flows downstream and is easier to manage as a result. The project continues, and the uncertainty of information that is passed on solidifies. Incorporating the insight of downstream functions earlier in the process reduces the likelihood and scope of change later in the process.

Outsourcing the design of a subsystem, while other subsystems and the top-level product are being designed, is an example of collaborating across the value chain. Two or more groups are simultaneously performing work that affects the others; each party is a producer and consumer of information. Data translations, design changes, and other forms of noise and impedance affect the flow of information in both directions. The activities are likely on a compressed schedule, with change occurring until the final deadline is reached.

10.4.1 Concurrent Engineering

Shina(1991) defines concurrent engineer (CE) as, "The direct involvement of all parts of an organization into the conceptualization, specification, development, manufacturing, and support of new products. The ultimate goal . . . is to produce products that meet customer expectation, with high quality and low cost." Manufacturability, serviceability, and the other "-ilities" are considered from the beginning, when the cost of change is lowest, by "the earliest possible integration of the overall company's knowledge, resources, and experience in design, development, marketing, manufacturing, [service], and sales." CE comprises a body of knowledge and practices—QFD, DFX, FMEA—that go beyond simply sharing information. But efficient and effective communication between functions responsible for different parts of the value chain is critical to its success.

When discussing CE, it is more useful to refer to a product realization process (PRP) rather than to a product development process. An integrated project management approach, which encompasses activities from concept development to production to preparations for servicing the product in the field, ensures that dependencies are understood and interests are aligned with the project and not individual departments. Thus, engineering is driven by the milestone to hit a sales window with a product that meets customer expectations, rather than by the need to check a box that it delivered a technical data package to production by a preset date.

Concurrent engineering is not a new concept; most manufacturing organizations at least pay it lip service in some form. It is helpful to contrast it with the apocryphal *throw it over the wall* serial approach illustrated in Fig. 10.3.

At first glance, referring to the communications model, it almost seems ideal: Well-defined, infrequent transfers of information across organizational boundaries—marketing, product development, and manufacturing. It is misleading because it is a static view, like watching a single frame of a video. What is missing, both from the picture and the process, is feedback and cooperation. Each phase produces a work

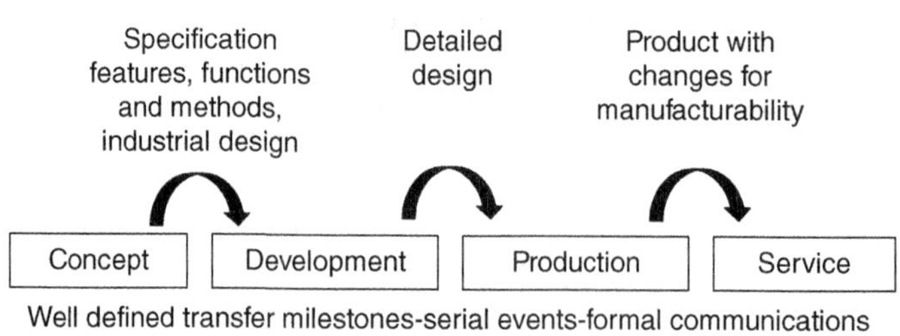

FIGURE 10.3 Serial project communications.

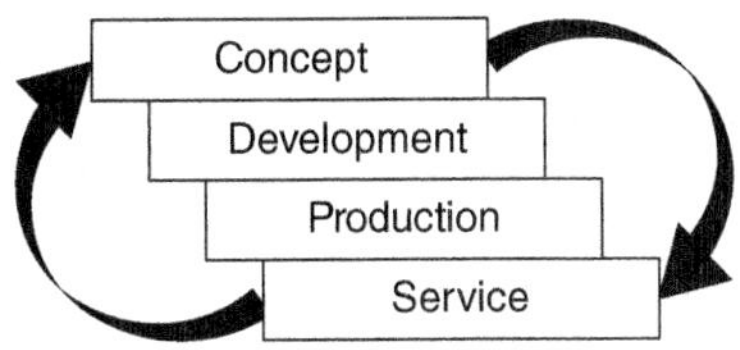

FIGURE 10.4 CE communications.

product without the benefit of input from those who must implement it in the subsequent phases, who then set about redesigning or reworking it to suit their needs to the extent they are allowed and able. Each redesign pushes out the schedule, inflates the cost of the development and the finished product, and reduces quality, as changes later in development are more expensive to implement and more likely to be work-arounds than sound design implementations. The true cost of this is only realized when the product is in the field, through repairs, replacements, loss of customer satisfaction, and a tarnished company reputation.

By comparison, the information flow in the CE process (Fig. 10.4) seems chaotic. The parallel execution of formerly sequential activities requires that work be done with incomplete and uncertain information. That information needs to be shared promptly across organizational boundaries—departments within an organization or external organizations such as customers and suppliers. Responsible persons in these organizations need to know who requires the information they have and when. Likewise, persons who need the information should know from whom to obtain it and from where. In addition to actionable information, the different functions need to know the status of parallel activities, availability of resources, and changing of priorities or direction.

CE requires more than just the effective and efficient communication of bits of information to conduct day-to-day activities. It requires the incorporation of core knowledge from different functions—marketing's knowledge of the customer, manufacturing's knowledge of what can be achieved by various production techniques, service's understanding of what makes for a quick repair—in the PRP as early as possible. CE uses a number of techniques to achieve these goals. Some require change at the organizational level and are beyond the scope of this book.

The formation of cross-functional teams helps achieve these two goals by bypassing the impedances that normally exist between departments and organizations. A transfer of people creates the transfer of institutional knowledge. And reduced physical and social distance facilitates day-to-day, real-time communication between disciplines.

The cross-functional team should be formed early in the PRP process. The keyword is "formed." It does no good if to be determined (TBD) is assigned to half the blocks on the project organizational chart; he would be a very busy person. Although each member is representing different departments, his or her responsibility is to the team and the product. Involvement in the collaborative development of the early concept helps to solidify this, and the integrated project planning codifies it.

If a function is not filled by a full-time team member, a specific person should still be designated as a resource. That person's responsibility to the team should be made clear, including the priority of the work, response times for requests, and participation in reviews and associated meetings. Establishing a regular pattern of communication and involvement—weekly team meetings and status reports, for example—will improve coordination by increasing predictability and reducing uncertainty (Jarvenpaa and Leidner, 1998). By the same token, the core team is obligated to keep part-time members involved in and apprised of the team's work, especially when it touches that person's area of concern.

Collocation of the core team members greatly improves communication, which is important at all stages but critical in the beginning of the project when trust and effective working relationships are born. It may not be possible for all members to be present at all times because of logistical or organizational reasons. So techniques such as partial collocation, front-loading travel, and traveling members can assist with this. Collocation and these variants are discussed in a later section.

The parallel-development inherent in CE often requires team members to perform work even though they do not have complete information. To avoid resulting costly rework/redesign loops later in the process, the team must become adept at sharing new information promptly as it becomes available, in a way that it does not overwhelm the recipients and become noise. The team should also make quick, incremental adjustments then prototype, validate, and test as it goes along.

Facilitating push and pull communications is one way to reduce noise. At the center of this effort is a structured database of project and product data. It must be well-organized, indexed, and accessible through a wide variety of tools. In addition to sharing information directly with team members in real-time when necessary, the sender also pushes information to the central database. When someone needs a particular piece of information, it can easily be pulled using search and other techniques. One of the greatest benefits of CE is realized when the same BOM and assembly documentation can be used for both design and production, with little (if any) modification. Sourcing can also use the working BOM to identify long lead components and work with suppliers.

10.4.2 Collaboration across the Value Chain

The term "concurrent engineering" is often used interchangeably with collaboration, but it is only a particular method of collaboration—one with specific and well-developed focus, principles, and techniques. CE is focused on collaboration along the value chain, with the intent of capturing and integrating recommendations and requirements of downstream steps into the design process as early as possible. It does not preclude other forms of collaboration, but nor do the tools and techniques of CE address the challenges or needs of other forms, in particular:

- Collaboration That Crosses Organizational Boundaries: In addition to capturing the input of downstream functions within an organization, CE captures the inputs of customers and key suppliers who are outside of the organization. Some downstream functions are also likely to be outsourced. So CE may cross organizational boundaries, but its tools do not address the related issues directly. These issues include separation by distance and time, technological barriers, cultural and organizational differences, security, and data compatibility.
- Collaboration Across the Value Chain (Fig. 10.5): The downstream collaboration partners in CE are typically consumers of information generated upstream; the very point of CE is to capture and consider these consumers' needs early in the

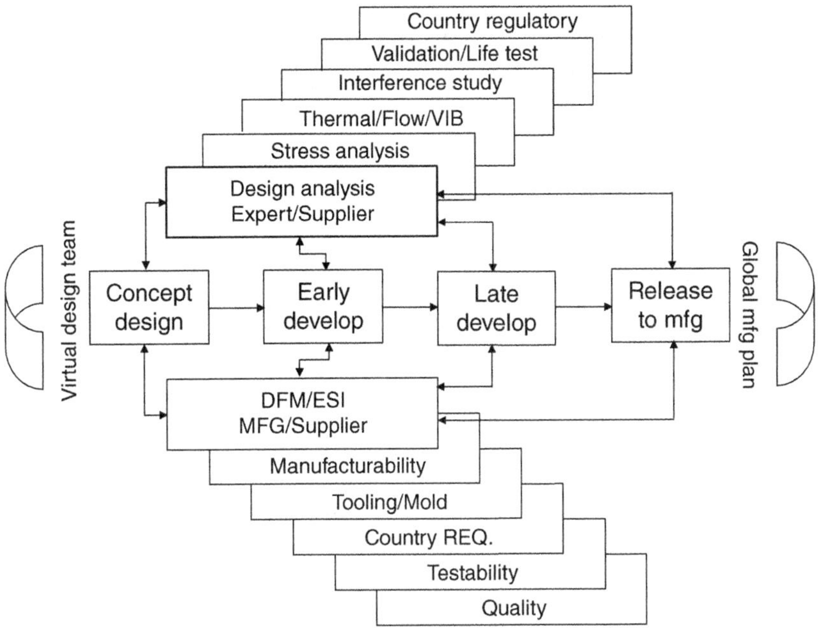

FIGURE 10.5 Collaborating across the value chain.

process to minimize the need for change when it reaches them. When collaborating across the value chain, partners are working on the same step and work product, acting on information from partners that is likely still in a state of flux. The flow of information and change may occur in both directions. The lead organization must mediate potentially conflicting requests.

Minimizing the interdependence of work packages eases this burden. One method to achieve this in product development is to organize the project work along the architectural lines of the product. Clearly defined work specifications and interfaces decouple the details of one team's work from the details of another's, allowing that team to work effectively in parallel. But it increases when the pieces are finally brought together; this is called integration risk. This technique is discussed in the section "Architecture as a Collaboration and Communication Tool."

Task interdependence is not always as direct as mechanical coupling and communication protocols. An example is an organization outsourcing the thermal analysis of a system and the design of a circuit board in the system to different specialists. Both the thermal analyst and the board designer are provided a power budget and space claim for the board. As development proceeds, the designer requests a change in board footprint to increase usable space and improve trace layout. The analyst concludes that either the power consumption must be reduced or the mounting area must be increased to reduce thermal resistance. The organization must mediate these potentially conflicting requests in real time and manage the change and its impact, internally and externally.

A formal change control procedure can mitigate this, but formality is not conducive to rapid development. Less formal, open communication channels can create visibility and allow differences to be resolved with less disruption. The change control procedure can be used to codify the final solution.

In developing the project communication plan, carefully consider the collaborations that will be required. What resources, technology, and training are needed? What impedances—whether organizational (e.g., need for approvals to share data), technological (e.g., navigating security systems between organizations), or cultural (e.g., language)—exist? What sources of noise will affect the integrity of the message? The project manager must provide those resources, plan for those difficulties, and monitor those systems throughout the project.

10.5 Collocated Teams

Collocation is an organizational technique in which key team members are brought together to share a physical workspace to accomplish a specific goal. The intent is to improve the chance of project success, which can be defined in terms of customer satisfaction, time to market, or any number of other measurable criteria, by improving communication, team dynamics, and oversight. Collocation is a key technique used in concurrent engineering and rapid product development. An ideal concept is a cross-functional team of development engineers and representatives of marketing, manufacturing, and other roles working in an open environment, centered on a collection of prototypes and tools, wall schedules and QFD charts, and other project artifacts.

The advantages of collocation are partly based on the idea that physical proximity promotes more frequent and effective communication, as discussed previously. Collocation also enables spontaneous, informal interactions that can be efficient and effective ways to build trust and common purpose; teaching or learning a skill that will directly or indirectly benefit a project; solving a problem; and sharing information. The background noise that exists in any workspace is more likely to be relevant to one's own work, allowing the worker to pick up on details that affect him or offer relevant information or assistance using the *cocktail party effect*, whereby one can detect important cues even when not focused directly on a conversation.

Psychological and social factors are at play in collocated teams. Proximity and frequent contact reinforces "social similarity, shared values, and expectations" and "increases anxiety pressure resulting from the possibility of failing to meet commitments" (Rad and Levin, 2003). Team members remain informed of each other's progress, activity, and disposition in informal and unobtrusive ways, improving coordination, ensuring people get assistance if required, and being reassured of mutual effort.

Zenun, Loureiro, and Araujo studied three pairs of similar new product development (NPD) efforts at Embraer Aerospace. Each pair consisted of one project executed by a collocated team and one project executed by a noncollocated team. In all cases, the collocated team outperformed in metrics related to budget and schedule. The number of product changes required by the noncollocated teams was greater, which may have contributed to the longer schedules. In one case, the collocated team also outperformed the other in terms of customer and business related factors, such as customer satisfaction and market share achieved; in the other two, performance was similar.

Collocation is not without cost, especially if it is being implemented for the first time. Costs that should be anticipated and considered in a cost/benefit analysis include:

- Remodeling an existing space or constructing new space
- Increased overhead: collocation typically uses space inefficiently, due in part to the temporary nature of projects and the transitional requirements of project starting and ending
- Moving personnel and equipment
- Disruption of the move and ongoing inconvenience to other's outside of the group
- Opportunity cost associated with the reduced availability of team members to consult on earlier projects or to support ongoing tasks of their functional group

Collocation also poses organizational and personnel challenges that should be addressed:

- Functional managers' fear of loss of control/political issues.
- Separation from employees' functional areas results in loss of technical development and currency: organizations can form groups focused on certain technical interests, tools, or processes with the intent of sharing knowledge and resolving difficulties. These can cross functional boundaries, such as a CAD user group that includes members of product development and manufacturing, with the added benefit of building trust and communication. They can also cross organizational boundaries, allowing different locations of the same company to share their latest advances in focal plane arrays, for example.
- Feelings of insecurity and uncertainty of employees leaving a *permanent* home for a temporary project environment. Uncertainty is a demoralizer. The organization plans ahead for the inevitable end of the project and communicate this to the team members. If the project is a pilot for collocated teams, how the team is treated after the project will have a significant impact on perception of future efforts.

As the size of the team grows, the benefits of collocation diminish. There are examples of large-scale collocation. Embraer Aerospace built a new building to house a product team of about 600 engineers formerly spread around the world. Chrysler built a center to house all of its development activities, including support functions such as model shops and test labs. More akin to traditional project teams, each of Chrysler's platform teams occupied a dedicated space within the building. But groupings of this size do not fit the model or derive the benefits of collocation as defined here. The location of the model shop within the same building as the platform team is more like a strategic move by a supplier to be close to its customer. Within these large collocated organizations, there likely are examples of more traditional collocation models (as described here)—with a representative of the model shop *living* with the platform team, for example.

10.5.1 The Collocation Environment

The collocation workspace must be large enough to accommodate the core team with room to spare for growth, part-time members, visiting members, and common work areas. The layout should decrease individual space and increase common space and conference rooms. It may be possible to carve out enough space from an existing cube farm so that team members can be near each other, but this is not ideal. The space should be contiguous with open common areas populated with project artifacts, such as prototypes and equipment; QFD charts and wall schedules and whiteboards that draw people in.

Place team members close together; a good metric is that they should be close enough to hear each other's telephone conversations. Shore (2010) suggests this as a basis for arranging the team, grouping people "according to the conversations they most need to overhear." It is also promotes impromptu information-sharing if they can engage in a conversation without getting up. This suggests proximity as well as reduced visual barriers, such as low dividers or workstations radiating out from an open center.

Collocation and especially open environments raise concerns about noise and distraction. There is a genuine difference between private offices with doors and a more open environment, but that is the desired effect in a project team setting. In fact, people with offices should be encouraged to keep their doors open.

Collocation-friendly arrangements do not necessarily generate any more buzz than rectilinear cube farms, and the buzz is more likely to be of interest and value to team members than talk of unrelated projects or sporting events. Project managers and program managers may have more frequent and distracting conversations; they can be situated at the periphery but still close enough to be in tune with and part of the ambient environment (Shore). Small, *quiet rooms* can be used for making personal calls or engaging in sensitive discussions; conference rooms can be used for extended and focused discussions such as design reviews to avoid disrupting others.

If possible, the collocation space should be out of the way and inconvenient for non-team members to reach. As discussed previously, proximity promotes increased frequency and quality of communication in direct and indirect ways. A corollary is that isolation and distance discourages outside interruptions and allows team members to focus on the project. With the prevalence of electronic communication, locating the team in a distant wing of the building or even across the street is not the barrier to communication it once was. But the same sociological factors that are at play in collocation—bonding, familiarity, out-of-sight/out-of-mind mentalities, informal and incidental contact—will discourage outsiders from relying on team members and instead will encourage the use of local resources.

The logical center of a team's activity may change as the project progresses. Development of a prototype may be done in a skunk works environment, but as the product moves into a pilot production run, it may make sense to relocate the team closer to manufacturing, if it is remote. This can be disruptive, time consuming, and costly, so analyze the cost/benefit of the move.

Collocated teams may require more space than noncollocated teams, with an attendant increase in overhead. Since projects have a finite life, project team members will be moving into and out of collocation spaces. The timing of one project winding down and another starting up will never be perfect, so a certain vacancy rate should be maintained to accommodate both. Alternatively, the new project could delay its move into the collocation space, but the initial, fluid stages of a project are when it can benefit most from collocation.

A better alternative may be for the existing project group, presumably further along and with established trust and working relationships, to disperse. But such a disruption could happen at a critical time. Temporary workspaces, such as large conference rooms with whiteboards, could be pressed into service at these early stages. In general, if an organization standardizes on a more open environment for all purposes, it is easier to repurpose space for collocation with less waste.

Projects have a finite life, and teams will form and disband. If team members are relocated from a formerly *permanent* home, they may experience anxiety about their future after the project is complete. The organization should have a plan for this eventuality and share it with team members. The organization may choose to preserve the team member's original home. Minimally, it should assure them they will all have a home.

10.5.2 Partial Collocation

The ideal of collocation can be difficult to achieve: collocation has definite costs of implementation; it can be hard to justify full-time representatives of all functions on the team. In addition, the full project team may be too large to accommodate, and few organizations have the luxury of completely controlling product development and production. Partial collocation is an attempt to realize as much of the benefit of collocation as possible within these constraints.

The goal should be to keep as much of the core team collocated and intact throughout the project as possible. But there will be resources who are considered too valuable to dedicate to a single project, functional representatives such as purchasing who would not be fully utilized at certain stages, and extended members (customers, key vendors, design partners) whose chief role is performed elsewhere. The project manager must identify critical tasks and periods in the life of the project for which the expense and effort to bring them together are justified. No such time is more critical than the beginning of a project. The team is forming at the same time that important decisions are being made. The trust, team cohesion, and working relationships that are established will be the foundation of communication and collaboration as the team disperses and begins working in parallel, separated by time and distance.

Consider the kickoff of a product development effort. Critical specification and architecture decisions will be made at this time that will become more difficult and expensive to change as the project progresses. Friction and mistrust between marketing and engineering can be high, impeding the reception of each other's message, including details on why a feature is critical or how major benefits of that feature could be attained with a simpler design. Engineering may be unaware of a key vendor's capabilities, which could resolve some of these differences. Lack of involvement in the process or a belief that the specification and architecture are unstable may lead some team members to delay work on a feature, anticipating that the specification will change and invalidate their work. The immediate and long-term social and communication benefits of collocation would be most profound at this time.

This focus on intense, up-front collocation makes it easier to justify the cost of travel for remote members to temporarily join the team. The value of their involvement in these decisions and of the relationships built will both pay dividends throughout the project, as the remote team members return to their locations and become connectors and enablers for the rest of their team.

Other periods and events that may justify the establishment of a larger collocation environment include periods of rapid change or uncertainty in the project or environment, critical design reviews and decision points, and the integration of independently developed components. Each instance is an opportunity to do useful work and strengthen team dynamics.

The core team at the lead organization should be collocated. As much as possible, teams at remote locations—on the other side of the building or the other side of the world—should be collocated. Depending on the degree required and practical, as determined by time, organizational boundaries, and other impedances, these teams may be autonomous or extensions of the main team. If the teams fall within the same organization and have responsibility for a function, component, or module within the boundary of the main design, it is generally best to treat them as you would a local resource and not as a separate project; the latter can lead to working at cross-purposes and impede communications. The techniques of virtual teams can be used in these cases.

If the full-time presence of certain team members, such as marketing or purchasing representatives, cannot be justified, provide a workspace and other resources for the time they are with the team. For periods of critical involvement, they should relocate to the team space. Regular team meetings, design reviews, and informal meetings are opportunities to strengthen working relationships and reinforce team commitment and should be held in the primary team area.

Bring the team together on a regular basis if manufacturing and engineering are in separate buildings and daily stand-up meetings are still practical. Hold them at a central location to the project at that phase, such as the main co-working space early in design, the engineering lab during prototyping and testing, or the manufacturing floor during a pilot build. The decision on location should be purposeful and not personal; reasons should be clear so as to avoid resentments.

10.6 Dispersed Teams

Globalization is driving outsourcing and collaborations that span political and geographical boundaries. With increasing distance between parties comes increasing differences in time and culture (language, norms, and assumptions), all complicating the communication process. Technology can mitigate some of these issues, but it can create problems itself, as incompatibilities and security layers impede communication.

What may first have been a response to competitive pressure is now a potential competitive advantage. For an organization that overcomes these challenges and extends the concept of the team beyond its walls, "Geographic location is no longer the primary context in which to define and pursue business opportunities in support of strategic goals and competitive objectives" (Rad and Levin, 2003). The organization can utilize the best resources for a project, independent of location.

The issues and techniques discussed mainly will be in the context of a large-scale dispersion. But what if a team is only dispersed between different floors of the same building? The benefits of collocation discussed in the previous section diminish rapidly with distance and the number of people involved. Addressing the challenges of dispersed project teams is an opportunity to look at all forms of collaboration with open eyes.

10.6.1 Dislocation

Communications between members of dispersed project teams cross one or more boundaries: spatial, temporal, organizational, cultural, and technological. Each of these represents an impedance, which increases with the frequency of communication. Thus project work that requires close cooperation has a high degree of uncertainty or is subject to rapid change suffers the most.

A project may be organized in a way that minimizes the amount and frequency of communication across these boundaries. In a product development effort, responsibility for components is often divided between locations along architectural lines. Interface control documents and change control procedures minimize the need to interact during development, and most of the interaction is around those well-defined items and procedures. The section "Architecture as a Collaboration Tool" describes this strategy in detail.

Technology is a complex case, used in some cases to bridge gaps and in others to bridge them. These applications and challenges are discussed in the section "Technology and Communication."

In a collocated environment, common space is usually dedicated to various project artifacts and assets, such as wall schedules and prototypes. Informal communication and familiarity promotes greater information sharing and team dynamics. A project website, often implemented in conjunction with the tools discussed above, can help bridge these gaps in a dispersed environment, providing such features as:

- Photo and video galleries of prototypes, tests, other artifacts
- Live schedules, specifications, and other project documents
- Social networking features—team member pages with photo, contact information, biographical information, and a stream of activities and updates
- A map of the world that indicates the location of various team members, the time at each location, and whether it is normal working hours

Unfortunately, the Internet cannot do the same for all physical assets. For example, to meet the project's schedule, it may be necessary to reproduce expensive test equipment or prototypes at multiple locations. Alternatively, human resources can travel to physical resources, such as to assist with the integration of a module they developed with the main system.

Software and electrical engineers working on a hardware product can do some testing and development via simulation, but eventually they will demand hardware. Software engineers could access remote prototypes over the Web; a technician on the prototype end should be available to assist. Electrical or electromechanical fixtures can mimic interfaces and interactions with missing components (see the section on architecture). Or the project could provide duplicate prototypes at the necessary sites. This last option is most complete but requires an investment in time and money to create and maintain the asset.

Collocation derives many of its advantages from the positive effect of proximity on communication. One of the virtues of daily stand-up meetings is that attendees are standing up; it promotes brevity and focus. And attendees are often literally face-to-face, gathered around the engineering lab, a prototype, or other project artifact. This experience cannot be fully replicated with dispersed teams.

But do not mistake techniques often associated with collocation as deriving all of their benefits from collocation or as applying only to collocated teams. Other advantages of these meetings are the daily rhythm of coordination and accountability they establish and the presence of nearly everyone involved, so that many issues can be immediately resolved. Daily remote meetings, with a level (or lack) of planning and formality that matches daily stand-up meetings, can provide these benefits. Examine practices that are effective in regular team environments, understand the benefits and how they are derived, then adapt them for use with a dispersed team.

10.6.2 Time Differences

A difference of several hours between a lead organization and a design partner can be an inconvenience, especially for meetings and other activities that require the joint presence or coordination of personnel at each site. Add a key vendor a few hours in the opposite direction to the mix, and there is only a narrow window during normal business hours that makes synchronous, live communication possible. Locate manufacturing near the customer on the opposite side of the world, and serious communication compromises must be made. These circumstances are not unusual in today's global environment.

The natural difference in energy levels at different times of the day should be considered. While it may be 9:00 AM and a time of peak energy and productivity locally, it may be late afternoon for a colleague in another office. And that is a small difference; consider coworkers on the other side of the globe. A simple resource can help scheduling, planning, and understanding: compile a list of team members by location and relative time zone. This could be part of the team directory, discussed later, or hosted on a project website. A dynamic document could display the current local time, not just the time zone. A graphical representation, such as a map of the world with team members' locations shown, communicates additional information such as potential cultural differences.

At the extreme, synchronous communication can only occur if one or both parties participate well outside of normal business hours. This off-time collaboration should be used sparingly because of the disruptive effect it can have on morale, health, and the rest of the team and workday. Additional planning is required to ensure the time is productively used and that off-time participants do not become frustrated. This is discussed in the section on remote meetings, but such consideration also apply to less formal off-time interaction.

The greater the time difference, the greater the need to rely on asynchronous communication. Conversations carried out by asynchronous communications are not smooth. Facts can change between when a message is sent and received, affecting its meaning or validity. Errors, omissions, or confusion take an additional cycle to resolve. The more dependent each resource is on the other, the greater the drag that latency will have on the project. Because of the care that is required, the project communication plan should include protocols for asynchronous communications.

Email is the most familiar asynchronous communication medium; its drawbacks were documented earlier. More structured equivalents such as discussion forums are a better option, which maintain a centralized, ordered record of the exchange. Task-specific forms of asynchronous communication are even more effective because they change the focus from a back-and-forth exchange to doing the work. Co-authoring a document is an example. Rather than emailing changes and notes back and forth, the document is stored in a collaborative workspace that permits simultaneous editing. Changes are tracked automatically and can be reviewed and approved. Notes can be inserted in the document itself.

10.6.3 Language and Cultural Differences

Language is the most obvious and significant way that communication can differ between cultures. The inability to communicate freely, without technology or an intermediary translating, contributes to more than lack of understanding. It can create social distance between team members and can mask other significant deficiencies.

If communication between non-native speakers will play an important role, consider it in the planning process. Identify the languages spoken, the roles involved, the information to be shared, and the paths the information will take. Staff the team with this in mind. In engineering projects, technical expertise will usually trump language skills. If the needed technical and language expertise align, problem solved. More likely, someone in another role who speaks the language could facilitate as part of his duties. Alternatively, arrangements should be made for a resource from outside of the project team to be available on an as-needed basis; the resource should know the technical terminology of the project in both languages. Offer training to team members who desire it or should have some level of competency in it.

Technology can play a role. Software can translate the written and spoken word. But it can be difficult to express complex ideas or delve into detail, making creative tasks and problem solving difficult. And the translation dictionaries may not support the technical terminology of the project.

Style is another way that communication can differ between cultures. One example of this difference is between low- and high-context cultures. In low-context (often Western) cultures, most of the meaning is carried in the explicit code, the words used. Expression of thoughts, feelings, and information is clear and comprehensive. In high-context (often Eastern) cultures, much of the meaning is transmitted through inference, nonverbal cues, and pauses, often filtered through such factors as the situation, the environment, and the relationship between the parties (Samovar, Porter, and McDaniel, 2012). This complicates in-person cross-cultural communication; the lack of media richness in remote communication is even more challenging. Team members on both sides of the divide should be aware of these potential differences, seek clarification and confirmation if necessary, and be sensitive of the unintended or incomplete messages they themselves may be conveying.

Asynchronous, written communication—used to replace or complement synchronous communication—has some advantages in situations involving differences in language or cultural style. The lack of media richness forces the intended meaning to be made explicit and reduced to words. Accents are eliminated and translation on the fly is unnecessary. And more time to compose and process messages can reduce language errors, miscommunication, and misinterpretation. In addition to improving the effectiveness of the communication, it may increase the perceived similarity of team members by filtering out the cultural differences (Jarvenpaa and Leidner, 1998). A written follow-up to a live interaction can confirm mutual understanding.

Culture can also affect the reception and interpretation. Team members' expectations regarding everything from the role of the project manager to status reporting and the definition of success can be influenced by their culture. The project manager must look beyond his own experience to understand how these may vary and affect team dynamics and the performance of work. Hofstede (2010) defines a number of dimensions by which cultures can be compared, including:

- Power Distance: "The extent to which the less powerful members of institutions and organizations within a country expect and accept that power is distributed unequally."
- Individual Versus Collectivism: The degree to which individuals are coupled, whether "everyone is expected to look after him- or herself and his or her immediate family" or people are "integrated into strong, cohesive in-groups, which . . . continue to protect them in exchange for . . . loyalty."

- Uncertainty Avoidance: "The extent to which the members of a culture feel threatened by ambiguous or unknown situations."
- Long-Term Versus Short-Term Orientation: "Fostering of virtues oriented toward future rewards" versus "fostering of virtues related to the past and the present."

These variables could affect how team members in different locations perceive the project manager as a coordinator or a leader, an equal or a superior. The project manager likely cannot change this perception, but understanding it may exist enables him to better manage the project and relationships.

Evaluating success is another example of how these differences could manifest. A team from a more present-focused culture may be satisfied with meeting the minimum functional requirements set forth in the product specification. A team with a long-term orientation may be additionally concerned with customers' experience of those features and the subjective quality assessment they make.

10.6.4 Remote Meetings

The ease with which a project manager can hold regular meetings or call for an impromptu gathering is one of the strengths of a collocated team. Daily stand-up meetings in the engineering lab allow team members to quickly and broadly disseminate information and coordinate the day's activities. Manufacturing and marketing teams can attend weekly project meetings. Close proximity and face-to-face communication eases the burden of preparation, both in the lead up to the meeting and in the meeting itself.

Meetings involving remote attendees, some of whom may be in another country, require a higher level of preparation and coordination to be effective. The agenda should be exceptionally tight, adequate notice must be given so participants can prepare, and the meeting must be well managed. This is important for efficiency and to communicate to those who may be sacrificing personal time that the meeting won't be a waste of time.

The project-communication plan should identify the preferred medium for remote meetings and ensure that the necessary technology, training, and support are provided at each site. The setup should be tested in advance of each meeting, a procedure for attempting to *reboot* after a technical failure should exist, and a fallback method should be available. The primary medium may be a teleconference with screen-sharing software that ensures all participants are viewing the same slide of a presentation at the same time. A fallback in case the network connection between sites fails could be to outfit each location with a projector for local sharing, distribute the presentation in advance, and establish a protocol for synchronizing pages and passing control of the presentation.

Scheduling a remote meeting can be complicated by different time zones, lack of knowledge about the remote organizations' norms and practices, and technology incompatibilities. Therefore, they require a more proactive approach than checking online schedules and sending a meeting invitation. Reach out to representatives at each location to discuss times that work given these factors; it may be necessary for the representative to check the schedules of local attendees. Only after confirming a time that works for required attendees, send out invitations with the agreed-upon time. Because it is easier to neglect to participate in a remote meeting, confirm attendance and technical details in the days leading up to the meeting.

Given the difficulties in conducting and controlling a meeting remotely and the unusual time demands the meeting may place on some participants, an agenda is a necessity. The attention of remote participants is more difficult to hold and monitor, so the agenda should be narrow and specific (Rad and Levin, 2003). Likewise, the invitation list should be selective and reflect the agenda.

The time allocated for the meeting is likely to be more limited and difficult to extend because of scheduling rooms and resources and because it is easier for remote participants to call an abrupt end. Allot time for each agenda item and carefully manage the meeting to ensure you cover all necessary points.

It is difficult to know who the participants in a remote meeting are, especially as they may enter and leave at different times. This uncertainty can confuse or even inhibit free discussion. The meeting should begin with attendee introductions; entrances and exits should be announced. Ideally, any communication software for web-based meetings will have a feature that indicates presence. A diagram of the attendees by location, who belongs to what group, and what the local time zones are can greatly improve meeting dynamics. Speakers should address people by name, to involve them more deeply, identify them for other remote participants, and remind others of their presence.

Remote meetings are more difficult for the facilitator to monitor and control and for participants to follow than conventional meetings. A live agenda that highlights the current topic and displays notes can assist both groups. Transitions from one topic to another should be made clear.

Remote meetings lack many of the nonverbal cues that are a benefit of in-person meetings, and participants may be hesitant to engage in task- and process-related communication. The facilitator should actively solicit and provide feedback to ensure that remote participants remain engaged and their needs are met. The use of back-channels such as chat windows should be encouraged to provide process-related feedback (such as about the quality of the connection) and asynchronous information. For example, if a Web resource is mentioned during the discussion, post a link to the resource in a chat window.

Converging on a decision is difficult in traditional meetings, even more so in a virtual one. Divergent activities, such as brainstorming alternatives, are more engaging and less confrontational. Convergence requires commitment and has consequences. The impedances, noise, and technological limitations of remote communications make sensitive give-and-take negotiations more difficult and also require more structure. A physical separation makes it easier to focus on the needs of the individual or local team over the needs of the team as a whole. The facilitator must exercise more formality and control to guide the process, if not the outcome. Feedback is even more important during this process, ensuring that individual issues are addressed and buy-in happens at each step.

10.6.5 Using Time Differences to Your Advantage

Time differences are typically seen as a challenge to collaboration and outsourcing. Once the investment in infrastructure and experience needed to collaborate on a global scale are in place, having workers in multiple time zones can actually help firms reduce project costs and meet project deadlines.

It is common practice for an engineer to begin a complex, automated analysis at the end of the day so the results are ready for the beginning of the next workday. It is not uncommon for that engineer to return the next day only to find out the analysis failed to complete or produced useless results. The engineer investigates the failure, corrects the setup, starts anew, and hopes for the best. A day of productivity is lost—more than a day in schedule time—because the results were needed to guide a key decision.

An alternative would be to use a skilled technician in another time zone to monitor the progress of the analysis. Providing the worker with a troubleshooting checklist would assist him in recovering from most common errors. If the analysis completes, the technician can begin a new analysis based on a punch list or rubric and prepare a report of the results of the first analysis. As with all project communications, formality and consistency in the process will enhance the overall efficiency and effectiveness.

Not all tasks are suitable for time-sharing. Smith and Reinertsen (1998) discuss a company that had design teams working on the same project in multiple locations—San Francisco, London, and Tokyo. At the end of each local workday, one team would hand off the electronic work package to the next in line. Rather than resulting in three times the productivity, the practice resulted in three redesigns each day.

The same heuristics used to determine the suitability of a collaboration process apply here. There is clearly too much interdependence when different engineers are working on the same task of the same design. Support tasks such as creating detail drawings from part models are ideally suited; they are well-defined, loosely coupled to the parent task, and typically performed by another individual or department. This fits the communication model well—with feedback in the form of finished drawings that the engineer or designer can redline to start a new cycle. The analysis task described above is less defined but can still be reduced to a procedure. Certain tasks may even require 24/7 support, such as monitoring tests and collecting data.

10.7 Technology and Communication

The effects of distance and time, cultural differences, and other impedances on communication between dispersed teams were addressed earlier. Technology is a more complex case. We use email, Web conferences, and collaborative authoring to level some barriers, while using firewalls, access control, and encryption to erect others. Additional technological barriers are unintended: incompatible communication protocols and file formats, unfamiliar interfaces, and unreliable connections. The project should develop a communication strategy that consolidates the complexity of technology, simplifying management and minimizing these difficulties.

The project could standardize on a suite of tools from the same source and/or built on a common platform. The interface and protocol for any tool will be the same across all locations, the interface will be similar across all tools, and the tools will interact more freely. The suite can be part of an extranet or hosted by a third party; either way, the problem of secure access only needs to be solved once, and permission management is done in one place.

Figure 10.6(a) shows a scenario in which distinct, best-of-breed applications are used to meet the communication needs of the project. Each requires a unique account; the user interfaces are inconsistent; and protocols, authentication, and APIs vary, making interaction between the applications difficult. Figure 10.6(b) shows the same services delivered as part of a common platform. The platform may be proprietary or open, supported by a vendor or self-implemented. But a single account provides access to all of the services, consistent interfaces ease learning, and tighter integration

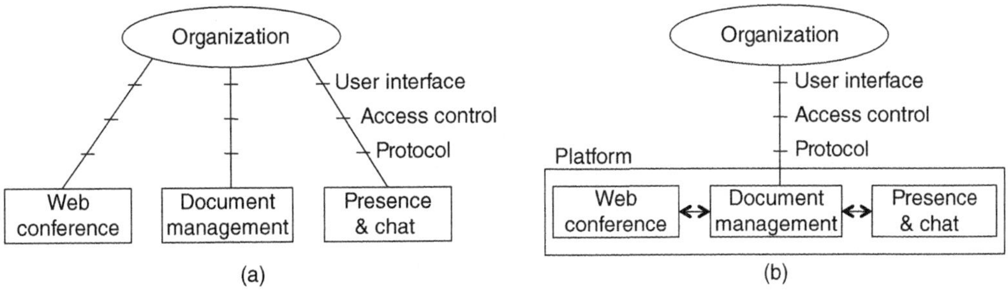

FIGURE 10.6 Containing complexity.

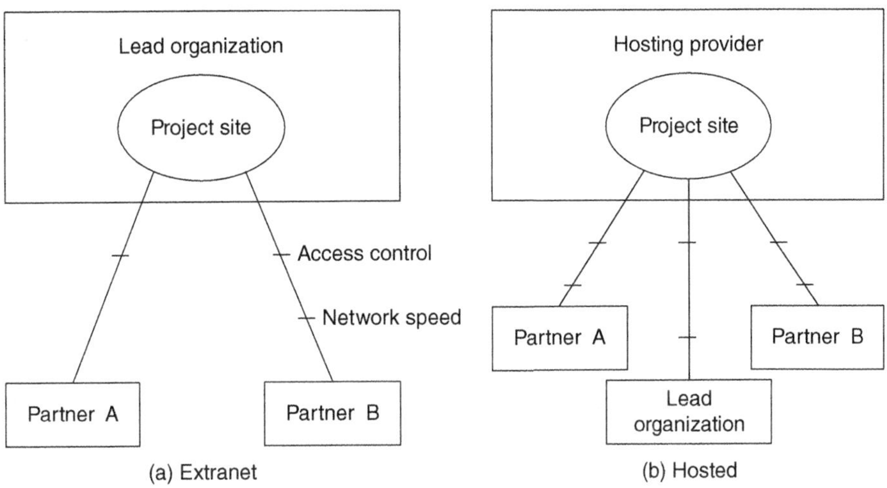

FIGURE 10.7 Extranet or hosted.

provides greater functionality such as reviewing a document from the library via Web conference.

If the chosen platform lacks a necessary capability, the project should standardize on a single solution for that capability, for all locations. Preference should be given to open solutions.

The platform implementation could be hosted by one of the organizations involved in the project, with access extended to partners via an extranet, as shown in Fig. 10.7(a), or it could be hosted by a third-party, as shown in Fig. 10.7(b), with all partners accessing it in the same way, albeit with one partner having greater permissions to manage the implementation. The extranet would be more costly to implement but offer a greater level of control. This does not necessarily mean it would be more secure because the burden of maintaining the site and applying security fixes would be on the organization. As do many firms that specialize, hosting firms often do a better job of delivering their core functions than a generalist organization. In the first case, there is likely a performance benefit for the organization that hosts the extranet, especially for functions that involve moving large amounts of data across the network.

The project communication plan should specify which technologies to use for specific purposes and provide for implementation and training. Two critical pieces of implementation are providing access to all who need it and testing that the infrastructure works, before either are needed in practice. Key operations, such as the sharing of proprietary CAD data with external partners, should be mapped out in detail so that all requirements are understood.

An implementation typically consists of the tools themselves; a set of databases, forms, and Web pages; and predefined user groups and associated permissions. Although the details vary, the structure is consistent from project to project. Many platform-based solutions enable the structure to be saved and reused for future projects. This reduces the cost of implementation and shortens the learning curve for future teams.

10.7.1 Project Websites

In a dispersed project, a project website can serve many of the same purposes as a common space in a collocated project, hosting photo and video galleries of prototypes, tests, and other artifacts; Gantt charts, schedules, and other project documents; and interactive features that reduce the social distance between team members who may never meet in person. Collocated projects can benefit as well, consolidating resources that would otherwise be spread about the network. A large part of the project website would be the tools discussed in the previous section; much of the rest will be implemented with the tools, such as creating a library with the document-management features. What remains may be as simple as a few Web pages to glue the information together.

The problem of providing security has already been discussed. A related issue is granting the appropriate level of access to specific data and functions on the project website to individuals and groups. A common tendency is to default to higher security and less access when there is a question; this should be avoided. Team members will only make so many attempts to find information they need before being discouraged by arbitrary security levels. People will make decisions without the necessary information and develop methods of bypassing the security, such as using uncontrolled tools and methods to store and access the information. (Pritchard, 2004) Understanding the security needs of the project is the subject of the next section.

The currency of information on the project website is another factor in determining its success. If the site is sporadically updated or contains outdated and incorrect information, team members will not turn to it as a resource. The project-communication plan should specify what information is to be published to the site, with what frequency or driven by what event. Technologies such as RSS (rich site summary) can allow team members to subscribe to updates to specific pages or sources on the site, so that they do not need to manually check for updates.

10.7.2 Security and Communication

Most projects will be working within an IT and security infrastructure defined or dictated by a parent organization. Unless a hosted solution is used (a decision that may itself be subject to approval), the project team must find a way to work within these confines. Each organization involved will have its own layers of security and bureaucracy, so concentrating complexity in one organization will minimize the effect of these barriers.

When in doubt, the tendency of many organizations is to implement higher security and greater restrictions on who may access what information. This hinders communication and encourages users to bypass security, perhaps developing parallel methods of storing and sharing information that are unknown to the organization and outside of its control, resulting in less security. For this reason, security decisions should not be ad hoc. The needs of the project should be understood, documented in the project communication plan, implemented, and enforced. Security training provided as part of the plan should include the why as well as the how.

Questions to consider include:

- What data needs to be protected
- How the data will be protected—technology, training, and procedures about handling, labeling, etc.
- Who needs what level of access to which data and ensuring they have the necessary tools and training
- Who is responsible for authorizations, when needed
- How is control and chain of custody maintained once information is *out of the vault*—email practices, encryption, etc.

Areas to consider include:

- Organizational security such as firewalls, access control
- Intellectual property such as filing invention disclosures to protect rights, nondisclosure agreements, marking of materials
- Export restrictions, such as International Traffic in Arms Regulations (ITAR) in the United States, which strictly limits what, how, and to whom defense-related information can be exported. The categorization of defense-related and the definition of *export* can be surprisingly broad.

10.7.3 Exchanging Engineering Product Data

All projects generate their share of data. But in nontechnical projects, that data is generally in the form of relatively simple, open, or de facto standards such as word-processing documents, spreadsheets, and presentations. Engineering projects, on the other hand, often generate volumes of complex data in closed, proprietary formats. There is no dominant, de facto standard for most types of engineering data.

This data needs to be shared along the value chain—from product development to manufacturing to service—and across the value chain, between design partners and analysts. Translation between formats typically results in some information—such as the design intent in a mechanical CAD model—being stripped out; it is not a lossless process. The losses are a form of noise, and the translation itself is an impedance.

Three-dimensional mechanical CAD is a typical example. It uses a sophisticated relational model, with spatial and temporal relationships between parts, assemblies, and drawings, which are represented as individual files. Temporal relationships refer to different versions of the same file: Assembly A that contains Part B, Version 1, is different than Assembly A that contains Part B, Version 2. Versions can also differ by internal data; Part B, Version 2, might different from Part B, Version 1, by the type, order, or parameters of features used to construct the geometry.

Translating CAD data will likely result in the loss of information, such as the features used to create the geometry of Part B; only the final shape will be transmitted. The identity of the files and associations between them will also likely be lost. If Assembly A, Version 1, and its component files are translated and provided to a vendor, followed by Assembly A, Version 2, with its updated Part B, each version is likely to be treated as unique.

Access is another issue to be resolved. The CAD data will typically be stored in an organization's product data management (PDM) or product lifecycle management (PLM) system, which understands the structural and temporal relationships among files and grants access based on a user's role or identity. Specific versions of files may be further categorized by lifecycle status, such as whether that version is obsolete for released to manufacturing (RTM). This status can be used to further refine access. If outside partners can access the PDM or PLM system, many problems are resolved. For example, outside vendors may be allowed to access that are RTM and that they are an approved vendor for.

If external access is not allowed, a hybrid method using automation and manual processes to integrate different services is required. Partners could be granted access to a document library. Cues inside the parent organization, such as the release of a new version to manufacturing, would flag a responsible person to copy the file to the document library. Pull-type notifications, such as an RSS subscription to the folder, could notify the vendor of the new version.

If design and/or manufacturing responsibility is divided among organizations with different product data management and/or CAD systems, processes outside of these systems must be developed for communicating what has changed and what the changes were and only exchanging the changed data. Avoid the situation where one party is uncertain what has changed, or even if there has been a change, but retransmits the data *for good measure.* Transfers of data should include a manifest that details the content of the data package, the relationship of the files, and what has changed between previous versions. This is especially important for mapping data between systems if an intermediate conversion is necessary.

Data conversion is an operation that is best pull-based rather than push-based. Figure 10.8 illustrates a scenario in which a customer shares the same CAD model with two different vendors. One vendor uses the same CAD system as the customer; the other uses an incompatible system. In Fig. 10.8(a), the customer publishes the native file and Vendor A retrieves it. The customer also converts the file to a neutral format, which Vendor B retrieves. In Fig. 10.8(b), the customer publishes only the native file, which both vendors retrieve; Vendor B converts the files itself.

The first scenario creates two different workflows for the same basic operation within the customer's organization. In the second scenario, each organization manages only a single workflow. The first scenario also creates uncontrolled data—the converted file—which also has any links to the original data severed. There is a break in custody until Vendor B creates a controlled version of it within its own data-management system. Finally, the file created by the conversion is likely to represent a lowest common denominator of common data, with the maximum amount of information stripped out. Vendor B could have tools that would convert the native file with greater fidelity, but the information is already lost.

Depending on the level of collaboration and interoperability required—viewing, redlining, one-way modification—conversion can be avoided altogether. Web-based and on-premises solutions that understand a variety of CAD formats can serve the necessary graphical representation for the purpose at hand and map markup or minor changes back to the original data sets. Some of these systems can also convert files on the fly, choosing the best target format based on the recipient's tools. They may also understand CAD versions and the relationships between files, serving complete data sets based on the permissions of the user.

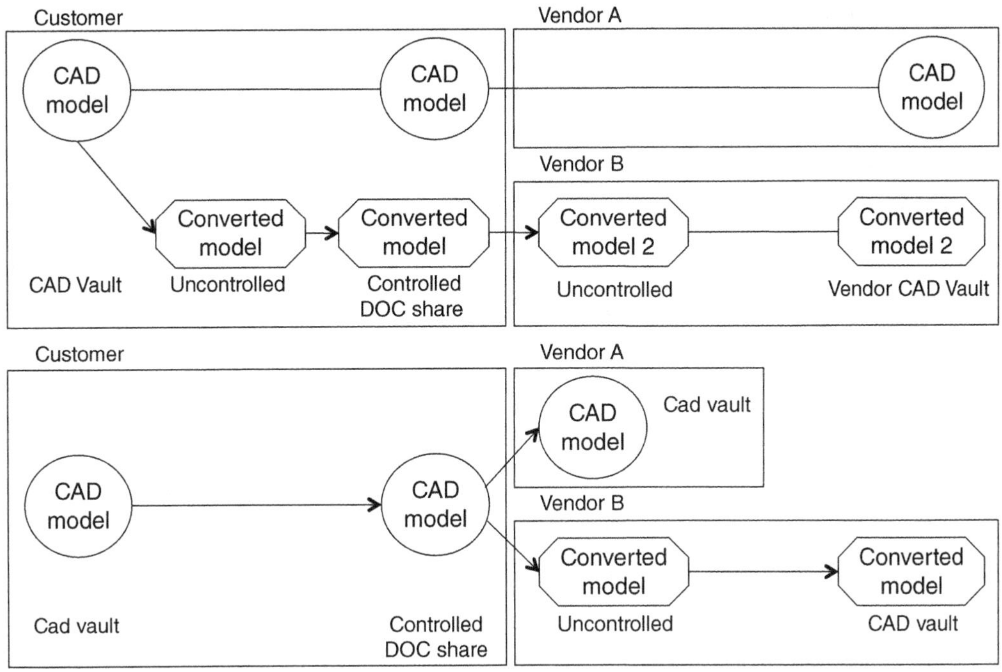

Figure 10.8 Exchanging CAD data.

10.8 Architecture as a Collaboration Tool

System architecture and interface specifications play an important role in any product development effort, but even more so when the development resources and responsibilities span organizational boundaries. Interfaces are typical failure points in products and the projects undertaken to develop them. When used properly, interfaces also can be points of communication, stability, and control.

Architecture is the assignment of the system's functions to various components and subsystems, the arrangement of those components and subsystems, and how they interact with each other. Ulrich (2000), in a mechanical context, refers to those components or subsystems as "chunks," but it is useful shorthand for describing the interaction of mechanical, electrical, and software components as well. If individual chunks perform a small set of functions without interacting with other chunks, except through well-defined interfaces, it is said to be a modular architecture. If the performance of a given function depends on the interaction of multiple chunks, and a given chunk tends to participate in multiple functions, it is said to be an integral architecture.

The tight coupling of an integral architecture tends to result in higher performance and lower cost (in volume), but requires similarly tight coupling in the design process. Both of these factors result in inflexibility and a high cost of change in terms of time and money. The robust interfaces in modular architectures increase cost and reduce performance, but maintain flexibility in design and implementation. As Smith and Reinertsen (1998, location 1890) state, "A good interface isolates design decisions in one module from those in others." Recall that communication impedances tend to occur at boundaries, so

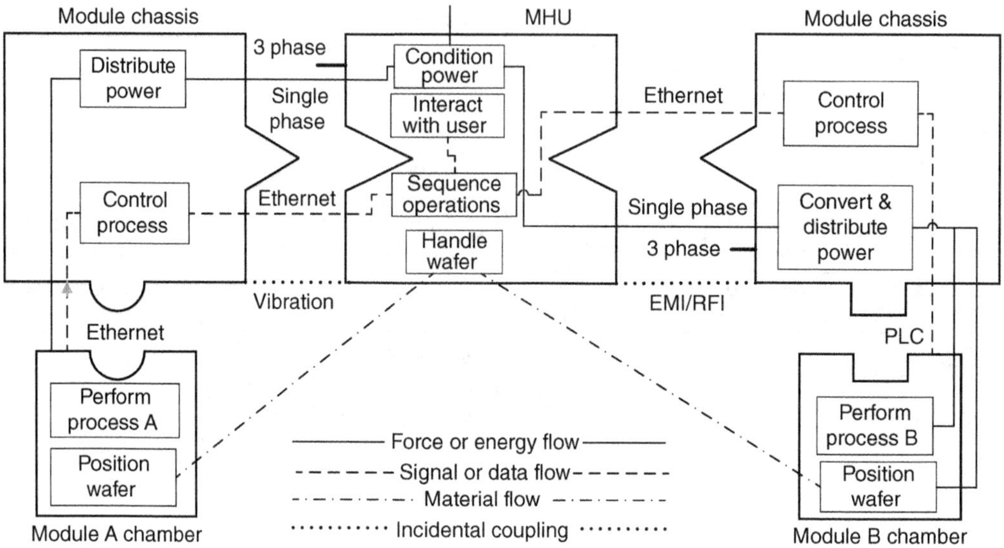

FIGURE 10.9 A system architecture.

these interfaces lend themselves naturally to be used as organizational boundaries by reducing the need for communications to cross them. Communication that is required tends to be narrowly focused on the interfaces themselves. This allows the development of subsystems in a modular architecture to proceed largely in parallel. This is especially powerful with dispersed teams with time and technology mismatches.

Figure 10.9 presents one possible architecture model for a semiconductor wafer processing system in which wafers are passed by a central MHU through a number of processes, each handled by a different module. The interlocking shapes on each element represent mechanical interfaces that secure and align mating components. This is an example of a bus-modular architecture, with modules A and B able to interface with either side of the MHU; other common architectures include slot-modular and sectional-modular. The connecting lines represent fundamental couplings—transfers of energy, data, and material related to the operation of the system—and incidental couplings—EMI/RFI, shock, and vibration as well as other unintended effects that must be considered.

10.8.1 Developing the Architecture

Architecture is a logical structure, not just a physical one. In a complex system, it can involve mechanics, electronics, and software, independently or operating in concert. In general, architectural decisions that involve software, electronics, or mechanics are increasingly difficult and expensive to change, in that order. Implementing more capability in software and electronics can yield more flexibility when managing the project.

Smith (2007) identifies five key architectural decisions:

1. System Boundary: Where does the system end and the rest of the world (ROW) begin in terms of its footprint, inputs, outputs, and interactions with the environment? Is the power conditioning part of the system or the function of a

separate supply? Will the system interact with the environment directly or through some intermediary? For example, a material handling system could operate on material directly, or it could be palletized.

2. Modularity: Will the system be highly integrated or modular? This is typically a trade-off of high-performance/low-cost versus flexibility. That flexibility enables more concurrent development and freedom in scheduling the project and distributing the work.
3. Location of Functions: In a modular architecture, functions typically will not cross boundaries. Minimizing the functions that each chunk performs further increases modularity, but not without additional costs as discussed above. And there are a number of reasons for grouping similar and even dissimilar functions, including geometric precision, vendor capabilities, and localizing change.
4. Interfaces: These provide support and coordination between modules of a system and between the system and the rest of the world. When project work is organized along modular lines, interfaces also allow work on both sides of the interface to be performed independently. To serve both purposes well, interfaces should use standards when possible, be stable, and be robust. The next section will discuss stability and change control. Robust interfaces are less likely to require change to accommodate errors or change in scope and handle unexpected conditions. Some characteristics of a robust interface include:
   - Operates acceptably with a wide range of input (voltage, frequency, air pressure)
   - Has a narrow range of output
   - Has spare signal paths to account for omissions or expansions
5. Flexibility: An architecture that maintains flexibility can speed time to market by allowing development to proceed before all market conditions, technology questions, or supply-chain issues are sorted out. A certain game console went to market with support for an external HD optical media drive before the battle for that platform dominance was settled; it also had built-in support for lower-quality streaming media. The optical media standard it chose lost the platform war, but ultimately optical media in general lost out to streaming. In the end, the console was not burdened with the high cost of an unwanted feature.

A subsequent section will discuss organizing the project along architectural lines to enhance communication and improve efficiently. One should be careful not to do the opposite—define the architecture to put available resources to work immediately. This can be particular problematic if a large project is staffed quickly before adequate planning and conceptual work is done. The boundaries should be based on technical and strategic factors.

10.8.2 Change Management and Architecture

Interfaces are key product documents as well as communication tools. They should be documented in formal, controlled ICD. These ICDs may take the form of text documents, 2D drawings, 3D models, or some combination.

Interface specifications should be developed with the involvement of all affected by the decisions, but the final specification can be owned by only one organization. There should be a project-level, inter-organization change process that notifies and involves all parties affected when there is a proposed interface change. It should provide a mechanism for parties that do not own the interface to request a change.

In the agile project management methodology, architecture is an ongoing activity that must be examined at every design iteration to account for uncertainty at the beginning of a project. Architecture evolves and can have a different meaning and purpose at different stages; at the early stages, a skeleton architecture framework may be enough to organize the people and work. But it should be subject to a change-control process that takes into account costs associated with changes, especially as specificity is added to the interfaces.

Highsmith (2010) notes that effective change management "is not a matter of control—who gets to make the decisions—as much as it is a matter of coordination—making sure the team understands the total impact of a change and which groups are affected." Inadequate architectural planning will make it difficult for projects to identify the scope of change and react accordingly. Highly coupled architectures will require frequent changes and coordination and may indicate a need to revisit the architecture.

10.8.3 Organizing around Architecture

Figure 10.10 presents two possible ways that the project could be organized around the architecture shown in Fig. 10.9. In this example, the lead organization's expertise is integration and providing an end-to-end solution to their customers. It is partnering with subcontractors A and B, who are best-in-class in their fields, to provide those capabilities. The first structure, in which the lead is responsible for all of the module chassis, may have economies of scale. Recall, however, that organizational boundaries are areas of high impedance. Bureaucracy, tool compatibility, and time-zone differences are some of the challenges that can separate two organizations trying to collaborate.

In either project structure, each subcontractor only has to design to a single external interface. But in the first structure, the lead must manage two different interfaces. In the second structure, both the lead and the subcontractors only need to collaborate around a single interface, which is the same interface for both subcontractors; investments in problem-solving and optimization for one also will benefit the other.

The interface between the module chassis and the MHU is largely standards-based, so it is likely well understood and relatively stable. The interfaces between the module chassis and each process chamber likely will be unique and involve areas of active

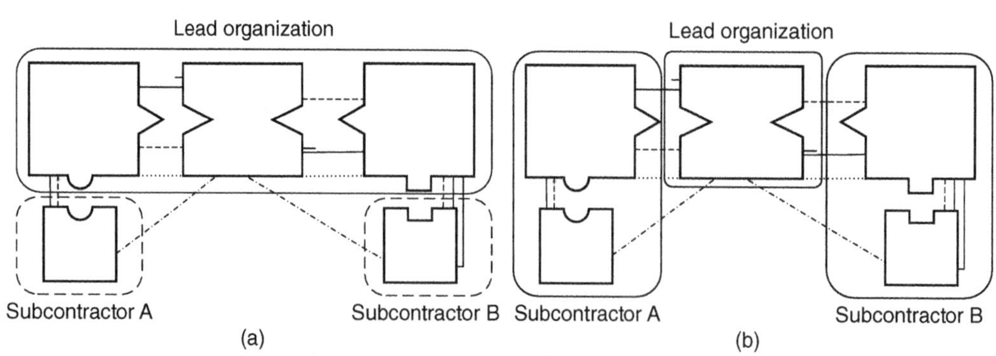

FIGURE 10.10 Organizing around architecture.

design with a high probability of change. The latter would require greater communication between the responsible parties. The effects of impedances increase with frequency; the second structure reduces the need for high-frequency communications to cross the high-impedance organizational boundaries.

The previous discussion has revolved around multiple organizations collaborating. But architecture and modularity is an important consideration even when a single organization is responsible for the design. Individual contributors and sub-teams can still benefit from the clarity and de-coupled decision making that a clean architecture provides.

10.8.4 Integration Risk

A risk when using modular architecture to organize projects, especially when different organizations have responsibility for different modules, is that certain problems will not be uncovered until they are brought together as a, hopefully, operating whole. This often occurs as some important deadline looms, without the direct assistance of each module's developer. This is an example of integration risk.

Robust interface specifications and change control procedures are intended to minimize integration problems, but omissions, assumptions, and other errors come into play. Common problems include:

- Communication mismatch (protocols, frequency, or noise)
- EMI/RFI
- Power requirements and sensitivity
- Mechanical fit and range of travel or adjustment
- Cosmetic mismatch
- Mirroring of interfaces (left-to-right, male-to-female)
- Static specifications (weight, center of gravity)
- Operating specifications (accuracy, repeatability)

Early and frequent integration, real or virtual, is a key to surfacing and correcting these issues before they become showstoppers. Integration can take different forms, depending on the type of project or module, logistics, and the stage of the project, such as:

- Software builds
- Software simulations and mockups
- Test and integration fixtures that replicate mechanical, electrical, fluidic/pneumatic, and other interfaces
- Exchanging geometry envelopes
- FEA, multi-physics, and motion simulations of combined model
- Maintaining common tracking documents, such as weight or error budgets
- Exchange of physical interface components

Frequency of integration should depend on the application. The purpose of designing to interfaces is to reduce the need for frequent communication that crosses impedances. Nightly builds are a common practice in software development, but the equivalent would be a burden for mechanical models.

Consider the mechanical interface that aligns the module chassis to the MHU in Fig. 10.9. The MHU and the interface are developed by the lead organization; the module design may be subcontracted out. A solid CAD model and a proxy representation of the module, which is supplied to the subcontractor, represent the interface. The subcontractor would incorporate the necessary details of the proxy model into its design, perhaps even recreating the details if the CAD systems are incompatible. Referring to the communication model, the subcontractor would provide feedback confirming that the message has been understood by sending the geometric envelope of its module to be integrated with the customer's, thereby closing the loop. This should be done early and then updated at fixed intervals or at significant points throughout the design.

Another source of integration problems are incidental couplings, such as EMI/RFI, thermal effects, and dynamic effects such as shock or vibration. These can be difficult to predict via analysis and other tools because the loose coupling between modules, literal and figurative, can result in nonlinear and indeterminate behavior. The project plan should identify these risks and include them in the mitigation plan.

One should not overemphasize the integration risk of a modular architecture over an integral one; in a sense, they are just different degrees of the same thing. A modular architecture has the advantage that many functions can be tested fully before integration, while it may not be possible to test most functions of an integral architecture until the last piece is ready.

10.9 The Project Communication Plan

The project manager's principal communication role is as planner and facilitator. The PM is responsible for identifying the communication requirements of the project, including by whom and how will they be performed; enabling those responsible with the technology and training to satisfy those requirements; and ensuring that the system functions and that the communications are made. The project communication plan is a tool to help the PM and the team achieve these goals.

The plan identifies stakeholders and their information requirements, the most efficient and effective means for communicating with them, the individuals responsible for specific communications, and the expectations (format, frequency, and distribution) for specific communications. It establishes protocols and formats for specific communications (meeting minutes, status reports, change requests) as well as communication forms (for example, the use of email for project-related communications, as discussed earlier). And it identifies the resources needed to enable these communications, including training and technology.

The project communication plan has three components:

1. Stakeholder registry and team directory
2. Communication protocols
3. Activities and resources

10.9.1 Stakeholder Registry and Team Directory

The communication plan should include a stakeholder registry and a team directory. The content and structure of both are similar; they differ only in their intended use, their audience, and the data fields stored for each entry. The fields in the stakeholder registry will be a superset of those in the directory, so a single database could be maintained for both purposes. In the following discussion, "directory" refers to both the team directory and the stakeholder registry; explicit references to the stakeholder registry apply only to it.

The directory could take the form of a simple list maintained as a spreadsheet, but it would be much more useful as a database that can be searched, sorted, and filtered. The team directory should bridge the gap that is often left when organizations with different information technology (IT) infrastructures collaborate, comprising internal and external team members, stakeholders, and other resources. Ideally, it would be accessible from inside or outside the lead organization. Alternatively, it could be maintained by the lead organization and published at regular intervals as a static document for external partners.

The security concerns for sharing data between organizations discussed earlier apply to the directory, whether it is a static document on a shared site, a Web page on the project site, or a Web application. Because of the sensitive and political nature of some of the information in the stakeholder registry, access control would be required even within the lead organization, limiting which entries or data fields could be viewed, depending on the user's permissions. This would be extended to all users outside of the organization. If the directory or stakeholder registry is a static document, abridged versions could be published for different audiences.

The existence of simple resources such as a phone directory is one reason for improved communication within collocated teams; the team directory addresses that for dispersed teams. Improved social dynamics—due to familiarity, informal interactions, and incidental relationships—is another. The directory can help bridge this gap as well by linking to social network-like profiles maintained by the team members themselves, including photographs, presence indicators, and personal information and status updates the member chooses to share.

Table 10.2 lists data that could be included in the directory. The country and time zone are particularly useful for projects that span those boundaries. If a database system or application is used and allows it, displaying the current time in the listing would be more powerful. Other visual cues such as color could be used to indicate the relative time of day.

10.9.2 Communication Protocols

The communication plan should catalog the specific communications and general communication tasks that need to be performed in the project and how they will be performed—by whom, frequency, distribution (specific recipients, internal or external, etc.), medium, structure and content, responsibilities of the receiver, and security (e.g., who may see it, stamp proprietary, do not export). Protocols for using specific media, such as email or Web conferencing, should be specified as well. The level of detail required for each entry may vary from a single line in a table to extensive description, as was discussed in the section on email. Typical information specified includes:

- Communication name
- Purpose or application of the communication
- Content and level of detail
- Message format: guidelines for subject line and replies of emails; headings for a status report
- Recipients or audience

Directory Fields
Name
Organization
Role
Contact methods and addresses
Preferred communications method
Best time to contact by method
Frequency of communication
Country and citizenship
Language(s)
Time zone and/or local time
Security notes and restrictions
Additional Stakeholder Fields
Primary contact/owner of the relationship
Secondary contact/owner of the relationship
Interests
Influence
Management strategy
Restrictions on contacting the stakeholder

Table 10.2 Directory Data Fields

- Producer/responsible person
- Time frame and frequency, e.g., at regular intervals, event based (e.g., preliminary design review), as the need arises (e.g., test reports)
- Medium, e.g., meeting or announcement
- Security issues, e.g., encrypt; stamp proprietary; do not export
- Data retention and archiving

Most communication needs and methods can be anticipated during the initiation and planning phases of the project. Brainstorm, interview, or survey stakeholders and key team members and review past project successes and failures. Consider inbound communications, internal communications, and outbound communications. Referring to the communication model, pay special attention to where communications encounter impedances and noise. Some of the methods may prove ineffective, and other needs could be uncovered as the project progresses, so monitoring communication effectiveness and updating the plan should be two of the communication-management activities, discussed next.

10.9.3 Activities and Resources

While project managers' responsibilities may include specific communications, their most important communication-related function is to facilitate the communication of others. Some of this communication is formal, such as meeting minutes or status reports. But the vast majority of it is the informal communication that is incidental to getting project work done.

The project manager must identify the resources needed to enable these communications. High-level needs can be drawn directly from the communication protocols already established. But determining lower-level details will require further review of past projects and interviews with team members and support personnel. Resources may include permissions and approvals from senior management, communication technologies, and training on technologies and protocols. Time and money must also be allocated for implementation and training.

Critical communication systems, such as sharing CAD data with external partners, should be tested before they are needed. Special attention should be paid to sources of impedance and noise, such as when communications cross organizational boundaries or CAD data is translated to another format. The health and effectiveness of systems and communications should be monitored throughout as well. The methods should be refined and supplemented as necessary.

10.9.4 Stakeholders

The *PMBOK* defines stakeholders broadly as, "Persons and organizations such as customers, sponsors, or the performing organization, and the public that are actively involved in the project, or whose interests may be positively or negatively affected by the execution or completion of the project." This can be broken down into a registry of individual stakeholders, but an intermediate categorization of three groups is more practical for planning purposes: Influences, active stakeholders, and passive stakeholders. The common trait of these groups is that their interests are affected by the project; their difference is the impact on the project.

Stakeholders should be evaluated based on their level of interest in the project and their ability to affect the outcome, through influence of direct action (their power). Develop a strategy to manage the relationship with each stakeholder accordingly. Figure 10.11, adapted from the *PMBOK*, maps the level of interest and power to general strategies.

Influencers can leverage direct power, such as budgetary or resource allocation, or indirect power, such as noncooperation or a position of respect in the organization, to help or hinder the goals of the project. Their needs typically come down to three primary questions:

- What is the information?
- In what form is that information?
- Under what circumstances does this information occur? Does it occur at scheduled times or in response to certain events or changing conditions?

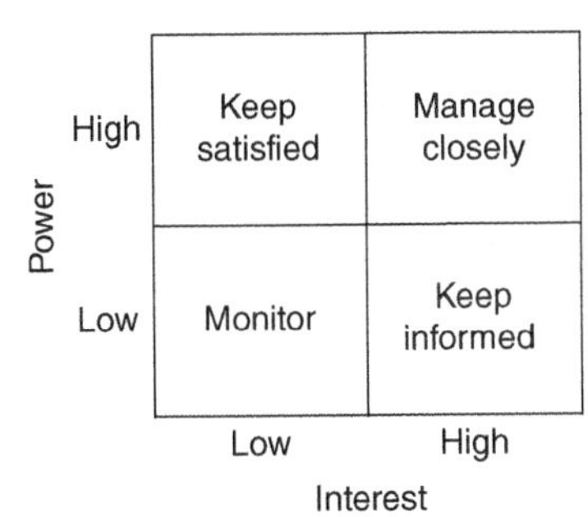

FIGURE 10.11 Stakeholder management strategy.

Because of the influencers' potential impact on the project in response, the project manager is not simply a passive conduit for information; he or she will be actively managing stakeholder expectations.

Active stakeholders perform the work of the project: engineers in the lead organization, design partners, key vendors. Active participants have these same needs as influencers to be aware of project goals, status of the project, and urgent news about the project. But not all of their information needs can be anticipated. It is more a question of what communication tools and processes they need to perform their work and what is expected of them to support their peers' work and the project manager's efforts to coordinate the project. The project can break down if there is a failure in either of these facets. The project manager needs to enable the communications and ensure expectations are met.

The work of the project will affect passive stakeholders; often these members are the reason the project exists in the first place, despite not having an active hand in the work. Think of these passive stakeholders as a constituency—voters who can determine the ultimate success or failure of a project long after the work is done. Surrogates with a more active role, such as the marketing department or focus groups, could represent the interests of passive stakeholders.

Depending on the nature of the project, the ultimate customer could be passive, such as the purchaser of a consumer electronic device, or have significant influence, such as a government agency tasked with selecting the winner of a competitively bid proposal.

References and Bibliography

Highsmith, J. A., *Agile Project Management: Creating Innovative Products*, 2nd ed., Kindle Edition, Addison-Wesley, Upper Saddle River, NJ, 2010.

Hofstede, G., *Cultures and Organizations: Software for the Mind*, 3rd ed., Kindle Edition, McGraw-Hill, New York, 2010.

Jarvenpaa, S. L., and D. E. Leidner, "Communication and Trust in Global Virtual Teams," *Journal of Computer-Mediated Communication*, 3.4: 1998; Web: May 2013. <http://jcmc.indiana.edu/vol3/issue4/jarvenpaa.html>.

Mittleman, D. D., R. O. Briggs, and J. F., Nunamaker, Jr., "Best Practices in Facilitating Remote Meetings: Some Notes from Initial Experiences," *Group Facilitation: A Research and Applications Journal*, 2:6–15, 2000.

Optiz, C., and J. Wimmers, "Challenges and Best Practices in International Projects," Thesis, *Fachhochschule Für Oekonomie & Management*, 2010.

Pritchard, C. L., "The Project Management Communications Toolkit," Artech House, Boston, MA, 2004.

Project Management Institute, *A Guide to the Project Management Body of Knowledge: PMBOK Guide*, 4th ed., Kindle Edition, Project Management Institute, Newtown Square, PA, 2008.

Rad, P. F., and G. Levin, *Achieving Project Management Success Using Virtual Teams*, J. Ross Pub., Boca Raton, FL, 2003. Compiled HTML help format.

Rechtin, E., and M. Maier, *The Art of Systems Architecting*, CRC, Boca Raton, FL, 1997.

Samovar, L. A., R. E. Porter, and E. R. McDaniel, *Communication Between Cultures*, [Independence]: Wadsworth Cengage Learning, Boston, 2012.

Shina, S. G., *Concurrent Engineering and Design for Manufacture of Electronics Products*, Van Nostrand Reinhold, New York, 1991.

Shore, J., "Sit Together," *James Shore: The Art of Agile Development*, New York, 2010; Web: 2013. http://www.jamesshore.com/Agile-Book/sit_together.html.

Smith, P. G., and D. G. Reinertsen, *Developing Products in Half the Time: New Rules, New Tools*, Kindle Edition, John Wiley & Sons, New York,1998.

Smith, P. G., *Flexible Product Development: Building Agility for Changing Markets*, Jossey-Bass, San Francisco, CA, 2007.

Ulrich, K. T., and S. D. Eppinger, *Product Design and Development*, Irwin/McGraw-Hill, Boston, MA, 2000.

Walsh, J. P., and N. G. Maloney, "Collaboration structure, communication media, and problems in scientific work teams," *Journal of Computer-Mediated Communication*, 12(2): article 19, 2010; Web: http://jcmc.indiana.edu/vol12/issue2/walsh.html.

Discussion Topics

1. List three communication-related responsibilities of the project manager. Give an example of each in practice.
2. List two communication difficulties associated with dispersed teams and a method to mitigate each.
3. Give an example of a stakeholder who could have a positive influence on a project and another who could have a negative influence on a project. How might you manage the relationship with each?

CHAPTER 11

Engineering Project and Product Costing

Keeping track of project and product costs are important aspects of the project manager's (PM) job. The project cost is estimated at the project inception and should be tracked via the earned value (EV) system discussed in Chap. 7. The most important element of keeping project costs on target is maintaining the project schedule. A project that is delayed means that the project team is getting paid beyond the project budget. Since the majority of expenditures in most high-technology projects are the engineering team salaries, a delayed project results in excessive project cost.

An accurate and up-to-date product cost estimate is important, since it has an impact directly in the ROI calculations for the project. The product cost estimate should be reviewed at each phase of the project cycle. This chapter will address various methodologies for cost estimating models for new products and contingency estimates for different levels of potential sales. The topics of this chapter discussion involve the following sections:

11.1 *Project and Product Cost Relationship with Lifecycle Stages.* In this section, the relationship of project and product costs is outlined for different stages of new product lifecycle development.

11.2 *New Product Cost Estimating Methodologies.* In this section, the basic elements of cost structure and new product cost estimating methods are outlined. Cost estimating models such as ABC costing and other technology-based cost models are discussed. Issues of quality and future sales and their impact on the product cost estimates are explored.

11.3 *New Product Cost Estimating Process.* In this section, the complete process for estimating new product cost is outlined, with provision for modifications of the cost strategy based on alternate sales volume estimates.

11.1 Project and Product Cost Relationship with Lifecycle Stages

The manufacturing costs of technology products are highly dependent on their lifecycle stage as outlined in Chap. 2. The following is a summary of the project and product cost and price profile for each stage.

11.1.1 The Start-Up Stage

The first stage is called start-up or market development, where the emphasis is on the performance of the product. Product development is marked by quick time to market, with minimum concern over manufacturing cost. Extensive inspection and test processes are used, resulting in additional manufacturing cost because of poor attention to design for manufacturing (DFM) issues as well as many engineering change orders (ECOs).

This stage is marked with high project costs, as much as 40 percent of the expected yearly sales volume, resulting in a high return on investment (ROI) of >18% or a return factor (RF) minimum of 15. The selling price of the product bears no relationship to the manufacturing costs. This is typically used when the customer's perceived value of the product benefits is high in terms of productivity or utility of the product. The company can charge whatever the market will bear but has to keep an eye on the time it will take the competition to develop competing products, using their track record of new product introductions. Part of this strategy would be the planning of a follow-on family of similar products. The PM needs to think forward to the next generation of products and the timing of their introduction into the market:

- How would the PM and marketing plan the rollout of follow-on products, and what time frame should pass before introducing the next-generation product?
- What is the track record of the competition, and how quickly will they manage to react and create competitive products?
- How much will patents and intellectual property issues delay or hamper the competition?
- What will the price-performance curve for the product family look like?
- Does the product offer additional sales in terms of consumables and additional content?
- When will the start-up stage end and the product lifecycle proceed to the next stage?

The new start-up product selling price can be determined by making a thorough analysis of the issues above. The product costs are not important at this stage, since the manufacturing costs constitute a much smaller percentage of the selling price.

Examples of these strategies for products in the start-up phase abound. Apple's iPAD is a good example of a product that came out at a fairly high price. The competition caught up quickly with lower-cost products, which forced Apple to travel down the price performance curve and offer a smaller physical version with lower manufacturing costs. Apple utilizes *premium pricing* strategy, which is to set the price high to reflect the exclusivity of the product. Other examples of this include Porsche and Rolex.

At the opposite end, the Amazon Kindle came at a low price compared with its manufacturing cost, but with the potential of additional revenue from providing customers with extra content in the form of e-book purchases. Amazon is using the Gillette *razor blade* model where the company distributed the razor for free and sold the razor blades, realizing a high markup.

Another classic example is the HP-35, the world's first pocket scientific calculator, which was marketed in the 1970s. The company had additional follow-on products for calculators in the financial market and even a watch calculator. While the HP-35s were snapped up originally at $395 each in 1972, many other start-ups as well as other large companies entered the market shortly afterward, and the price plummeted to less than $10 each in less than a few years. Needless to say, calculators are a commodity market now, and HP obsoleted the HP-35 three years later.

11.1.2 The Growth Stage

The competition in this stage is in the differentiating aspects of the product features. These include the customers' long-term cost of ownership such as quality, warranty, field service, and the growth of ancillary products. Product development during the growth stage is characterized by the focus on introducing the DFM guidelines of capabilities and constraints and beginning to concentrate on manufacturing as a strategic weapon to emphasize low cost and high quality. Working with suppliers is increased and major parts of the product manufacturing are outsourced.

In this growth stage, the project costs begin to level off in the range of 10 to 25 percent of the expected yearly sales volume, resulting in a high ROI (>12 percent) or RF minimum of 10. The financial indicators can be more accurately estimated since the market potential is more realized. There is increased emphasis on controlling manufacturing costs in order to lower the sales price. The use of manufacturing tools such as Six Sigma and DFM are more prominent in order to control product quality and cost. Examples would be the competing technology and characteristics of products such as Kindle and iPAD for e-readers, and video game consoles (Microsoft Xbox, SONY PlayStation, and Nintendo Game Cube), which use alternate microprocessors and graphics chips from AMD and Intel.

11.1.3 The Maturity Stage

This phase is characterized by the emergence of a dominant technology or technique for the product design and manufacturing. At the same time, the relative growth of the market is low, proportional to the growth of population or customer base. The competitive emphasis in this stage is on price and quality, as the dominant technology does not allow too much variation on the basic design or manufacturing processes of the product. Product development in this phase is focused on continued improvement in manufacturing process costs by improving yields and reducing test times through the quality tools of control charts, continuous quality improvement, and robust processes. The suppliers for the product are involved early and often for DFM purposes, and the design process is more robust through the use of analysis and simulation tools.

In this maturity stage, the project costs become critical, in the range of 4 to 8 percent of the expected yearly sales volume, resulting in a medium ROI (8 to 12 percent) or RF minimum of 5. The use of the supply chain is critical in terms of sharing the development costs of new features in the dominant standards or technology. Several possible outcomes in the project/product price would follow, since the market is changing, and the competition in price and performance becomes heightened:

- The company does not realize that the maturity stage of its core products requires a different profile in project and product costs. The company continues to develop products using project and product data based on the ROI for the start-up or growth products. This leads to financial losses, and the company could become bankrupt or forced to be sold to a competing company. The high-technology history is full of dominant companies in their field that were not able to thrive as the market matured. An example would be those in the minicomputer market (DEC and Prime), workstations (Sun Microsystems and Apollo), and personal computer markets (Compac and IBM PCs). As new disruptive technologies and revolutionary products such as laptops and tablet computers were developed to compete for their traditional customers, their market share was eroded over time, and the companies ceased to exist.

- Alternatively, as a company reaches the limits of technology improvements in its core products, the company cannot continue to operate within the price performance curves. The higher performance is not easily available (or requires expensive new technology development), and the lower-cost competitors occupy the lower performance curve. The company is stuck at its level of price and performance and rapidly becomes less profitable as the lower-cost providers take away market share. An example would be a company providing diagnostic medical products using ultrasound technology, where the next performance upgrade would be MRI technology.
- The company might decide that it cannot internally compete in the design of new products at lower project and product costs, and it will begin to rely on technology acquisition by buying up companies with advanced technology or manufacturing expertise. The company might concurrently reduce its development resources and show greater profitability in the short term. HP is an example of a company that went from internal innovation to outsourced products and technologies in its non-core printer business. Cisco Systems is another example of a company that relied on this technology-acquisition model in order to greatly expand its business from its Internet-routing core to other related products with great initial success.
- A new company operating in the maturity market might decide to use a *penetrating pricing* strategy to set a lower price for its products to increase market share. Once that is accomplished, and its customer loyalty base is increased, the company can increase the price. Examples of this include Hyundai and Kia in the auto industry.

It is important for the PM of a new product in the maturity stage to consider all of the above issues in carefully managing the new project and product costs as well as product price.

11.1.4 The Final Stage

This final stage can take one of two forms: either the product will decline as the need for it is overwhelmed by new technology, or it will develop into a commodity. The emergence of standards of use, manufacturing, interconnection, and quality will make the price and, hence, manufacturing costs the only competitive leverage. The revenue per unit decreases rapidly, and the products will essentially be interchangeable from one manufacturer to another, with high customer expectations of quality and reliability. The manufacturing costs are extremely important, as they are the only differentiator in the market. The manufacturing of commodity products exhibit several characteristics centered on reducing costs:

- The supply chain becomes extremely relevant in the design of the new commodity products. The chain becomes the innovation leader in the market, with product enhancement technologies within the acceptable standard design. Safety and comfort innovations in the auto industry flow from its supply chain. Examples would be action suspension, dynamic braking, and electronic stability and traction-control systems. The supply chain can thus spread out the technology developments and manufacturing tooling costs among several commodity companies. Examples of trends in commodity companies are as follows:

- The company essentially supplies the product packaging, styling and marketing strategy. It might engage end-to-end design services and a manufacturing supply chain for developing and producing new products. In this manner, the cost of development projects can be significantly reduced. The irony is that original communications companies such as ATT (then Lucent) developed land-based communication products that were designed for reliability to last 40 years. The company was suddenly thrust into designing products for the mobile market with a life cycle of less than one year. Some of the household long lifecycle consumer products companies such as washing machines and dryers turned to global electronic suppliers for designing and manufacturing their products with electronic controls, even though these suppliers usually design for much shorter lifecycles in the communications and computer industries. This helped transition the industry into higher-technology products without investing in the technology or training their engineering staff.
- In turn, the supply chain might consolidate to better manage costs and communications with global suppliers and OEM customers. The industry might switch to a tier system of suppliers, similar to that used by the auto industry, in order to control the design and manufacturing processes better. The industry and its supply chain might agree to strict standards for design and manufacturing processes such as QS-9000 (ISO/TS 16949) for the auto industry.

The role of the supply chain is critical for the project and product cost and price in the commodity stage. The PM should consider the balance of the supply chain versus internal competencies in deciding on managing the new product development project.

11.2 New Product Cost Estimating Methodologies

Product cost is based on several factors that add to the total product manufacturing costs. These costs are then augmented by the general corporate administration and technology development costs that are then used to calculate profits. The product and project costs are related as follows:

1. Direct Manufacturing Costs. These manufacturing costs are directly chargeable to purchase labor and materials used in manufacturing the product. Labor rates are governed by the prevailing wages in the local area, which is a direct result of the average cost of living. They are also determined by minimum wage laws. The amount of labor can be greatly influenced by the use of early supplier involvement (ESI) and DFM techniques in the design stage to lower the mechanical motions required for locating parts and assembly. They also can be significantly lowered by the use of automation in manufacturing. The use of ROI, coupled with a good marketing forecast is required to justify the need for high levels of automation.

In high-technology products, the amount of labor is usually a small part of the manufacturing cost, especially for electronic products that have inherent standard processes, which allow for a large degree of automation. A successful ESI/DFM strategy and good use of standard operations can reduce the labor costs of new products to less than 3 percent of the selling price, depending on the new product development lifecycle stage. Labor costs are not well defined in the early phases of the product development, and techniques have been developed to account for more accurate estimates of the labor used in the final product and will be shown later in this chapter.

Materials costs are more defined in the early stages in the development cycle, with the design team having to create a bill of material list (BOM). This is necessitated by the long lead times required to develop or purchase the materials as well as qualifying them for use in the product to ensure high reliability. The use of common or standard parts is encouraged to reduce the cost of purchasing and qualifying the parts. As mentioned earlier in Chap. 6, the use of the supply chain can help in reducing material costs by increasing the material leverage available to the supplier.

2. Direct Overhead Costs. These are financial burdens placed on top of the labor and materials that account for the additional shared manufacturing costs of the products. These costs include salaries of managers and supervisors in production, costs for space, heating and lighting and for supporting the procurement of the product such as purchasing, stockroom, and production consumables. Other costs also include product quality and support costs such as incoming inspection, failure analysis, inventory carrying charges, scrap, obsolescence costs, manufacturing, and quality-engineering support.

   Care must be exercised to manage the overhead rates. Traditionally, the rates were levied solely against labor, which tended to focus undue attention to labor costs. With standardization and automation of manufacturing, the overhead burden was being distributed over a much smaller labor base, resulting in large overhead labor rates. This led PMs to seek more outsourced parts and assemblies to reduce in-house labor content of parts, and shift cost from labor to materials. Currently, there are blended or distinct overhead rates placed separately on labor and materials.

3. General and Administrative (G&A) Costs. These costs are not directly tied to the direct manufacturing of the product, but involve corporate overhead and other functions directly related to the planning, development, selling, and marketing of the product. The corporate profit is directly related to the revenue derived from the selling prices of its products subtracted by the direct manufacturing and G&A costs. Profits can be manipulated in the short term by lowering the development costs or by reducing customer services such as marketing, sales, and product support costs. These short-term strategies can have devastating long-term negative effects, reducing new product investment and the stream of new products. Conversely, keeping high development costs as the market matures also can lead to reducing profits in the long term.

An example of a product costing structure is shown in Table 11.1. In this example, two overhead rates are applied, one for material (25 percent) and another for labor (100 percent). In addition, a G&A overhead cost is applied at 20 percent, leading to the

Source	Direct Cost	Direct Overhead	G&A Overhead	Total Cost
Product Selling Price (100 units)				140K$
Labor	$25,000	$25,000 (100%)		50K$
Material	$40,000	$10,000 (25%)		50K$
Total Direct				100K$
Total Loaded Cost			$20,000 (20%)	120K$
Profit From Selling 100 units				20K$ (14.3%)

Table 11.1 Example of New Product Cost Structure

total loaded cost of 100 units to equal $120,000. Using an example of 100 units, selling at $140,000, the corporate profit is equal to $20,000 or 14.3 percent. It can be seen that any modification of the overhead rates could have a substantial effect on the profit derived from selling the product.

The overhead structure can be devastating in terms of applying unreasonable pricing to ordinary items, such as common spare parts or accessories for the product. A scandalous $600 was the price of a military airplane toilet seat cover in the 1980s. It arose from the simple calculations that any item accrued by the military supplier had the overhead applied against it for a limited-lifecycle product. There are many examples in high-technology products as well. In the author's experience, a razor to be used for shaving the skin of a patient as an accessory for a medical product was priced at $30. The marketing manager said, "No customer will ever buy the razor from us; instead they will purchase it from the pharmacy for $1, but we priced it according to the FDA."

Applying the overhead rates directly on top of labor and material using this common accounting system has led to these imbalances in the pricing of certain items and products. The burden of overhead can be unfairly applied to different products, some requiring more than their share of overhead resources within the company. This has led to the development of differing nonaccounting methods for determining true product cost, such as activity-based costing (ABC).

11.2.1 Activity-based Costing

ABC is a product cost estimating system that attempts to determine the true costs of new products. In addition, it can be used as a viable cost-estimating tool for new products. This estimate can be used during the development phases of new products and updated frequently to reflect the current estimated cost based on the latest design information. An accurate cost estimate is an important part of the design phase review process and can be used to ensure the project and product goals are being met.

ABC addresses the traditional shortcomings of cost accounting determination based on overhead rates. The rates are applied based on all activities/facilities in manufacturing, whether or not the product uses these activities. There are no specific penalties or rewards for the correct level of work-in-process inventory, test, or quality level for individual products. These are usually the outcome of good DFM and product design strategies. All products are assumed to consume the same level of resources in

	True Product Cost		Cost Accounting Overhead Derived Cost	
	High Quality	Low Quality	High Quality	Low Quality
Labor	$150	$50	$150	$50
Material	$200	$200	$200	$200
Quality Costs	$50	$150		
Other Overhead	$200	$200		
Overhead (300%)			$450	$150
Total Cost	$600	$600	$800	$400 = $1200

TABLE 11.2 Comparing True Cost Versus Traditional Cost Accounting

manufacturing, such as manufacturing engineering or quality. In this manner, a PM that strives to use all of the cost and quality optimization tools in their new project development is penalized by the burden of non optimized design projects.

An example of a scenario for comparing true versus traditional product cost accounting is given in Table 11.2. It compares two types of products; one is high-quality using Six Sigma design, requiring much smaller quality resources. The other is a low-quality design requiring a lot of testing, rework, and scrap. The true costs of both products are the same, considering that the low-quality product consumes much more resources from the quality system than the high-quality product. If we assume that the other overhead costs are the same for both products and combine the overhead quality costs with other overhead costs, then the overhead rate is 300 percent. This is based on the division of the total overhead ($600) by the total labor consumed by the two products ($200). The cost-accounting system applies this overhead rate to both labor contents in the two products, resulting in an inaccurate (by a factor of 33%) assessment of the product cost. The total cost for the two products remain the same for either cost estimate at $1200.

ABC techniques involve the use of a cost driver model. The driver is a factor generated for each resource in the manufacturing area in terms of types and amount expended per unit output measure. The cost driver is the primary measure of the activity in each area. More than one cost driver can be developed per activity. The cost driver can be calculated initially and updated frequently. The update should have a time cycle less than the new product development cycle to keep the model current with manufacturing wage scales and process changes. A quarterly update is usually used.

An example of a manufacturing department that can be modeled using ABC is purchasing. The activities the department generates include purchasing orders (POs) for product materials and suppliers, expediting shortages, and managing the stockroom. The output of the purchasing departments is the number of POs that they submit to their suppliers. The POs can be separated into non product-specific POs for maintenance and process support as well as the POs needed for the parts in the new products. Thus, two cost drivers can be generated, one for non product and another for new product costs. The contribution of the purchasing department to the new product cost is equal to the number of parts in the product multiplied by the cost of each PO to purchase these parts. A more complex model could incorporate an additional cost driver by separating the

1.	Total # of pieces/lot
2.	Material thickness and type
3.	Length in x direction
4.	Length in y direction
5.	# of nonstandard tools used
6.	# of hits/piece
7.	# of fold angles
8.	# of folds/piece
9.	# of pems and/or rivets per piece
10.	Paint yes/no
11.	Type of paint finish
12.	Class of paint finish
13.	# of silkscreen colors

Table 11.3 Sheet-Metal Cost Driver Model Construction

unique part of the new product versus multi-use parts used in many products. Another level of refinement could be the use of cost drivers for the different levels of costs of parts that require different stocking strategies. This is called A, B, and C part inventory categories, where the cost for an A category part is high and is ordered strictly as required for use in products. The PO cost driver for category B parts is dependent on the frequency of ordering. Category B parts are ordered as needed for maintaining a predetermined inventory level to limit the number of POs, while the C parts are ordered in bulk. Thus up to three or even six levels of cost driver factors could be calculated, though most companies opt for a single level of purchased-parts cost driver.

A typical cost for a fabricated part for sheet metal is shown in Table 11.3. It illustrates that many attributes of the sheet-metal design can be added to the cost model. The cost driver calculations can help the design engineer understand the different elements making up the cost of a sheet-metal part. In a sense, the cost model could be used as a DFM analysis to see if the cost can be reduced.

The ABC model can be used for estimating the cost of new products at any time during the design phase and updated with each successive phase. For a good estimate, with 10 to 15 percent accuracy, up to four to six cost drivers could be used, with a fixed sales estimate from marketing. For a more accurate estimate of 5 to 10 percent accuracy, many additional cost drivers might be required. In addition, several sales volume scenarios might have to be explored to estimate the accuracy of the model based on different automation levels corresponding to the different sales.

While some PMs might question the validity of using ABC to augment the accounting-overhead-based traditional costing systems, the benefits of ABC might outweigh the extra cost of maintaining the ABC system. The following are the potential benefits of using the ABC system:

- Provides for a more accurate costing system in the early stages of the project development phase, eliminating unpleasant cost surprise upon release of the product to manufacturing

- Acts as a possible DFM guide to alert the design engineers to the constituents of cost in their parts and encourage them to use less complex part geometries, manufacturing processes, or reuse parts already in other products
- Serves as alert system to convey to the design engineers any changes in the cost structure or automation in certain parts of manufacturing that they can take advantage of
- Gives the design engineers a comparative cost estimate using different processes in manufacturing to enhance process choice decisions

ABC can be used in projects for new product development, but it does not replace the official cost-accounting system for the company. The company auditors will only acknowledge the traditional cost accounting system, since it is based on the accumulations of costs and overheads in yearly periods.

In summary, the traditional cost-accounting system conforms to the generally accepted accounting principles (GAAP) while the ABC estimating system is not accepted by the GAAP. The traditional costing system assigns manufacturing overhead to manufacturing volume produced with an overhead rate that is driven by labor or material costs or both. The traditional accounting system fails to properly account for the amount of support activities and other nonmanufacturing costs needed to produce the new product.

ABC is a supplemental accounting activity within the project acting as a DFM guide for the PM to make sound decisions based on the cost of design activities as well as an early estimator of ultimate manufacturing costs. This system gives a greater understanding of overhead costs of indirect labor and thus addresses the shortcoming of traditional cost accounting. The overhead rates in the traditional costing system are applied to all activities/facilities in manufacturing, regardless of whether the new product uses these activities. The ABC system is clearly more beneficial to the PM than traditional costing, since the PM can allocate project resources according to real overhead costs and select the best design methodologies for lowering the cost of the new product.

11.2.2 ABC for Electronic Products

Typically, printed circuit boards (PCBs) account for 90 percent of the total material costs of an electronic product. Developing PCBs cost models can vary depending on the accuracy level needed and the electronic product lifecycle development stages. Consumer product's prices are sensitive to cost variation, while new technology products are less sensitive.

PCB assembly is performed on equipment with different capabilities and constraints in automation, speed, component technology, and quality. The Association Connecting Electronics Industries (IPC) has developed classifications for three different types of PCB assemblies. A combination of 10 different machine categories can be used to produce these types in differing permutations. The task of assembling a PCB can be achieved by many different alternatives of equipment and processes, some of which are overlapping in function, yet different in quality and productivity. This renders the cost analysis of PCBs very difficult and time-consuming. Several surveys have been taken to estimate the cost distribution of PCBs, some by makers of automation equipment, to show the cost distribution. The data can then be used to justify the ROI for purchasing manufacturing automation. Figure 11.1 is a cost distribution of a typical PCB. This is for direct costs only and does not include G&A overhead costs.

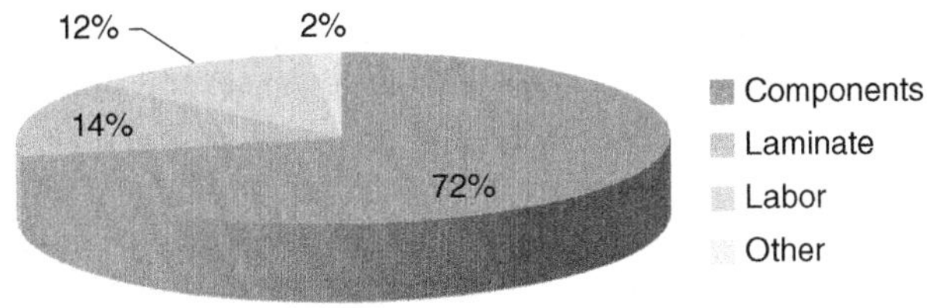

FIGURE 11.1 Cost distributions for a typical PCB.

It can be seen from Fig. 11.1 that the primary cost contributors (72 percent) for PCBs are the components. The next contributor is the laminate or the unloaded (raw) PCB cost at 14 percent. It is closely followed by the cost of labor at 12 percent. The remaining 2 percent cost contributors include various elements such as maintenance, consumables, floor space, utilities, and quality costs. Quality costs include scrap, rework, and replacement material. Given that the total PCB material, including components and laminate, account for three-quarters of the total PCB cost, most ABC drivers for PCB include a material-based system.

Three levels of PCB cost systems are in common use in the electronics industry, depending on accuracy and resources available to manage the cost system:

1. The Materials Based Cost Estimating System. Since material costs account for the majority of PCB costs, all other costs are calculated as a percentage of the material used in the bill of materials (BOM). Cost parameters are shown in Table 11.4 for costs for a PCB with analog and digital circuitry designs. The ranges are based on querying four different PCB manufacturers (used in the supplier-selection example in Chap. 6). The nonrecoverable expenses (NRE) are based on the

Direct Costs	**Digital**	**Analog**	**Range/Comments**
Raw Material	$500	$1000	+/− 5%
Markup (Material and Labor Overhead)	$50	$100	5 to 20% of material
Labor	$50	$100	$30/hour or 10% of material
General & Administration	$50	$100	8 to 10% of material
Earnings Before Taxes	$50	$100	7 to 15% of material
Total	$700	$1400	Material Multiplier 30 to 50%
NRE	**Digital**	**Analog**	
Stencils	$800 (laser)	$500 (etch)	Laser for Fine Pitch 1 Each Side
Programming/ Document	$1000	$1000	
Total NRE	$2600	$2000	

TABLE 11.4 Costs Parameters for Typical PCB with Analog and Digital Circuitry Designs

tooling necessary for making stencils for applying the circuit images on the PCBs and the test setup. Test costs are based on several iterations (spins) of the PCB design in the development project.

There is a refinement in the cost model, based on whether the components are for an analog or digital design. The range column is based on four quotes delivered based on the cost model for the four different alternative suppliers. Several interesting conclusions can be drawn from Table 11.4 for the PCB cost parameters:

- The material cost range is based on the material purchase leverage applied by the four alternatives, as was discussed in Chap. 6. This range was low because of the standardized components.
- The material markup includes the material and labor overhead which ranged as much as 5 to 20% of raw material cost.
- Direct assembly labor was calculated as approximately 10 percent of the material costs at $30/hour (including overhead).
- The G&A costs were closely ranged from 8 to 10 percent of material costs, reflecting lean supplier organizations.
- Earnings before taxes ranged widely, depending on the supplier size and willingness to compete for the business of the new product.
- The NRE expenses included the cost of making the stencil for the PCB image transfer in the manufacturing process. There are requirements for difference stencil drilling technologies in analog versus digital PCBs (etch versus laser). The programming cost is for the automation software development in the PCB line.

The test costs for PCBs are included in the Table 11.5. They include the cost of developing the test methodology and the associated cost of debugging and programming. Two types of tests are used: in-circuit test (ICT) for eliminating workmanship defects, which requires a test fixture, and a final (functional) test. The final test is divided into the design and debugging of stages and the qualification and approval of the test by the QA department. The assumptions for Table 11.5 are PCB test-development costs, which include the salaries of the manufacturing staff at in-circuit test at $75/hour, final electrical test at $50/hour, supporting engineers at $3000/week, and construction of the test fixture at in-circuit test at $6/pin. Total estimate tests per PCB at production quantities for the communications market was assumed at $30 to $60/PCB. It assumes two revisions (spins) per test fixture and a six-week period of debugging the test fixture and methodology. It takes about $65,000 to develop the cost for a single PCB with combined analog and digital components.

2. The Technology PCB Cost Driver Model. In this model, component technology and process methodology selection are used in calculating the costs of the PCBs. A single cost driver is assigned to each PCB production or assembly step, including a driver for each type of component technology used. Through hole (TH) and surface mount technology (SMT) are the most commonly used components. Additional complex component types such as ball grid array (BGA) and chip on chip can have their own entry in the cost driver model for PCBs.

ICT Costs	# of Spins	Cost/Spin	Total Cost
ICT Fixture ($6 per test pin)	2 (spins)	$10,000	$20,000
Engineer for ICT Setup	1.5 (weeks)	$3000	$4500
Tooling NRE	one-time	$5000	$5000
Subtotal PCB ICT Test			$29,500
Final Test Costs	**# of Weeks**	**Cost/Engineer Week**	
Engineer for Final Test/ Debug	6	$3000	$18,000
Incremental QA Test Approval	6	$3000	$18,000
Subtotal PCB Final Test			$36,000
Total PCB Test Development			$65,500

TABLE **11.5** PCB Test Development Costs

Each driver is a representation of the cost of the assembly function in the ABC system. The $/rate per driver is based on the total costs for each manufacturing step for a certain period of time, typically every quarter, divided by the number of driver units processed by the manufacturing department for that period. All costs specific to the assembly step should be used in the driver rate, including costs for factory space, consumables, energy, supervisory and technical support, and equipment maintenance and calibration. The number of cost drivers used to determine the PCB costs will vary with the accuracy needed. A 15 percent accurate model can be constructed with four drivers, while a high-accuracy system of 5 percent error must be made with at least 15 drivers. To be effective in reducing costs of new products, the cost driver $ rates should be recalculated periodically (quarterly) at a faster rate than the new product development cycle. Modifiers can be used for tooling holder setup ("h"), batch size ("b"), and quality loss ("l"). Examples are given below.

A PCB technology cost model is shown for a typical PCB board in Table 11.6. The model lists all of the assembly operations required to produce the PCB. The setup modifier "h" for machine loaded TH components is based on the number of PCBs that can be accommodated in the machine work holder. If it takes 25 seconds to load each PCB into the machine, at an estimated $20/hour loaded labor rate, this amounts to (25/3600)*20 = $0.14 additional cost for every PCB loaded on the machine. Since a PCB contains an average of 100 components, this cost represents an addition of $0.0014/component. This cost is halved when the work holder used can accommodate two PCBs at the same time.

The batch modifier "b" is based on the batch size or the number of PCBs in each job dispatched to the manufacturing floor. For each new job, the machine has to be unloaded from the previous job's components and then loaded with the programming information and the components for the new job. For TH components, an axial insertion machine changeover time is 25 minutes. At the $20/hour labor rate, this amounts to (25/60)*20 = $8.33 additional cost for every PCB job loaded on the machine. For a typical job of 100 PCBs and 100 components to be machine loaded, this modifier represents an additional cost of $0.00083 per component. This cost is inversely proportional with batch sizes.

The PCB technology cost model shown in Table 11.6 is a composite of high-technology electronic companies in the consumer, instrument, medical, and defense electronic companies. These tend to be of lower volume and, hence, of many batches (b) and lower automation (h) in PCB assembly. For very high-volume companies, the batch "b" and tooling holder "h" modifiers are not used, since their PCB manufacturing includes a high level of automation in the setup as well as production and, therefore, the set-up time is reduced to zero. Sometimes, they have dedicated automated lines to a very high-volume product.

3. Quality Based Technology Cost Model. The cost driver in this model is augmented by the quality performance of each assembly step. The quality performance is derived from defect prediction based on Six Sigma analysis of

$ Modifiers		**Material**			
Process Function	**Driver Units**	**Rate**	**Setup**	**Batch**	**Loss (l)**
Component procurement	Material $		.032		
Material storage & delivery	# of part numbers	.022			.0005
Hand load TH	# of insertions	.120			.0005
Machine load TH	# of insertions	.012	.14/h	8.33/b	.0003
Hand load TH post solder	# of insertions	.24			.0005
Solder and wash for TH	# of PCBs/ holder	3.56			
Hand load SMT	# of placements	.036			.0005
Machine load SMT	# of placements	.024	.08	3.15	.0002
Special operations	# of placements	.360			.0010
High-accuracy SMT	# of placements	.075	.10	5.00	.0002
Solder paste and reflow	# of PCBs/ holder	10.00			
Test and repair	# of components	.035		1.00	
Visual inspection	# of components	.030			
h = work holder load/unload cost; b = batch size setup cost; l = quality loss/component					

Table 11.6 PCB Technology and Quality Cost Model

the design. Alternatively, it could be based on the quality historical performance of the assembly step. However, historical performance based quality data does not encourage PMs to increase the quality of PCBs in the new development project using DFM techniques, since they should set higher quality targets. For all cost drivers for each PCB assembly step, parts per million (ppm) defect rates are calculated based on actual machine operations. Parts per million rates are determined from periodically dividing defects generated by all of the driver unit output for each process. The defects generated are then summed up to the product level, and a test strategy is developed for the most efficient removal of these defects.

The material quality modifier or quality loss (l) is based on the defect rate of the assembly process. The misloaded component rate for TH axial insertion is typically 2000 ppm, or 1 in 500 components. At the \$0.15 typical cost of a through-hole component, the quality loss is 0.15/500 = \$0.0003/component.

The quality based cost model assumes a defect rate for each manufacturing operation, based on historical data. The quality loss modifier is derived from the defect rate and calculated for defective material and process costs. A more detailed model could be constructed from adding up all of the defects extracted in final test for a more accurate test and repair driver. The advantages of this model are to predict the number of failures in the final PCB test as well as the number of test equipment and technicians needed to operate them. This will allow for better planning to manage the test and repair function. A complexity factor could also be added to distinguish between the different types of densities in the PCBs.

An example is given in Table 11.7 for PCB fabrications (raw PCBs) complexity factors and reject rate calculations. The different steps for fabricating a PCB are given, with a reject rate for each step, based on complexity. The reject rate could be added up to a total reject rate per raw PCB. The reject rate can then become the basis for the test and repair driver. In addition, the total *Cpk* for the raw board could be calculated in a manner similar to the method using standard normal distribution (SND) curves discussed in Chap. 1.

Process	Low	Medium	High (Complexity)
Image transfer	.02	.03	.04
Copper etching	.01	.02	.02
Lamination	.005	.01	.01
Drilling	.004	.01	.01
Metallization	.04	.05	.05
Solder mask	.02	.03	.03
Total reject rate	0.099	0.15	0.16
$f(z)$	0.0495	0.075	0.08
z	1.65	1.44	1.4
$Cpk = Z/3$	0.55	0.48	0.47

TABLE 11.7 PCB Fabrication Complexity Model and Reject Rate Calculations

The reject rate is assumed to be double-sided, leading to a $f(z) = \frac{1}{2}$ of the total reject rate. A z value is returned from the SND table and is divided by three to determine the *Cpk*.

11.2.3 ABC Summary and Variance from Classical Cost Accounting

A major difference between traditional cost accounting and ABC is that the overhead is not generally applied in ABC. Overhead is applied only as it actually pertains to the resources expended in producing the product.

ABC accounting is advantageous to the PM because it allows for a more accurate cost estimate at the beginning of the project. This could result in decreasing the probabilities for schedule and budget variances in the later stages of the project. ABC also can be used to show design engineers the cost distribution in their portion of the design (depending on how many cost drivers the PM decides to have). This can be used as a way to determine where the design can save money by using less complex geometries (DFM) or less expensive parts, etc. The advantages of ABC over classical costing are:

1. ABC addresses the frequency issues of up-to-date GAAP. GAAP refers to the standard framework of guidelines for financial accounting, generally known as accounting standards or standard accounting practice. ABC is usually updated quarterly while GAAP cost accounting is updated yearly. ABC is more effective in capturing the increased efficiency in manufacturing.
2. ABC has diminished the overhead-rate issues as discussed earlier. There is no decision to be made on subcontracting based on labor versus material overhead.
3. ABC gives a more accurate cost estimate at the beginning of the project, thereby decreasing the chances for schedule and budget variances in the later stages of the project.

The ABC method's disadvantage is that it requires substantial resources to implement and maintain. It can potentially overlook some costs. It does not conform to GAAP standard, necessitating the use of two different accounting systems at the same time.

11.3 New Product Cost Estimating Process

An accurate cost estimate for high-technology products is dependent on many factors:

1. Development schedule realization. The new product cost estimate should improve as a new product moves closer to production. Different cost models discussed earlier in the chapter can be used for the cost estimate. The timing of the product introduction might influence the sales forecast, especially if there is new technology incorporated in the design. The cost estimate plans should include provisions for aggressive (50 percent probability of success) as well as standard new product introduction schedules (90 percent probability).
2. The sales forecast should be as accurate as possible. The marketing department should include up and down sales potential, competitive analysis, and price-performance curve strategies. These would help in selecting the optimum manufacturing strategy in equipment and tooling and hence determine the appropriate cost structure of the product.

3. NRE expenses should be quantified, including tooling and capital equipment acquisition costs. A determination should be made whether some of those costs could be shared with other products or resources in the form of a cost center that allocates an overhead or burden rate onto other products that use the NRE tools and equipment. A depreciation schedule and methodology, using straight line (SL) should be agreed upon. Typically three to seven years of depreciation for electronics-based capital equipment are used, depending on the types of equipment. Longer schedules might be available for more permanent structures such as buildings to house the equipment (25 to 40 years). The accounting department should be the best source for determining the depreciation method. An ROI should be performed on all capital equipment purchases to ensure compliance to the company's financial target.
4. The BOM should be as complete and up to date as possible. It should include provisions for options, raw materials, and selected suppliers. Non identified suppliers should be investigated and estimates of material costs as well as reliability studies initiated. There should be a material cost reduction program for developing lower-cost material alternatives to the current BOM. These materials may be substituted when newer technology is available or when lower-specification materials might offer comparable performance in the product. A good target for such a program is 3 to 5 percent cost reduction per quarter after release to manufacturing. Material volume discount schedules should be available and readily incorporated with the forecast into the cost structure.
5. The product routing scheme should be reasonably developed. The routing includes all of the manufacturing operations or steps necessary to fabricate, assemble, inspect, and test the product. A determination should be made whether intermediate steps in the product assembly should be treated as line fabrication items with no inventory control points or subassemblies. It is always desirable to have the minimum level of assembly to reduce assembly time and cost as well as lower inventory requirements.
6. The direct labor needed to produce, assemble, inspect, and test the product should be accumulated for each manufacturing step. The amount of labor needed is dependent on other factors, such as outsourcing, which converts possible in-house labor into purchased materials. The use of tooling and the level of automation and production volume based on the marketing forecast can also have an influence on the amount of labor needed to make the product.
7. The overhead rate should include provisions for equipment and workspace allocations, along with special requirements for energy and environmental considerations. It should take into account any special skills needed to manufacture and technical support of the product such as process, quality, and manufacturing engineers and their managers. As materials might contribute significantly to product cost, as well as increased trends toward outsourcing, several overhead rates can be applied. These include one for material and another for labor. Material overhead should also include costs for material warehousing, obsolescence, scrap, and purchasing and inventory control.
8. The quality plan for the product, including the costs of expected yield, rework, scrap, inspection, and test should be identified. The defects imparted by the

raw materials suppliers should be added to the defects inherent in the design as well as those incurred in production. A test strategy is then developed for the optimum removal of these defects, including where to perform intermediate as well as final test of the product. The number of test stations needed to remove assembly and process defects should be planned and test technicians trained on the use of the equipment.

9. G&A costs including royalties paid to corporate R&D investments, profit margins and provisions for taxes and reinvestment should be calculated as overhead to the product direct costs.
10. Start-up Costs. These should include costs for design revisions, equipment, and tool debugging and support costs for additional technical and material support during the prototype (alpha) and production (beta) phases of the product.

11.3.1 Determination of Costs and Tracking Tools for New Product Development

After collecting background information, several tracking tools can be used in order to make product marketing or financial decisions affecting the cost of the product. Some of these tools are as follows:

1. The return factor (RF) of the product, which is the total profit (sales revenues – manufacturing costs) returned by the product during its lifecycle divided by the development costs. This return factor should be compatible with the historical trends of the product family and its competitors. In addition, ROI analysis should be performed and reviewed each quarter. The ROI is determined by the time adjusted present value of the return factor. It should be in the range mentioned earlier in this chapter depending on the product lifecycle stages. Both RF and ROI were discussed in Chap. 5.
2. Cost history of the project and product. The costs of the product can be identified in terms of the individual components of cost including labor, material, overhead, depreciation on capital, NRE tooling, and quality and administration costs. These costs can be tracked over the design as well as the production phases of the product to show impact of design changes and investment in automation. The profile for the different elements of the project cost structure should be reviewed at each design phase to make sure the development project is still within its original targets of profit and ROI. Figure 11.2 is an example of a product cost structure as it proceeds through it development cycle for an electro-mechanical consumer product. It is based on a modified depreciation structure of 25 percent in year one, 50 percent in year two, and 25 percent in year three of the product lifecycle. The differences in the project profile during the different phases are due to design changes, sales forecasts, and start-up costs.
3. Volume sensitivity of the product. Depending on forecast accuracy and upside potential, several levels of automation and manufacturing strategies can be used to estimate product costs. These alternatives are shown in the pie charts of Fig. 11.3, representing typical cost percentages of a consumer electronic product based on several levels of sales forecast. From these charts, automation and

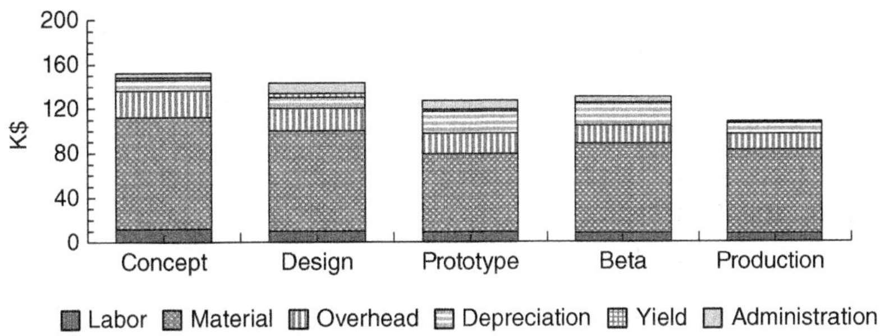

Figure 11.2 Product cost structure versus development phases.

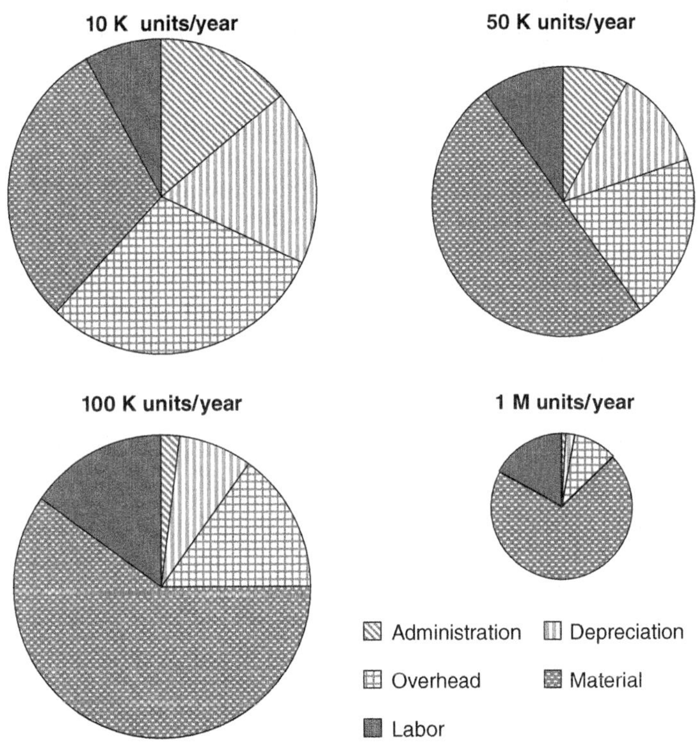

Figure 11.3 Product cost structure versus sales volume.

outsourcing plans for the new product could be developed in order to obtain the optimum cost through carefully investigating the automation and outsourcing levels desired for manufacturing the product.

11.4 Conclusions

A good understanding is needed to evaluate the relationship of high-technology product cost to other factors such as marketing and forecast, material and process selection, the level of automation of manufacturing equipment, and in designing process methods and flows. Cost determination, estimation, tracking, and control are varied, depending on the lifecycle phase of the product and the market it competes in. Several types of cost-estimating systems have been developed and are used as accurate estimates of the product cost through the project development stages.

References and Bibliography

Banks, S., "After ISO There is ABC," *Surface Mount Technology Magazine*, p. 25, August 1993.

Dunlop, R., " Design for Profit," *Proc. National Electronic Packaging and Production Conference*, NEPCON West, Anaheim, CA, pp. 139–147, 1992.

Institute for Interconnecting and Packaging Electronic Circuits (IPC) *Guidelines for Printed Board Component Mounting*, Lincolnwood, IL, IPC-CM-770, 1997.

Hobbs, D., *Lean Manufacturing Implementation: A Complete Execution Manual for Any Size Manufacturer*, J. Ross Publishing, Inc., Plantation, FL, 2003.

Monk, E., and B. Wagner, *Concepts in Enterprise Resource Planning*, 3rd ed., Course Technology Cengage Learning, Independence, KY, 2008.

Myerson, P., *Lean Supply Chain and Logistics Management*, McGraw-Hill Professional, New York, 2012.

Nicholas, J., *Lean Production for Competitive Advantage: A Comprehensive Guide to Lean Methodologies and Management Practices (Resource Management)*, Productivity Press, New York, 2010.

Ostwald, P., and T. McLaren, *Cost Analysis and Estimating for Engineering and Management*, Prentice Hall, Pearson Higher Education, Upper Saddle River, NJ, 2004.

Reid, D., and N. Sanders, *Operations Management*, 4th ed., John Wiley & Sons, New York, 2009.

Saigal A., and S. Shina, "An Algorithm for Selecting the Electronic Design Implementation in Printed Circuit Board Fabrication Based on Cost Factors," *Proc. ASME Winter Annual Meeting*, Atlanta, GA, pp. 757–769, 1996.

Saigal A., and S. Shina, "A Design Quality Based Cost Model for New Electronic Systems and Products," *Journal of Materials*, 50 (4), April 1998, 29–33.

Saigal, A., and S. Shina, "Technology Cost Modeling for the Manufacture of Printed Circuit Boards in New Electronic Products. *Journal of Manufacturing Science and Engineering*, 120(2), May 1998, pp. 368–375.

Wilson, L., *How To Implement Lean Manufacturing*, McGraw-Hill Professional, New York, 2009.

Discussion Topics

1. Discuss three differences between the traditional cost-accounting system and ABC. Use your own experience in industry or from reading this chapter. Quantify your opinion of the three differences as to whether they are beneficial or counterproductive for the PM.

2. Discuss the need for single or multiple overhead rates in direct costs of a product. Give two reasons why you would select overhead on labor or material or both in a company.

3. The chapter featured discussions about the relationship of product cost to product price, using an example of start-up products such as the iPAD or the HP-35 and the pricing strategy between the Apple iPAD and the Amazon Kindle. Give two examples from historical sources or your own experience as to why there were such differences and how and when were they used?

Problem

9.1. Write a generalized view of a price structure for a consumer product such as a CD player, computer monitor, or laser printer. Trace the technology development versus prices for the product, as it traversed from start-up into the maturity mode using the Web. Indicate if there are any differences in price from competing manufacturers or sellers such as the *big box* stores or Internet product providers. How is the price differential, if any, justified?

CHAPTER 12

Building and Managing Teams

As the global marketplace becomes ever more competitive, the ability of firms to utilize technology as a competitive advantage is fast disappearing. As the cost of a sophisticated technological infrastructure continues to fall, all companies large and small have an equal opportunity to access the global marketplace. The traditional barriers are being minimized.

The key performance difference maker would seem to be the ability of competing organizations to assemble competent teams, execute on the given tasks, and solve problems to produce the requisite product or service to the specifications and delight of the customer. Whether that customer is internal or external is no longer a point of differentiation given today's age of fluid marketplaces, companies are no longer obligated to accept mediocre supplier or vendor performance simply because of intraorganizational favoritism. This de-coupling of internal buying preferences, places even more pressure on businesses and organizations to perform—to successfully execute on their challenges with problem-solving teams.

In this chapter, we will look at the people, knowledge, skills, and behaviors necessary to consistently lead project teams and organizations to high performance and mission success.

There is a clear recipe for success in the development and management of teams and projects. In fact, taking the right steps in the formulation of projects and teams can often determine whether there is even the likelihood of success for any new initiative under consideration. With these concerns in mind, we offer the following considerations for project and organizational success.

First, we lead off with a discussion on understanding the value of cross-functional teams and how to get the most productivity from them. Second, we look at the major events and activities that are the key to the execution of successful projects. Finally, we take a close look at the team leader and the pivotal role that he or she plays in leading, and often, driving the team toward success.

The chapter closes by offering several case studies in order to give readers an opportunity to appreciate the potential impact of operationalizing, or putting into practice, the

This chapter was written by Ralph E. Jordan.

insights shared in this chapter. The discussion questions, at the very end, will serve to stimulate readers more fully to digest the key learning associated with understanding and gaining mastery over the all-important "people" aspects of project management.

12.1 Teams versus Groups: What's the Difference?

Let us begin by clearly delineating the difference between a group or organization of people and a team. So often the terms are used interchangeably, as if they were in fact synonymous. This mischaracterization often leads to overlooking the most important difference between the two. A group of people working together in order to accomplish a particular mission or task is not necessarily a team. A team has the following three characteristics:

Shared Work Product: A team is composed of people working toward a common objective or product, and an individual member of a team cannot, in and of his or herself, accomplish that objective. Team members need to necessarily depend upon another to get the job done.

Interdependent Tasks: Team members have different interrelated roles and responsibilities. Often these responsibilities require different skills and/or competencies in order for the role to be fulfilled. Once again, team members need to necessarily depend upon one another to get the job done.

Common Rules, Policies, and Boundaries: Team members will have committed to one another that they will obey certain rules, interact in a certain fashion, and be bound by together by certain limitations to accomplish their objective.

A group, however, while sharing some of these attributes, does not have that tightly bound interdependence that sublimates individual accomplishment to team success. In particular, group performance can be projected as a simple calculation of the sum of the individual member contributions. Furthermore, and most importantly, team performance or productivity can be more than the sum of the individual member's contributions. That additional dynamic or output of productivity—synergy—is a function of a team achieving high performance as members commit to each other and excel together. Groups are incapable of realizing this dynamic.

12.1.1 When Are Teams Needed?

Getting any number of people to come together as a team for some finite period of time requires a real investment of time and energy. The investment only makes sense if the task is complicated, a great deal of creativity is required, or there are a number of different functional or geographic organizations that will be impacted by the work of the team. In the workplace, teams are often formed in order to accomplish one of the following objectives:

Develop or Deliver a Particular Product or Service: A team may be formed to enable an organization to design and develop a new product or service for the business-to-business or business-to-consumer marketplace. This would result in enabling the organization to access additional prospects and revenue opportunities and as a consequence sustaining or growing the business.

Solve Problems or Explore Potential Opportunities: A team may be formed to investigate issues of importance to the leadership of the organization. When a variety of viewpoints must be considered, or multi-functional expertise is required to negotiate opportunities or resolve problems of sufficient size or complexity, a project team may be just what's needed.

Make Process Improvements: Lean Six Sigma process improvement teams are standard fare among the best of global firms. Typically, these cross-functional teams utilize a formal methodology and tool set to continually improve the effectiveness and efficiency of critical-to-quality major business processes.

Whether we're talking about long-term, organizationally-oriented teams, major program teams, or ad hoc project teams, they all share the common characteristic of capitalizing on the differences among team members to craft the richest and most robust solutions to the mission at hand. In fact, the differences among team members make it challenging to get the members to work together effectively, and yet, those same differences are the most valuable characteristics of a team.

12.1.2 Differences: The Team Advantage

Teams are often composed of individuals with different complementary expertise (hardware, software, networking engineering), representatives of different impacted stakeholder functional organizations (research and design, product marketing, sales, customer support) and/or representatives of different impacted geographic locations (North America, South America, Middle East, Africa, Europe, Asia Pacific). These individuals are expected to contribute to the teams' execution on the mission, while serving in the interest of their respective organizations. In addition to these organizationally-fostered type differences, each of us as individuals are unique. We have our own identity, cultural, and cognitive differences that we bring with us everywhere we go, including the world of work.

Identity Differences: We typically think of identity differences as those issues that are either readily apparent or generally well-known; such as religion, race, gender, age, marital status, ethnicity, and physical and mental challenges. In actuality, this never-ending list goes on and on to include socio-economic status, citizenship, occupation, education, sexual orientation, birth order, and others. These identity differences have a lot to do with how we view the world and how the world tends to label us.

Cultural Differences: Cultural differences tend to be a function of where and how we are raised; the accepted belief systems, mores, values, and rituals of the society within which we grow up. While often intermingled with religious differences, culture, in and of itself, plays a huge role in how we see ourselves, how we see others, and how others see us.

Cognitive Differences: These differences are more individual-oriented. How we think, how and what we feel, and how we individually orient ourselves to the world are included in this latter category. Included are such issues as how we deal with conflict, emotional intelligence, IQ, and how we think and feel about the world around us. Personality traits such as whether we are extroverted or introverted, deal in the here and now or focus on tomorrow, make decisions being purely objective or with relationships in mind, and whether or not we are planners.

In short, we also differ in how we approach problem-solving; based upon how we take in and process information. When we sort by attribute, the attributes we choose as most important will vary by our individual experiences. We all have different perspectives on the same issue or viewpoints on how we see the world. Finally, our heuristics, or how we infer cause and effect, will differ based upon how we think the world works.

Each of Us Is Unique: All of these characteristics, attributes, traits, and tendencies, when coupled with our unique experiences, are what make each and every one of us unique. We can look at the same thing and interpret and categorize it differently; when it comes to problems and opportunities, we can infer different cause-and-effect relationships. This is the real value of teams. When the issues are complex, and creativity is desired, a well-functioning team of diverse individuals is much more likely to exceed expectations than any single, well-meaning, individual.

12.1.2.1 Summary

- In assessing the benefits of identity, cultural, and cognitive differences, we consider the issues of problem-solving and prediction. They include much of what teams are asked to do; come up with alternative approaches, evaluate potential courses of action, and choose a viable solution.
- People with identity and cultural differences may think alike. However, people belonging to the same race, age, gender, religion, and socio-economic class might see things differently.
- Identity and cultural differences may contribute to better outcomes only if the task is appropriately complex and in need of inspiration and creativity (if we are developing an innovative solution), but these differences may not have much added value if the task does not require problem-solving or prediction.
- If well managed, identity and cultural differences create benefits, provided they are made manifest in cognitive differences and provided the task is one in which those differences matter.
- Organizations and teams that solve problems should seek out people with diverse experiences, training, and identities that translate into diverse perspectives and heuristics.
- In addition, the benefits of diversity also apply within individuals. People should acquire diverse sets of perspectives and heuristics so that they can make larger contributions.
- In conclusion, race, age, gender, religion, and ethnicity matters, but so do our experiences: what we learned from our parents and elders, what we learned from, and while we are with friends, our lessons from school and education, our vacations, and unique mix of experiences that form our life influence our ever-changing collection of cognitive tools.

12.1.3 Selecting and Launching Teams: A Recipe for Success

Several lessons have been learned overtime as they relate to the successful development and launch of project teams. These lessons can be organized into ten issues that must be carefully considered in order to provide a team with a strong opportunity to be successful in meeting its challenge. Let us now take an in-depth look at each of these ten critical success factors in order to further understand what it takes to launch a project team that has a much higher likelihood of achieving success.

1. Capable Sponsorship

   When an organization is faced with a potential opportunity or perceived problem of sufficient magnitude and/or complexity so as to command the attention of its most senior leadership, there may well be a call for assembling its best and brightest into a project team. While the challenge at hand will command the attention of most of the leadership, one from among them should be designated as the "prime sponsor" of the project. As the prime sponsor, it is that executive's responsibility to make sure that the project is developed and launched in a manner that gives it every opportunity to be successful. The selected executive should be of sufficient stature and creditability so as to command the respect and best efforts of his or her colleagues at the executive level. That sponsor is responsible for making sure the project and team have access to those decision-making leaders throughout the term of the project. The sponsor must assist the team by fostering a supportive political environment within which the team can go about the investigative, developmental, and implementation portions of the project with the timely access to information and resources they will need to meet their objectives. Throughout the course of the initiative, the project team and its leadership are dependent on that executive, the prime sponsor, to provide the guidance and the *"political air cover"* needed to drive major change in an organization.

2. Clear Mission and Sense of Urgency

   In order for a project to be worthy of any significant investment of time and resources, it must be important to the organization; the leadership and key players must fully understand that importance and, as a consequence, attach a clear and compelling sense of urgency to the project's successful completion. Thus, it is critically important that the senior leadership, its selected prime sponsor, the project team leader, and the project team members all have clear and firm understanding of the exact reason for being part of the project team's mission.

   It is strongly urged that the so-called mission statement be clearly articulated and captured; it must convey what is to be accomplished by the project team, and why it is of such high importance to the future of the organization. In today's fast-moving and ever-changing world of reality, change is a way of life, but even so, that mission statement must always be up-to-date, representing a set target, and based upon a clear and compelling sense of urgency.

   Creating and maintaining a sense of urgency is critical to getting any organization(s) to really consider changing behaviors, habits, or course of action. And so it is with projects. In order for projects to become more fully supported and the potential changes more readily accepted, there must be a clearly articulated and fully understood reason on why change, and all of the uncertainty and discomfort associated with change, is a better option than continuing on the same path. If there is no real sense of urgency associated with the work of the project team, the project will soon lose its ability to command senior management's attention and the access to information and resources needed to complete the assignment.

3. Stakeholder Commitment and Representation

   As previously stated, teams are often composed of individuals with different complementary expertise, representatives of different impacted stakeholder functional organizations, and/or representatives of different impacted geographic locations. All of the organizations that have an immediate vested interest in the work of the team, or the outcome of the initiative should be represented on the team, or at the very least identified and represented as members of an extended team, so as to control the viability of the team dynamics.

The inclusion and commitment of these stakeholder groups is critical to the success of the project. Those who are impacted and do not feel they have been respected—as evidenced by an invitation to be members of the project team—are seemingly bound to stand in opposition to the later agreed-upon outcome or final product.

Project teams beyond 10 to 15 people can become unwieldy and difficult to manage to high performance because of lessening of intensity in personal relations and trust among members. Thus, it may make sense to utilize a tiered involvement approach where you form a primary team with representatives of the more highly involved stakeholder groups as members. This primary team's efforts can be complemented by the formation of an extended team made up of representatives of stakeholder groups not nearly as impacted by the work of the team or the outcome of the initiative.

4. Effective Team Leadership

While we discuss more fully (in a later section of this chapter) the desired knowledge, behaviors, and skills desirous of the critical-to-success effective project team leader, it is most appropriate that we discuss the responsibilities of the project team leader within the context of this section our recipe for success in these top ten factors.

The choice of a team leader is certainly critical to the success of any project initiative. The team leader is clearly the lynchpin—pulling it all together—and leading the team forward through the project management process. In order to do this successfully, the project team leader must accept several critical responsibilities, while remaining fully accountable to the following constituencies: (1) the organization's senior leadership; (2) the project or prime sponsor; (3) impacted stakeholder groups; and (4) project team members. The team leader responsibilities include:

- Orchestrate the project management process
- Create and manage a trusting and inclusive team environment
- Foster a participative and commitment-building culture
- Manage communications with all stakeholders, and in particular, senior leadership
- Stay on mission, while maintaining a sense of urgency
- Pursue excellence through the intelligent use of resources, and effective decision-making processes based upon the best available facts of the situation
- Recognize and reward the behaviors and contributions that model the way to project success
- Lead in the celebrations as major milestones are accomplished on the way toward a successful initiative

In order to fulfill these many responsibilities, it is important that the team leader be afforded the time and resources needed to put forth his or her best efforts toward project success.

5. Team Members' Relationships and Contract

As soon as is practical, the newly assembled team members should be inspired and enabled to begin to lay a foundation of personal relationships, and figure out how they can best work together, given their jobs and competing day-to-day responsibilities to their respective stakeholder organizations.

The investment of time in laying this foundation of personal relationships is an effort to enable team members to get to know one another as people and colleagues—beyond the immediate task or project at hand. This investment in time and energy is crucial; it yields the bedrock foundation upon which members can grow to trust one another over time—*trust* that any and all criticism of ideas are meant to be constructive and toward improving the richness of thought, and excellence of decisions toward a successful project. The development of trust among team members is what will enable each to give and receive constructive criticism in stride and develop confidence in each other's commitment toward project success.

The creation of the "team member contract" is aimed at getting the members of the team to commit to how they will hold each other accountable for working together as it relates to: (1) The values that should be of primary importance to the team; (2) How they will behave toward one another as a reflection of those values; and (3) What will be the logistical arrangements for the conducting of team-related business. The resulting contract should have been developed with the whole group's participation and agreement. The document should be of sufficient depth and breadth to keep team members accountable to one another for being good teammates—it must be rigorously yet respectfully enforced by all members.

6. Defined Boundaries

   In order for the project team to meet the expectations of senior leadership by proposing a solution that is effective and doable, its leader and members must be clearly aware of the boundaries within which they may propose a solution. If there are pre-existing limitations with regard to scope, legal, regulatory, financial, time, or other considerations why a potential solution might be untenable or unacceptable, the project team must be made aware of such limitations or restrictions as soon as possible.

7. Shared Vision of Success

   It is important that project team members develop a clear and compelling vision of the desired characteristics and attributes of the successful conclusion to the project. Whether it's the features, function, finish, or cost of the new product; the quality, breadth, depth, accessibility, or cost of a new service; or, the speed, quality, reduced waste or increased capacity of the transformed process—consensus among team members on how that new *thing* will meet the *narrowly defined* needs of each of the stakeholder groups is an important milestone of achievement toward project completion. It is the development and adherence to this *shared vision* that serves as the driving inspiration for launching the group onto the path toward team development and cohesion.

8. Access to Information

   As we will discuss in more depth in subsequent sections of this chapter, the guiding principles for successful team projects are threefold: (1) Trusting team members; (2) Focusing on customer success; and (3) Making fact-based decisions. In this section, we are concerned with the team having access to the data, information, and knowledge it will need to fully understand what is known about the situation in order to make decisions based upon the facts.

The project sponsor, senior leadership, and stakeholder organizations all have a vested interested in making sure that the team has access to the best information available, at all times, throughout the project. Making sure the team is given access to key personnel for information sharing, interviews, surveys, and focus groups are all an investment in the team gaining a complete understanding of the situation from all points of view—teams must be provided with access to the facts. When this access is coupled with a use of the tools for effective decision-making, team members are enabled to develop and implement solutions that meet senior leadership expectations.

9. Access to Resources

   When an organization's senior management commits to a project worthy of assembling a functional team, the management must also recognize there will probably be a need for additional resource and financial support as the project moves through the start-up, investigatory, developmental, and implementation phases of the project. Whatever limitations senior management is aware of need to be shared with the project team as soon as possible.

   The team's request for financial resources must be supported by a clear and compelling return on investment (ROI) case. Thus, allowing senior management to intelligently select among proffered solutions.

10. Ability to Perform

   Lean Six Sigma, as a total quality-management program and philosophy, offers a complete set of benchmarking, data collection, analysis, and decision-making tools and strategies to help project teams move through every phase of the project. The more experience and competence the leader and team members have with these tools and aids, the less likely they are to get *hung up* at any point in the project and the less likely they are to degenerate into unresolvable conflict. Organizations would do well to invest in training and development of key personnel in the tried and tested components of the Lean Six Sigma tools and techniques. Many of the best companies have Lean Six Sigma-qualified *resident experts* available to coach project teams as they execute their mission.

   Lean Six Sigma (formerly total quality management) is the current title for a formal approach to business process improvement through the use of employee teams trained in the application of scientific approach to removing unnecessary steps from a process and/or reducing variation at any specific step in a process. The removal of any steps in a process (steps that do not add value to the customer) is said to have made the process more *lean*. The reducing of variation occurring at any particular step in a process is focused on reducing the incidence rework or of a defective product being created at that step. The concept of Six Sigma suggests that no more than one defect per 3.4 million opportunities is allowable for the creation of a defective outcome (99.9996 percent defect-free production). The Lean Six Sigma concept is considered to be a product/service delivery philosophy or program that features a formal strategy of employee involvement through inspiration, empowerment, and enablement; with a formal methodology for problem-solving, decision-making, data collection and analysis, and ROI calculation in a data-based approach to process improvement. Participants in project teams engaged in these process improvement initiatives are often certified for their respective levels of expertise in the application of Lean Six Sigma tools and techniques.

12.1.4 Team Dynamics: The Four Phases

The transformation of an assembled group of colleagues into a high-performance work team is something that must be purposefully and intentionally managed. There are four predictable stages of development the group must go through. These stages are as follows:

1. Coming together
2. Learning about each other
3. Acceptance
4. High performance

This team dynamic is seldom static or long-lasting. Changes in leadership, mission, or membership will typically send the team repeatedly spiraling through this development process. Let us examine each of these phases, in turn, to better understand how to recognize each phase; and, more importantly, how the skilled leader can accelerate a team's passage through each in order to return to high performance as soon as possible. We are not suggesting that a team attempt to bypass any of the phases to high performance. Passage through each phase is an important developmental step; instead, the application of skills can enable the team to accelerate through the phases—(re)gaining high performance sooner and stay there longer.

1. Coming Together

Characteristics

This initial phase is a natural outgrowth of members being selected to join a team. Initially proud at having been selected, members are excited and optimistic. Each anticipates the opportunity to show what he or she can do—eager to validate their manager's choice in selecting them to represent their respective organizations on the team. At the same time, there is fear and anxiety as to whether the others selected are indeed up to the task and can be counted on to do their share of the work. While there is a tentative, initial attachment to the team, there are also suspicions and doubt. Each member silently asks him or herself, *why me, what's in it for me, what's expected of me?*

Behaviors

As members of this new team attempt to define the task, there are expressions of a need for more information and clarity. Members are waiting to be told what to do next. There are complaints about problems that may not even be germane to the task, as members attempt to learn acceptable team member behavior—as each member explores and probes to determine "who's who" on the team. If a leader has been named, all eyes are on that leader to see what he or she wants to do next—the unasked question is whether this newly selected leader is capable.

Leadership Requirements

The leader must immediately begin to build trust and confidence among team members. He or she must help members get to know one another, gaining a better understanding of why they were selected and who they are beyond a job title in the world of work. The leader must provide the information and the initial structure. The new members need clear direction and both a sense of purpose and a sense of urgency.

2. Learning About Each Other

Characteristics

This second phase is a function of team members' initial attempts at meeting and working together. Each member is striving to be accepted and liked. There is an abundance of conflict avoidance all of the attendant anxiety which typically accompanies such behavior. Each silently questions the wisdom of those selected for membership on this project team. Some team members may even begin to psychologically withdraw and consider ways to get out of the team assignment without harming their careers.

Behaviors

We begin to see some defensiveness and arguing among team members as they begin to compete with one another for position on the team's *pecking order*. As a result of the conflict that is looming, unrealistic goals are established as members attempt to impress one another.

Leadership Requirements

The team leader must resolve issues of power and authority. Ground rules must be established in order to guide team member behavior. Decision-making processes must be agreed upon, and an all-inclusive environment must be fostered as the leader pulls in everyone for equal participation. Questions about who's in charge and will I have influence are answered by the leader's demonstrated behavior.

3. Acceptance

Characteristics

The third phase is characterized by a growing sense of cohesion, acceptance of team membership, and relief that everything is going to work out fine. Team members are beginning to enjoy working together, and certain friendships are emerging. Confidence in each others' commitment on team success is growing.

Behaviors

Team members are consciously maintaining the ground rules and respecting the boundaries. Laughter and jokes can be heard, and there is a growing sense of harmony. Team members begin confiding in one another and are able to gently and constructively criticize one another.

Leadership Requirements

The team leader begins to make collaborative work the norm. Team members are encouraged to follow the ground rules as respect for personality and skill-set differences are proudly recognized, and seen as strengths to be strategically employed.

4. High Performance

Characteristics

Team members' primary loyalties shift to the team's success. Members begin to anticipate who should perform what tasks based upon an understanding of each other's strengths and challenges. Among members, there is a general satisfaction with the team's performance, thus far, and growing confidence that the team has what it takes to be successful in its mission.

Behaviors

Team members are able to anticipate each other's needs and are on the look-out for resources and information to help each other complete assignments. Team members gravitate toward each other for social settings and public recognize each other's contributions to team progress.

Leadership Requirements

The team leader must monitor work progress; and, design and lead external recognition and celebrations for achievement. The team leader must advocate for team members within their respective organizations and manage individual development.

12.1.4.1 Team Dynamics Summary

The team dynamics are an ever-changing phenomenon that requires the full attention of the team leader. Managing team member completion of assignments, participation and sense of being included, the team leader must be ever vigilant as the owner of both the duration, and the highs and lows, of the stages team dynamics cycle.

12.1.5 Roles and Responsibilities

Several roles are key for the development of a successful project team. The effective performance of individuals in each of these roles can inspire and enable project teams to that high-performance level-where the output of the team exceeds the sum of the individual inputs, i.e., synergy. The key project roles are team sponsor; leadership team; stakeholder organizations; quality management coach; team leader; team members; and ultimately, the owner of the new product or process.

1. Team Sponsor

   The team sponsor is the member of the senior staff or leadership team given or accepting the charge to solve the problem or explore the opportunity. The sponsor is the person who creates the project team and selects the team leader and members.

   Responsibilities:

   - The team sponsor, after consultation with the leadership team, defines the project.
   - The team sponsor is responsible for creating and maintaining an environment within which the project team can be successful.
   - The team sponsor, as a member of the organization's leadership team, is responsible for acting as the liaison between the project team and the leadership team.
   - The team sponsor assures that the team gains access to the resources and information needed for the project team to proceed and ultimately succeed.

2. Leadership Team

   The leadership team is composed of the senior leaders of the organization, ultimately responsible to the owners for the performance of the organization in pursuit of its mission. They are ultimately responsible for authorizing or commissioning the project team.

   Responsibilities:

   - The leadership team is responsible for creating and maintaining the sense of urgency throughout the organization that will inspire support for the project.
   - The leadership team must make itself available to host formal reviews and decisions at the key milestones of the project.

- The leadership team is responsible for providing rewards and recognition for team members for meritorious performance.
- The leadership team is responsible for providing the timely access to information and resources needed by the team in pursuit of its mission.
- The leadership team is responsible for managing an environment within which the team can be successful.
- The leadership team reviews all project accomplishments and major proposals in order to monitor the strengths and challenges of the team's approach to mission fulfillment.

3. Stakeholder Organizations

   Stakeholders are those organizations that have a vested interested or *stake* in the outcome of the project initiative, or will be significantly impacted by the results of the initiative.

   Responsibilities:

   - Depending on the level of engagement or dependency on the outcome of the initiative, stakeholder organizations must provide resources and information to assist the project team in its deliberations and initiative development.
   - Stakeholder organizations must negotiate future performance requirements with the project team.
   - Stakeholder organizations must negotiate future management system requirements with the project team.

4. Quality Management (Lean Six Sigma) Coach

   The quality management coach is the organization's expert on the application of Lean Six Sigma strategies, tools, and techniques to project team and/or problem-solving initiatives.

   Responsibilities:

   - The coach is responsible for the preparation and development of Lean Six Sigma-competent project teams and, tangentially, a workforce that is steeped in the philosophy of Lean Six Sigma.
   - The coach is responsible for the tactical execution of a program to inculcate the organization with a culture of continuous improvement.
   - The coach is responsible for project teams learning how to utilize the extensive Lean Six Sigma toolkit for exploring opportunities and solving problems.
   - The coach advises the team leader on managing team dynamics.

5. Team Leader

   The team leader is selected to guide, enable, and inspire the team through the project management process. The team leader is the lynchpin for the tactical execution of the initiative.

   Responsibilities:

   - The team leader is responsible for the orchestration of team activities.
   - The team leader serves as the contact point for project communications.
   - The team leader works with the project sponsor to resolve related progress problems.
   - The team leader manages team dynamics.

- The team leader coordinates activities among the project sponsor, coach, stakeholders, and team members.
- The team leader keeps official team records.

6. Team Members

   Team members are representatives of the various stakeholder organizations that will be significantly impacted by the initiative's outcome.

   Responsibilities:

   - Team members contribute fully to the project.
   - Team members represent the performance requirements of their respective stakeholder organizations.
   - Team members represent the management system requirements of their respective stakeholder organizations.
   - Team members share their knowledge and experience.
   - Team members participate in meetings and discussions.
   - Team members listen to others and stay open to ideas.
   - Team members assist the team leader with managing meetings, documentation, discussions, etc.
   - Team members carry out assignments between meetings.
   - Team members keep their respective managers up-to-date on teams progress and challenges.

7. Process Owner

   The process owner is the individual responsible for the on-going performance of the newly created product/process resulting from the initiative.

   Responsibilities:

   - As a member of the team, the process owner has all of those attendant responsibilities.
   - The process owner must be able to commit to the performance requirements.
   - The process owner must commit to the management system requirements.

12.2 Managing Events and Activities

12.2.1 Managing Meetings

The business meetings of the project team, and the formal review meetings between the project team and the organization's senior leadership, are both critically important events that all involved use to gauge the progress of the initiative and its likelihood of success. These meetings also offer a glimpse into the effectiveness of the team itself, even for members of the team. These meetings serve as a window into the initiative and its challenges; they also provide insight into the dynamics of the team and to what extent members are committed to team success. That these meetings be informative, effective, and even compelling is important toward:

- Inspiring and enabling team members and reviewers to experience progress in the initiative; critically important with regard to maintaining a sense of urgency for the planned change to be fully embraced
- Allowing team members an opportunity to show off their skills and talents as they offer results of their completed assignments

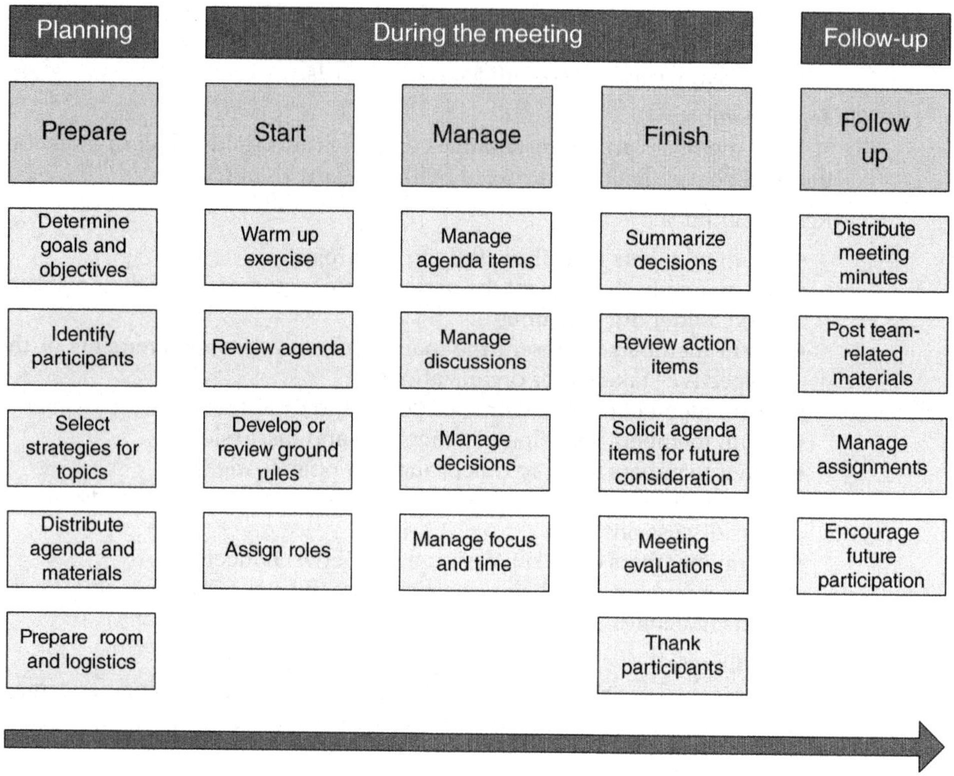

FIGURE 12.1 Meeting management process.

- Enable team members to gain access to the resources they need to continue to drive the project forward
- Provide reviewers with the wherewithal to make intelligent decisions at the critical junctions of the project

These meeting are so important to all involved so as to be worthy of an in-depth discussion in this section. It is summarized in Fig. 12.1, The Meetings Management Process.

In fact, the importance of being able to consistently host excellent meetings is better viewed as a meeting management process, rather than as a singular event. What happens between meetings will often dictate the effectiveness of the upcoming meeting event. Given that fact, let's take a look at the meeting management process, with its three distinct phases (planning, meeting, and follow-up) and how they work together to drive progress on projects and other initiatives.

1. Planning Phase: Before the Meeting

   The planning phase is about positioning the organization to do all of the right things to set the meeting up for success. It includes making sure that all of the right people are in the right place, on time, and prepared to act knowledgeably and contribute to the success of the initiative.

i. Stakeholder Analysis: Who to Invite

Stakeholder organizations are those functional or geographic organizations that have a vested interest in the initiative. What they do or how they do it will be significantly impacted by the outcome of change. If it is of critical importance that they be represented at the meeting to help shape the initiative, they should be contacted in advance so that the meeting can be scheduled around the availability of a representative of that organization, at the appropriate level for substantive contribution and making decisions.

ii. Scheduling and Location

The meeting should be on the calendars of the critically important attendees at least two weeks ahead of the meeting time and be held in a place that, if at all possible, makes it easy for them to attend. If the purpose of the meeting is to launch an initiative, the focus of the meeting should be heavily focused on establishing a foundation of personal relationships.

iii. Agenda and Background Materials

The agenda is the action plan for meeting, including the following:

- Meeting activities and time allotments
- Topics and discussion goals
- Roles and responsibilities
- Required preparation materials and actions

The agenda and associated background materials should be in the hands of all participators at least one week in advance of the meeting. An example of an agenda is given in Fig. 12.2.

iv. Meeting Accommodations

If a significant number of participants are coming from remote locations, the facilitator of the meeting will typically assure that someone assume host responsibilities so as to ensure the comfort and proper accommodations for all *out-of-town* guests. The meeting facilities should be properly sized and outfitted to meet the needs as set out in the agenda. A coordinated walk-through with the facilities manager should be conducted prior to the meeting. Logistical issues that become distractions must be avoided. As the host/facilitator/team leader, you must assume responsibility for the *whole meeting experience*.

2. Event Phase: During the Meeting

i. Meeting Roles and Responsibilities

In order to better manage participation and the business of the meeting, the host should consider assigning roles and responsibilities to several participants, in effect, forming a *meeting management team*. The purpose of this team is to: (1) make sure that meeting goals and objectives are met in a timely fashion; (2) ensure that progress, accomplishments, and key decisions made during the meeting are captured for the project archives and future dissemination; and (3) at the end of the meeting, all attendees leave feeling they have participated in a valuable experience, have been respected, and given every opportunity to voice their opinions and make appropriate contributions. That meeting leadership team is composed of the following:

a. Host. The host is responsible for managing all of the environmental, physical, and logistical elements related to the meeting event and experience.

Date: 30 March 2013 Time: 2:30–4:00 PM Location: PA 409	AGENDA		Team Goal: Advisement process improvements Meeting objective: Team launch	
Clock time	Topic	Lead	Method	Outcome
2:30–2:45 PM	Project definition	Project sponsor	Presentation w/Q&A	Problem statement
2:45–3:15 PM	Participants introductions	Team leader	Roundtable	Project team
3:15–3:45 PM	Team contract	Team leader	Brainstorming to consensus	Team contract
3:45–4:00 PM	Meeting evaluation	Team leader	Roundtable	Improved design

Figure 12.2 Meeting agenda example.

This role can be of particular importance when some or all attendees must travel from long distances or participate in the meeting via the use of technology. As the host, accommodations, amenities, and social considerations are all within the realm of responsibilities. For the host, maximizing the experience, while limiting distractions, should be the goal.

b. Facilitator. The meeting facilitator is responsible for the execution of the agenda in a way that meets the meeting goals and objectives, holds all participants accountable to living out the previously established *ground rules*, and creates and maintains an inclusive and collegial environment. The facilitator moves through the agenda one item at a time; facilitating appropriately paced discussions; managing focused and inclusive participation; selecting and executing appropriate decision-making strategies; and, ultimately soliciting the participants' evaluation of the meeting.

c. Timekeeper. Working closely with the facilitator, the timekeeper is responsible for advising on the management of time as it relates to agenda items. It is the timekeeper's responsibility to assist the facilitator in managing the pace of the meeting and keeping to the schedule of activities as set forth in the meeting agenda.

d. Scribe. Working with flip-chart paper and markers, the scribe has on display a visual record of the progress and decisions as the participants work through the agenda. These recordings are often referred to in order to keep the meeting on track and minimize *back tracking*.

e. Note-Taker. The note-taker is also keeping track of the proceedings of the meeting, but more for historical, and sometimes legal, purposes. These minutes or official records of the meeting are used to appraise stakeholders not in attendance of the major accomplishments of the meeting. Since they do represent the official records of the meeting, these minutes must be validated for completeness and accuracy by attendees before they are accepted or certified.

ii. Review the Agenda

The agenda is effectively a contract between the host/facilitator and the attendees with regards to the goals, objectives, and meeting process. As the plan for the meeting, we typically begin with a preview of the agenda, which ideally has been sent to invitees along with the meeting notification at least one or two weeks prior to the meeting.

iii. Ground Rules

The establishment or review of previously agreed-upon ground rules is an important component of meeting start-up. The ground rules represent a contract among team members and/or participants as to how they will work together in conducting team and meeting business. Typically covered by the ground rules are such topics as the following:

- Attendance: What are our expectations for attendance at meetings?
- Punctuality: What are our expectations for being on time for meetings?
- Disagreements among participants: How will we manage personal conflict between team members?
- Participation: What are our expectations with regard to member participation in discussions during meetings?
- Interruptions: How will we handle external and internal distractions to the flow of the meeting?
- Basic conversational courtesies: How will we treat each other when we are engaged in dialog?
- Confidentiality: How will we handle protecting anonymity and privacy among team members in order to encourage open and honest deliberations and discussions?
- Assignments: How will we handle work and information-gathering assignments in anticipation of conducting future meeting business?
- Breaks: How often and for how long will we take refreshment breaks during the meeting?
- Rotation of responsibilities: Who will perform which roles for leadership of the meeting(s), the facilitator, timekeeper, scribe, note-taker?
- Meeting place and time: How often, how long, and when and where will we meet?

iv. Warm-up Exercise

The purpose of the warm-up exercise is get all participants into the flow of collaborating while continually investing in the development of personal knowledge and relations among team members. These exercises normally run about a minute per participant, are engaging, invoke laughter, and are not work-related. Many are competitive and have an underlying theme of team-building, leadership development, and/or problem-solving.

v. Managing Discussions

a. Listening

Listening is the most important aspect of discussions, communications, and leadership. It is through listening that we come to understand what is being communicated. To whatever extent possible, listening involves more than simply understanding the meaning of the words being spoken. Really listening includes being attuned to the emotions and behaviors (body language) associated with the spoken word. Yes, words really matter, but so does the emotional context within the words delivered. Invaluable insight can be gained into the real needs and desires behind the words by fully focusing on the communication. If we train ourselves to fully listen to those who speak with us, instead of racing ahead of the speaker because we can hear and process information much faster than anyone can typically talk, we will be better able to understand what they intend for us to know and feel. We substantially increase our opportunity to connect with that person with whom we are engaged. Listening is the key to understanding the needs of others.

b.Active Listening

Actively listening goes a step further. This entails all of the laser-like focus on the speaker that we previously discussed, and importantly, making sure the speaker knows and feels he or she has your full and undivided attention. Active listening is an important recognition that being wholly attentive is a sign of respect—people do need to feel respected. By actively listening, we let people know that we respect them and care about what they have to say and how they feel—we connect as humans.

- Give the speaker our full attention
- Concentrate on understanding the speaker
- Listen for meaning behind the words
- Pay attention to the speaker's feelings
- Follow speaker's thinking
- Capture central ideas
- Demonstrate that we are listening by asking questions
- Translate central ideas into our own words; paraphrase
- Check the speaker for accuracy
- Relate what the speaker is saying to others' ideas

vi. Managing Meeting Behaviors

The facilitator has the distinct responsibility for managing the meeting room environment; making sure that environment is conducive to all attendees feeling engaged and respected. The facilitator must be attuned to the participation and discussion dynamics so as to keep the team feeling positive and moving forward through the agenda toward achieving the meeting's goals and objectives. In order to do so, the facilitator must not allow the attendees to take shortcuts for the sake of *saving time*. Time-saving techniques seldom work; team performance and cohesion are eventually compromised unless the team's ground rules have been clearly articulated, previously agreed upon, and consistently enforced by all members in general—and the facilitator in particular.

In managing meeting behaviors, it is important that you keep the five principles for sensitive conversations in mind:

1. Be in the right emotional frame of mind, take the time to do this to the best of your ability, or wait until later, after you have calmed down.
2. Criticize others in the way they would have you criticize them—gently and tentatively, focusing only on the specific behavior.
3. Take time to do it right, remember what you really want from the interaction—a stronger relationship with no more misbehavior
4. Make the other person feel safe (stressing the positive aspects and importance of the relationship) before you criticize; speak tentatively, seeking additional information.
5. Point out only the seemingly unacceptable behavior; never shame, blame, or attack the individual.

Managing discussions and decision-making is the primary focus of the facilitator. He or she sets the table for substantive dialogue even before the meeting starts by making sure the invitees have received whatever background materials might be appropriate to better prepare for discussions, prior to the start of the meeting. As the team moves through the agenda, it is the facilitator who opens the discussion by (1) stating the topic; (2) giving background information, including why it is important to the team; (3) explaining the intended outcome of the discussions; (4) suggesting methods for the discussion to proceed; and (5) initiating the discussion with a question. The facilitator manages the discussion, with an eye toward equal participation, by giving everyone a chance to comment, drawing reluctant participants into the discussion, corralling digressions and tangents, asking for points of clarification, and summarizing as discussions wind down. With one eye always keeping track of time, the facilitator brings the discussion to a close, and if appropriate, launches the decision-making procedure.

vii. Making Decisions

Managing the decision-making process is at least as important as managing discussions with regard to nurturing inclusiveness as a core value. While there are several methods for making decisions within a team, the preferred method is for the team to work toward achieving consensus on all major decisions. Achieving consensus requires that all team members be willing to:

- Invest the time it takes to be heard and feel respected for their diverse viewpoints
- Be encouraged to fully participate, actively seeking out different opinions
- Listening carefully, searching for alternatives, asking questions
- Keeping an open mind, thinking creatively, and not simply arguing for one's own point of view

The advantage of investing the extra time and energy to achieve a consensus among members is in keeping the team intact, without the divisiveness of having winners and losers. Achieving consensus does not mean a unanimous vote, nor does it mean that everybody gets exactly what they want. It means that everyone understands the decision, can live with it, and can explain why it is the best decision for the team. Yes, achieving consensus requires skills in communication, listening, conflict resolution, and facilitation. The pay-off is in enabling members to build off of each other's ideas and make decisions together. No member(s) feeling "left behind" because their side lost in the voting process

viii. Evaluations

At the end of each meeting, it is important that time be reserved for each of the attendees to comment on how they thought the meeting process was managed and to speak to the way that future meetings could be improved. This process of collecting immediate feedback serves to provide the facilitator with valuable insight as to how his or her *customers* feel about the service provided; it shows the customers or attendees that their time and opinions are valued and will be respected in future participation.

3. Follow-up Phase: After the Meeting

The follow-up phase of the meeting management cycle is the often-overlooked critical component of successful project management. When consistently well executed, the project moves smoothly through to a favorable outcome, and all members of the team feel proud of the results and are fully vested in the outcome. This phase is composed of three activities that enable a team leader to distinguish his or her self among team members and other stakeholders alike. The three activities are: (1) publishing the meeting minutes; (2) managing meeting attendance; and (3) managing team member assignments.

i. Meeting Minutes

The immediate publication and distribution of accurate and complete minutes of the meeting are an important symbol of the professionalism of the project environment. The minutes provide an important communication vehicle to remind attendees of accomplishments, commitments, and to set expectations for the coming next steps in the project. For those stakeholders not in attendance, the minutes serve the same purpose, and in addition, keep them abreast of the dynamics associated with the project. When properly constituted, these minutes will clearly and crisply articulate:

- The time and location of the meeting
- Who attended
- Who was expected and not in attendance
- What major issues were discussed and the conclusion of those discussions
- What were the major decisions
- What are the major commitments, or who will do what by when
- What are the next steps
- Time and location of the next meeting

An example of meeting minutes is given in Fig. 12.3.

ii. Managing Attendance

The team leader is ultimately responsible for the development and maintenance of an inclusive project environment that enables and inspires the project team to execute in an efficient and effective way. Critical to that enablement is the consistent participation of those expected to be in attendance. The leader

Meeting Minutes		
Date	Time	Location
Host/Facilitator Contact Information:		
Attendees:	Scribe:	Note taker:
Agenda Item No. 1		
Discussion Leader	Discussion Outcome	Action Items
Agenda Item No. 2		
Discussion Leader	Discussion Outcome	Action Items
Agenda Item No. 3		
Discussion Leader	Discussion Outcome	Action Items

FIGURE 12.3 Meeting minutes example.

must do whatever is necessary to keep those involved that need to be involved. To the extent that certain of those individuals may need special attention or assistance to fulfill that participation commitment, it is the leader's responsibility to provide what is needed.

iii. Managing Assignments

The team leader must be active in managing the fulfillment of commitments by offering gentle reminders, offers of assistance, and support to those who are expected to complete assignments between meetings. The team leader engagement with members between meetings encourages the member to fulfill his or her commitments. The one-on-one engagement serves as a powerful and persuasive reinforcement toward inspiring team members to meet expectations. These contacts show the members that the team leader respects and needs their participation, as they are gently yet firmly reminded of their importance to the likelihood of the project being successful.

12.3 Leading People and Managing Performance

Fundamentally, leadership defines influence in a relationship. The group's selection of a leader is derived from the consensus perception of who can best help the group achieve its primary objective. The leader's influence may be a function of the group's perception of his or her skills and abilities, knowledge of a particularly relevant subject matter, previously demonstrated success having dealt with a similar challenge, or overall dominance of the group. As the situation, challenge, goal, or membership changes, the group may opt for a different leader—leadership is situational and can be fleeting.

Good leadership makes a significant difference in a group or organization's performance. Depending upon the extent, an organization's performance is a function of collaborative knowledge and creativity, the performance difference can be as much as 15 to 25 percent.

While managers are appointed by the owners to get the work done via planning, organizing, directing, staffing, controlling, innovating, and recording; leaders focus on the people in the organization. They know who their people are, what they know, what they feel, where they've been, what they've done, and/or what they can do and be. Leaders know and grow their people. They inspire and involve them. They recognize and reward them. Real leaders are servants to the needs of the people in the organization; they inspire and enable them to be the best they can be for the sake of the organization's mission objectives.

Let's take a more in-depth look at what leaders must do for their organizations; let us explore the responsibilities of organizational leaders.

12.3.1 Leadership Responsibilities

For these twenty (20) responsibilities, there are behaviors that enable leaders to drive a distinctive culture throughout their team.

1. Share a Clear and Compelling Vision
   Leaders need to galvanize the members of the team around the mission, or the reason for the team's existence. Given this mission, what does he or she see for the team at some point in the future? What are the leader's aspirations for the team?
2. Define the Mission and Share the Vision
   Let everyone in and around the team know why you exist as a team, where you are going, and the important role they play in helping you all get there.
3. Proclaim for All to Hear
   Communicate your vision for the team's success and the core values you hold dear. Make sure all team members are able to articulate the mission, vision, and values of the team.
4. Listen Actively for Criticism
   Listen for feedback from the members and stakeholders; seek criticism and embrace opportunities to be a better leader.
5. Carve the Path and Lead the Way
   Leaders lead. They tell members where they are going, and how they are going to get there together. In everything they say and do, they make it abundantly clear what is important with regard to roles and responsibilities, performance, values, and culture.
6. Live Out the Values
   Leaders do more than give directions; they know that group members watch everything they do. Leaders are always on stage, so they live out team's values all of the time.

7. Reward and Recognize Behaviors
Leaders purposefully and intentionally drive culture throughout the team. They reward and recognize exemplary behaviors and performance. They offer correction and guidance to those in need of improvement.
8. Demand Accountability
Leaders drive a culture of trust and accountability. People are given an opportunity to show what they can do and are held accountable for their behaviors and performance.
9. Capitalize on Differences and Manage Change
Leaders recognize the value of differences—identity, cultural, and cognitive differences. Leaders know that those differences are the source of why and how teams typically outperform any single individual on complex problems and when creative solutions are required.
10. Embrace and Celebrate Differences
Leaders do more than tolerate differences among group members; they openly embrace and celebrate those differences, for all to see.
11. Create a Culture of Inclusiveness and Fairness
Leaders seek diversity; they intentionally create a culture that encourages personal relationships and yet demand that everyone be treated fairly.
12. Guide and Direct Innovation and Change
Leaders enlist everybody in their efforts to continuously improve the effectiveness of their team. They inspire and enable product/process improvement initiatives.
13. Engage the Hearts and Souls in the Culture
Leaders know the importance of keeping everyone fully in engaged and vested in the success of the team. They know that if they can help team members to see their individual success directly tied to the team's success, it will be reflected in the productivity of the team.
14. Know and Grow the People
Leaders know their people. They recognize the intrinsic need for them to feel they are growing in the mastery of their craft. Leaders invest in the continuous development of all team members—they help them get better.
15. Give All a Sense of Purpose
Leaders understand the importance of team members being proud of the team's mission and their performance in fulfilling that mission. Leaders track progress and celebrate accomplishments. They maintain a sense of urgency throughout the project.
16. Allow Them to Show What They Can Do
Leaders understand the importance of team members feeling respected and trusted. They give team members every opportunity to show their talents and that they can be trusted.
17. Meet the Need for Knowledge, Resources, and Tools
Leaders understand their responsibility to provide team members with an environment conducive to teamwork and high performance; this includes providing team members with the knowledge, skills, resources, and tools they need to perform at a high level and continuously improve.

18. Be a Strong Advocate
    Leaders understand that the team is dependent upon the leader to be the face of the team and, an effective advocate to get the team access to the information and resources it needs to perform at a high level.
19. Be a Good Steward and Colleague
    Leaders understand the importance of their mission and the trust that has been bestowed upon them by upper management. They are judicious in the use of resources and strive for positive relationships all around the team—for the sake of the team.
20. Be a Servant to Team's Needs
    Leaders see themselves as a servant to the needs of the team and individual team members, making sure that each team member is enabled and inspired to be at his or her best.

12.3.2 Motivating Team Members

Over the last few years, we have come to better appreciate the so-called knowledge worker with his or her unique characteristics. The knowledge that once an individual's basic financial needs are satisfied, the primary motivators come from within—the individual's own internal satisfiers, they are said to be more intrinsic in nature. By that I mean, unlike the production worker who receives more money by working harder (extrinsic motivation) and producing more output; for the so-called knowledge worker we have found these intrinsic motivators to be comprised of mission, recognition, and trust.

Mission As social beings, we want to be proud of what we do for a living and would like to believe that we contribute to making the world, or at least our corner of the world, a better place. The mission, or reason for being, of the organization on whose behalf we ply our craft and make our livelihood is an important factor in how we want to see ourselves contributing to the betterment of our society. This sense of belonging and contributing to an organization that is *bigger* than us is foundationally motivating to us. We are motivated by our perception that we somehow contribute to making things better for others.

Recognition for Accomplishments In addition to a sense of mission, we are also motivated by feelings of accomplishment; an internal belief or external recognition that we have a mastery of our craft. We want to know that we are good at what we do and are appreciated for it. Whether derived from internal or self-satisfaction, or the recognition of those we respect, being excellent at what we do for a living is a huge source of motivation—indeed inspiration, for each of us.

Trust It is important for us to know that people rely on us to do what we say we are going to do. It is a sign that we are a respected contributing member of our social group or society. To be "trusted" is an important validation of how well we are thought of and appreciated we can be trusted to fulfill our obligations, to exhibit our mastery of our craft, and to do our part as a member of the team—any and all of these attest to our sense of self-worth as a member of society.

Summary Mission, trust, and recognition for accomplishments; while these intrinsic motivators are within an individual's own capacity to satisfy, a well-versed team leader makes sure to foster a team environment that lends itself to, and facilitates the self-fulfillment of such core needs. In other words, it is the team leader's responsibility to foster an environment within which one can be motivated. While, as we have previously stated, money and other extrinsic rewards may not, in the long run, increase the output or creativity of the knowledge worker, the lack of money can be a huge de-motivator. When the lack of money is seen as an indicator or metaphor for a lack of trust and respect, especially among peers and other colleagues, it can be a powerful de-motivator.

12.3.3 Team Communications

Communication is how we share information, knowledge, and coordinate actions. Dictionaries tend to define it as a process by which these exchanges take place via a shared understanding of words, signs, symbols, and behaviors.

Given that a team is a group of individuals dependent upon one another to achieve a common goal, with defined roles and responsibilities, and bound by common rules, we know that effective communications among members are essential to enable the close coordination required for them to succeed. If that communication is to be open and honest and focused upon helping each other continuously improve as a seemingly unified organism, then it must be built upon a commonly held belief that we are all focused solely upon that one goal. That fundamental belief that we are in this together, win together, or lose together, must be steeped in a deep and abiding trust in each other as members of the same team—all for one and one for all! This can only happen when every member of the team, led by the team leader, intentionally and purposefully works to foster an environment conducive to open and honest, helping communications.

Personal Relationships as a Foundation In order for team members to grow to trust in one another, the team leader must intentionally foster the development of personal relationships among team members. These personal relationships, social in nature, provide a foundation that can lead to individuals feeling that they are members of the same pack—feelings that they are in it together, with a growing affinity for one another as they learn more about each other, begin to empathize with one another, as they learn they share the same wants and needs for themselves and their families, as they learn about their sameness as persons, even as they may have differences at work.

Trust Trust tends to develop over time. It is a function of team members consistently living up to their own set of expectations; individuals committed to the team, doing what they say they are going to do, when they say they are going to do it, to the best of their ability, over and over again. When a member exhibits this behavior so consistently that you can begin to predict their performance, we have the beginnings of trust. Once trust in commitment to team performance exists among team members, there exists a team environment within which the opportunity for open and honest criticism among members can be embraced for the sake of the betterment of team performance. Team members get better, and team performance is improved when members can openly and honestly, gently, critique one another—all for the sake of improving team performance.

Commitments to Goal, Values, and Rules Upon this foundation, or more likely, even as it is being built, the team sponsor and team members come to agreement on the team goals or reason for creation; the values they hold dear and how those values are translated into behaviors, such as how they will treat each other, and how and when they will work together. It is important they all agree on the fundamental purpose of the team and these other instrumental team issues.

Open and Honest Communications Open and honest communications is the core requirement for teams to become effective. Since teams are typically composed of individuals of different geographic or functional knowledge and responsibilities, the members are dependent upon learning from each other in order to solve problems or craft solutions that meet the needs and desires of the whole organization. That each of these stakeholder groups, as represented by the respective team members, have their issues and concerns properly figured into the equation, is a function of effective communications. Open and honest communications is the highly prized outgrowth that we seek from our considerable investment in the development of our platform for effective team building of: personal relationships; commitment to goal, values, and rules; and trusting team members.

Communications Model Communications is a dynamic process initiated when there is a desire to share or exchange ideas, thoughts, or feelings. In doing so, it is critically important for the communicator to clearly understand what it is that they want the other(s) to know, what do they want them to do, and how do they want them to feel as a consequence of the communication. Clearly understanding these three things—and being able to empathize and adapt in anticipation of how the intended receiver may interpret them—before attempting to communicate will help one go a long way toward successful messaging. This ability to empathize with the receiver(s) before initiating the communication is a function of being able to appreciate the receiver's background, beliefs, values, and language skills; and, the importance of all toward the receiver's ability to comprehend the message.

Ideally the sender's communications will be clear, concise, cogent, complete, consistent, and, above all, correct—thus greatly enhancing the likely effectiveness of the communications.

Sharing these thoughts can be done through writing, speaking, gestures, sign language, or even body language. In order for the communication to be successful (the receiver understands the sender's intended message), the parties involved must have a common understanding of the meanings of the words, tone, gestures, and body language employed. Given all of these components, it is easy to see why face-to-face communications should be utilized for communications whenever possible. In addition to enabling the receiver to gather supporting information from all of these nonverbal sources (tones, gestures, and body language), face-to-face communications provide an opportunity for the receiver to ask clarifying questions and for the initiator, or sender, to seek confirmation as to the receipt of the intended message.

Communication Means Other means of communications can only be employed if it is not reasonable to communicate face-to-face. Thus, in the hierarchy of project communications, face-to-face is best, video or teleconferencing is second, and any written form of communications is the least desired for other than the creation of a more permanent record of communications.

12.3.4 Managing Conflict

The real strength of teams is more than simply a multiplication of the number of individuals brought to a task; it is instead in the differences attendant to the way the team has been composed. Whether the differences are a function of (1) Geographic regions represented and their unique considerations; (2) Functional organizations represented with their own areas of concerns and responsibilities; (3) Cultural or identity differences with their particular physical, psychological, and social preferences; (4) Cognitive differences in the way they relate to the world or process information and make decisions; or (5) Individual differences in that each of us have our own unique worldview that is a function of what we believe to be true about the world we live in and all that it contains.

The point is that because of all of these possible differences, no two of us will ever see everything in exactly the same way. There will be conflicting viewpoints with regard to goals, objectives, methods, and means. The ability to embrace and capitalize on all of these differences, and meld the members into a high-performance team, working in a seemingly unified way to achieve success, is the true measure of the effectiveness of the team's leadership. A summary of conflict types, possible causes, and resolution strategies is given in Table 12.1.

1. Types of Conflict

   When conflict is limited to a focus on discovering the best solution toward achieving the team's agreed-upon goals and objectives or when the conflict is respectful of the team's agreed-upon roles, responsibilities, behaviors, and boundaries, it can be said to be "constructive conflict." This constructive conflict is the real power of the team, and must be sought after and embraced; this is the team leader's core mission and responsibility. When conflict is personal and/or disruptive, when it violates agreed-upon considerations, when it leaves other team members feeling disrespected or excluded, it can be said to be "destructive conflict." This destructive conflict cannot be tolerated, and must be thoughtfully and skillfully addressed by the leadership. If not addressed, as soon as is

Type of Conflict	**Possible Causes**	**Resolution Strategy**
Facts	Conflicting Interpretation Uncertain direction Inadequate information	Agree on criteria and find
Process or Methodology	Alternative apporaches Alternate emphasis or Resources	Discuss alternative and agree on criteria
Responsiblities	Undefined or confused ownership Lack of buy-in or accountability Structural, Procedural, or social impediments	Clarify and renegotiate
Objectives	Lack of clarity Conflicting goals Inadequate resources	Redefine or refer to prevously agreed upon priorities
Values	Different motivation or beliefs Different morals or ethics	Acknowledge difference and either find common ground

Table 12.1 Summary of Conflict Types, Causes, and Resolution

reasonably possible, every time it happens the resentment will fester, and eventually the camaraderie and inclusive team environment, essential to the achievement of high performance, will be negatively impacted, and in the worst-case scenario, perhaps destroyed. The distinction between constructive and destructive conflict is of critical importance. Team leadership must be ever vigilant and attentive, encouraging and nurturing the former, while jealously guarding against the latter with a "zero tolerance" policy.

Destructive Behaviors The team leader has to take the lead in managing team member behaviors. Developing a team environment where there are good behavior expectations and accountability is a key reason why it is important to invest in the development of a solid foundation based upon personal relationships and trust before diving into the actual work content of the project. Second, the early investment in a Team Charter, composed of a clear problem statement and strong team member contract, will make it much easier to keep team members in line throughout the length of the project. However, even with this strong foundation in place, it is inevitable that there will be instances of team-member misbehavior with which the team leader must be prepared to deal.

Most importantly, bad behavior cannot be tolerated, nor can the leader simply sit back and hope that it won't be repeated. If not addressed, the behavior most often gets repeated, and eventually begins to destroy relationships and the overall team environment. The kinds of destructive behavior that the team leader must be ever watchful for include the following:

- *Tardiness.* Team member promptness and attendance is a cornerstone of being a good teammate. It is crucial that meeting time and place be set in a way that is considerate of team member's willingness and ability to be on in attendance on time, all of the time. Once such an agreement has been achieved, all team members must be held to the agreement. The lack of attendance and/or promptness is typically a harbinger of additional behavior problems.
- *Broken Promises.* When a team member fails to live up to a commitment, the leader must confront that person and remind him or her of the importance to team success for all members to live up to their commitments and to make themselves available to assist in the future. The leader can act to prevent these incidences by working with members between meetings to live up to their commitments.
- *Dominating Participants.* Team members that tend to dominate discussions or to pursue adoption of their ideas with an unacceptable amount of vigor, need to be reminded of the importance of the team hearing from other divergent viewpoints. The use of a timekeeper can be of great assistance in keeping these efforts to manage participation equally.
- *Reluctance to Participate.* Recognizing a reluctance of certain members to participate in meetings can, in and of itself, go a long way toward enabling those members to gain confidence. Follow-up private conversations in an effort to diagnose the problem together can further inspire fuller participation.
- *Groupthink.* One of the biggest impediments to high performance is a function of team members simply agreeing with one another for the sake of not risking conflict. These superficial behaviors happen due to a lack of trust among members that

disagreement and conflict is all aimed toward genuinely seeking a better decision for the sake of team accomplishment. Groupthink can and must be prevented by the team leader. The key component for the prevention of groupthink is trust. Trust among members to feel free to critique ideas and constructively criticize any decisions that aren't based upon facts and data is all for the betterment of team performance.

- *Rush to Accomplishment*. Even though a team's mission should be caste with some sense of urgency, the team must adhere to a formal problem-solving or opportunity-investigation methodology. The use of a formal methodology precludes taking shortcuts and rushing to finish the task without exploring alternative approaches, and when appropriate, making decisions based on collected and analyzed data.
- *Attribution Failures.* In the execution of a project employing a Six Sigma-type formal methodology, it is critical that all decisions be based upon the collection and analysis of data or properly attributed facts as put forth by a creditable authority. Failure to do so, in every case, is not acceptable, and all members must be held accountable to that standard.
- *Discounting Team Members.* Team members that tend to either talk over or discount the value offered by certain other teammates can be particularly troubling. The emotional and psychological damage they can inadvertently inflict on the offended team member cannot be overestimated. In the most severe cases, both the offender and the offended will need private counseling. Any behavior that devalues any other member of the team is unacceptable. A core principle of the team concept has to be that it be "inclusive" in order to gain the full benefit of the team's diversity.
- *Digressions and Tangents.* Team members that tend to take discussions off-track or want to rehash previous discussions or decisions can have a serious impact on team progress. The use of a timekeeper, meeting minutes, and an agenda can be effective deterrents to such behavior. A gentle reminder from the team leader may also be needed to make the team member aware of such behavior may also be necessary.
- *Feuding Team Members.* Team members that bring a hostile relationship onto the team from previous encounters, or whose relationship turns contentious over the course of the project, can be a real threat to the health of the team's work environment. The leader must intercede to the extent that the issues bubble over into the team business. The leader may see a need to sit down with both parties together to facilitate commitment to a working relationship that does not negatively impact the team.

2. Conflict Management Styles

   As unique individuals, we react differently when faced with conflict. In fact, each of us tends to revert back toward a default tendency. It is this tendency that we call our "conflict mode." Think of it as comparable to left-hand/right-hand favorability; other things being equal, what comes most naturally. The four conflict modes most widely cited are as follows (see Fig. 12.4):

- *Avoiders.* These are people who would just as soon not engage in conflict. As such, they tend to be nonconfrontational and less assertive and declarative

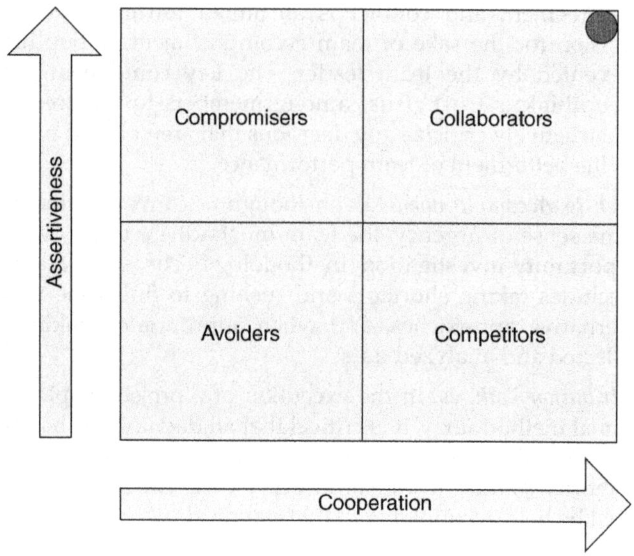

FIGURE 12.4 Conflict modes.

when asked to offer an opinion. The challenge for avoiders when in leadership positions is that they may allow bad behavior among team members to linger and fester, in hopes the bad behavior will not be repeated, thus enabling the leader to avoid the needed confrontation. Unfortunately, if a team member is not confronted when exhibiting bad behavior, and asked to cease and desist, a tacit acceptance is implied, and the behavior will typically be repeated to the detriment of the team working environment.

- *Competitors.* These are people who quickly become bent on competing and winning any confrontation. As such, they are prone to see conflict as an opportunity to impose their will. The challenge for competitors is not to get so caught up in the prospect of winning or losing, as to lose sight of what is the best decision for the sake of increasing the team's likelihood of achieving success. Competitors can be sidetracked by becoming more concerned with the score rather than the merits of the conflict.
- *Compromisers.* These types are more concerned with quickly bringing the conflict to closure, thus ending the confrontation. All too often, such haste to resolution can lead to the adoption of a sub-optimal solution. The challenge for compromisers is to strive for the best possible solutions, even though it may take more time and require an additional investment in team energy.
- *Collaborators.* People who tend to have collaboration as a default approach to conflict and are naturally inclined to have the vision and patience to seek out the optimal solutions to resolving conflict. They tend not to be distracted by concerns with regard to winning or losing these confrontations, but instead are able to maintain focus on the team's previously agreed-upon goals and objectives. They are often able to resist the urge to *cave-in* to the first possible

solutions, while continuing to seek-out the optimal solution. The challenge for collaborators is to keep the rest of the team from rushing to judgment.

While each of us is fully capable of employing any of the four modes, depending on the situation, and revert to varying levels of cooperation and assertiveness, we tend to default to favoring one mode in particular. In better understanding our default approach to conflict, we are better able to manage our challenges and accept differences in others.

3. Conflict Intervention

Open and honest criticism can lead to conversations that build substance into a relationship. Open and honest criticism when delivered in a constructive and helping way, can enable participants to identify where there is agreement and where there is disagreement.

Once these areas of contention have been identified and clarified, the parties can determine the reason for the conflict. Typically, these conflicts revolve around facts, processes, responsibilities, objectives, or values. Regardless, once brought to light, they are resolvable.

It is important to note that the deeper the type of conflict (downward toward *values*) and the longer-lasting (time left unresolved) the conflict, the more difficult it will be to resolve.

Determining whether or not confrontation between members requires intervention by the team leader is primarily a judgment call; a call that can be important to the team's performance. Determining whether confrontation is constructive can lead to an enlivened team environment in search of the best decision for the team's sake; if the confrontation is potentially destructive, it has potential to leave some team members devalued or frustrated. If in the team leader's judgment the conflict may be of the latter type then the situation must be promptly analyzed to determine what to do about it and when.

Among the questions on the checklist are the following:

- *The Right Issue.* Is the nature of this conflict constructive or destructive? Have the previously agreed-upon team accomplishments or behaviors been violated?
- *The Right Time.* Are the emotions of those involved elevated to an extent that it would not be a good time to intervene and attempt to gain dialogue?
- *The Right Place.* Are too many observers present for us to achieve meaningful dialogue without distortions and interruptions? Are all of the right people present for us to fully resolve this issue in a meaningful way?
- *The Right Person.* Am I the right person to call this person's attention to the issue at hand? Do I have the authority, responsibility, and credibility to command this person's attention to this issue? Given the situation, how would I like to be approached on this issue? Given what I know about this person that I will engage, how would they like to be approached?

4. Intervention Behaviors

Intervention and crucial conversation skills are certainly among the most important tools that a team leader needs to develop for his or her toolset. Fortunately, they are learned skills that can be acquired and continually improved with practice. Keeping the following ten (10) considerations in mind will provide you with a solid foundation upon which to practice and refine your conflict intervention skills.

a. Be Supportive: It is important that the team leader not initiate the discussion by shaming, blaming, or attacking the individual who may be in need of correction. Instead, the importance of the individual to the team should precede any query into the team member's questionable behavior.

b. Be Tentative: The leader's voice tone, body language, and words should all be oriented toward seeking to learn more about the situation. No shame, no blame, and no attack—ever. The leader must stick with the known or observable facts of the situation and refrain from assigning intent to anyone. No shame, no blame, and no attack—ever.

c. Be Clear: The leader needs to specific and clear in recalling the incident or behavior in question. The language employed should be easily understood and directed at the behavior only.

d. Be Sincere: It is important the leader makes it clear that any violation of previously agreed upon ground rules is a serious issue. The leader must, above all, be authentic in everything said and every feeling conveyed.

e. Listen Actively: As always, active listening is a core requirement to team-member engagement. It is particularly important in reinforcing the leader's respect and value for the team member whose behavior is being questioned.

f. Feedback Constructively: During the conversational engagement, the leader should utilize open-ended questions in order to learn more about the incident and understand the feelings during this conversation. Remember that criticism is most easily digested when offered up in a clear and concise fashion between reinforcement and appreciation—a sort of "goodwill sandwich."

g. Act Curious: The leader should attempt to understand why this good and able team member has possibly behaved in a way that is not a benefit to the team. There are no villains here; instead, only good team members perhaps behaving badly. Why?

h. Stay Focused: It is important that neither party be distracted during this conversation. The focus and intention must remain on resolving the issue at hand. If during the course of the conversation, an issue is raised that is either more time-sensitive, or of more importance to team welfare, then the leader must make a conscious and formal decision to *bookmark* the original issue and focus on the newly uncovered issue.

i. Be Patient: Every one-on-one conversation with a team member is an opportunity to further the bond among team members. The leader should be prepared to make the time available for that bonding to be nurtured as this conversation begins to wind down and to learn more about each other as people.

j. Offer Assistance: It is important that the leader continually offers his or her availability to provide assistance or to do a leader's part to resolve the issue and keep the team on track for high performance. We are in this together.

12.4 Our Project Team Leadership Summary

1. Be Transparent and Lead By Example: Live-out, proclaim, and sell your vision and values; be a positive and inspirational influence; become well-known for open and honest decision-making.
2. Set Clear Expectations Together: Create a behavioral contract and hold each accountable for compliance. Set targets and milestones along the way.
3. Be Inclusive in All Matters: Seek-out and embrace organizational, cultural, identity, and cognitive diversity. Invest time to gain consensus agreement on decisions—avoid voting when at all possible.
4. Hear First and Listen Actively: Listen actively to all around you; they will all appreciate it. As the leader, be last to offer your opinion, what you think tilts the scale—hear from others and learn along the way—first.
5. Seek and Embrace Criticism: Encourage constructive criticism for all that you do—feedback is a critical component to getting better. When seeking feedback, ask open-ended questions to gain thoughtful appraisals.
6. Maintain Positive Relationships and Delegate Tasks: Continually invest in the development and maintenance of positive relationships everywhere. Focus on inspiring and enabling the team members, and let them take care of the work.
7. Plan, Perform, and Measure Results: Develop agreed-upon project goals and objectives; manage to your plans and measure all results. Benchmark existing conditions, envision desired future state, and chart path to success.
8. Enable and Grow Talent: Provide team members with a positive and supportive project environment within which they can excel. Be a servant to team-member needs for opportunities, knowledge, skills, and tools to show their best work.
9. Celebrate Success and Behaviors: Recognize the importance of progress and achievement—take time to celebrate team accomplishments. Reward and recognize team members that exhibit the values and behaviors needed for the team to achieve high performance.

References and Bibliography

Bens, I., *Facilitating with Ease! Core Skills for Facilitators, Team Leaders and Members, Managers, Consultants, and Trainers*, 3rd ed., Wiley Jossey-Bass, New York, 2012.

Bradberry, T., and J. Greaves, *Emotional Intelligence 2.0*, TalentSmart; Har/Dol En edition, New York, 2009.

DuBrin, A., *Leadership: Research Findings, Practice, and Skills*, 6th ed., South-Western/Cengage Learning, Mason, OH, 2010.

Heath, C., and D. Heath, *Switch: How to Change Things When Change is Hard*, Crown Business, New York, 2010.

Kouzes, J., and B. Posner, *The Leadership Challenge: How to Make Extraordinary Things Happen in Organizations*, 5th ed., Wiley Jossey-Bass, New York, 2012.

Page, S., *Difference: How the Power of Diversity Creates Better Groups, Firms, Schools, and Societies*, Princeton University Press, Princeton, NJ, 2008.

Pande, P., R. Neuman, and R. Cavanagh, *The Six Sigma Way Team Field Book*, McGraw-Hill, New York, 2002.

Patterson, K., J. Grenny, R. McMillan, and A. Switzler, *Crucial Confrontations: Tools for Resolving Broken Promises, Violated Expectations, and Bad Behavior*, New York, McGraw-Hill, 2002.

Patterson K., J. Grenny, R. McMillan, and A. Switzler, *Crucial Conversations: Tools for Talking When Stakes Are High*, McGraw-Hill, New York, 2002.

Patterson K., J. Grenny, R. McMillan, and A. Switzler, *Influencer: The Power to Change Anything*. McGraw-Hill, New York, 2002.

Peoples, D., *Presentations Plus: David Peoples' Proven Techniques*, revised 2nd ed., Wiley, New York, 1992.

Pink, D., *Drive: The Surprising Truth About What Motivates Us*, Riverhead Books, New York, 2009.

Schein, E. H., *Organizational Culture and Leadership*, 4th ed., Wiley Jossey-Bass, New York, 2010.

Scholes, P., B. C. Joiner, and B. Streibel, *The Team Handbook*, 3rd ed., Oriel Inc., Madison, WI, 2003.

Discussion Topics

1. Why is it important to consider all stakeholder organizations when forming a project team?
2. Why is it important to gain agreement among all sponsors and team members on a written problem or opportunity statement?
3. What is the sponsor's role and what are the associated responsibilities?
4. Why take the time for team members to learn about each other's personal lives?
5. What are the responsibilities of the team leader, and what are the most desirable skills and behaviors?
6. What is the value of employing a formal problem-solving methodology and making all of the team's decisions based upon data and proven facts?
7. Why not vote on everything?
8. Why do most teams fail?
9. Why is it important to identify interim milestones and to celebrate their attainment?
10. What are a team member's primary responsibilities and most desirable traits and behaviors?
11. Why is it important for the whole team to participate in the development of a team contract?
12. What is the value of disagreement among team members?
13. How can the team leader recognize groupthink? What should he or she do about it?
14. Once basic needs for compensation have been met, how do you motivate the knowledge worker?

15. Why is it important for the team leader to adopt a *zero tolerance* policy for each team member not living up to the behaviors agreed upon in the team contract?
16. What is the value of utilizing flip-chart paper and a scribe at all team meetings?
17. What is the value of having a detailed agenda published before each meeting and publishing minutes soon after each meeting?
18. Why is the Six Sigma approach to solving problems, including tools and techniques, so important to team success?
19. Why is it important for the team leader to understand team dynamics?
20. Why is it so important for the team leader to employ active listening technique?

APPENDIX

ROI Tables

i% n Yrs	1%	2%	3%	4%	5%	6%	7%	8%	9%	10%	11%	12%	13%	14%	15%	16%	17%	18%	19%	20%
1	1.01	1.02	1.03	1.04	1.05	1.06	1.07	1.08	1.09	1.1	1.11	1.12	1.13	1.14	1.15	1.16	1.17	1.18	1.19	1.2
2	1.02	1.04	1.06	1.08	1.1	1.12	1.14	1.17	1.19	1.21	1.23	1.25	1.28	1.3	1.32	1.35	1.37	1.39	1.42	1.44
3	1.03	1.06	1.09	1.12	1.16	1.19	1.23	1.26	1.3	1.33	1.37	1.4	1.44	1.48	1.52	1.56	1.6	1.64	1.69	1.73
4	1.04	1.08	1.13	1.17	1.22	1.26	1.31	1.36	1.41	1.46	1.52	1.57	1.63	1.69	1.75	1.81	1.87	1.94	2.01	2.07
5	1.05	1.1	1.16	1.22	1.28	1.34	1.4	1.47	1.54	1.61	1.69	1.76	1.84	1.93	2.01	2.1	2.19	2.29	2.39	2.49
6	1.06	1.13	1.19	1.27	1.34	1.42	1.5	1.59	1.68	1.77	1.87	1.97	2.08	2.19	2.31	2.44	2.57	2.7	2.84	2.99
7	1.07	1.15	1.23	1.32	1.41	1.5	1.61	1.71	1.83	1.95	2.08	2.21	2.35	2.5	2.66	2.83	3	3.19	3.38	3.58
8	1.08	1.17	1.27	1.37	1.48	1.59	1.72	1.85	1.99	2.14	2.3	2.48	2.66	2.85	3.06	3.28	3.51	3.76	4.02	4.3
9	1.09	1.2	1.3	1.42	1.55	1.69	1.84	2	2.17	2.36	2.56	2.77	3	3.25	3.52	3.8	4.11	4.44	4.79	5.16
10	1.1	1.22	1.34	1.48	1.63	1.79	1.97	2.16	2.37	2.59	2.84	3.11	3.39	3.71	4.05	4.41	4.81	5.23	5.69	6.19
11	1.12	1.24	1.38	1.54	1.71	1.9	2.1	2.33	2.58	2.85	3.15	3.48	3.84	4.23	4.65	5.12	5.62	6.18	6.78	7.43
12	1.13	1.27	1.43	1.6	1.8	2.01	2.25	2.52	2.81	3.14	3.5	3.9	4.33	4.82	5.35	5.94	6.58	7.29	8.06	8.92
13	1.14	1.29	1.47	1.67	1.89	2.13	2.41	2.72	3.07	3.45	3.88	4.36	4.9	5.49	6.15	6.89	7.7	8.6	9.6	10.7
14	1.15	1.32	1.51	1.73	1.98	2.26	2.58	2.94	3.34	3.8	4.31	4.89	5.53	6.26	7.08	7.99	9.01	10.15	11.42	12.84
15	1.16	1.35	1.56	1.8	2.08	2.4	2.76	3.17	3.64	4.18	4.78	5.47	6.25	7.14	8.14	9.27	10.54	11.97	13.59	15.41
16	1.17	1.37	1.6	1.87	2.18	2.54	2.95	3.43	3.97	4.59	5.31	6.13	7.07	8.14	9.36	10.75	12.33	14.13	16.17	18.49
17	1.18	1.4	1.65	1.95	2.29	2.69	3.16	3.7	4.33	5.05	5.9	6.87	7.99	9.28	10.76	12.47	14.43	16.67	19.24	22.19
18	1.2	1.43	1.7	2.03	2.41	2.85	3.38	4	4.72	5.56	6.54	7.69	9.02	10.58	12.38	14.46	16.88	19.67	22.9	26.62
19	1.21	1.46	1.75	2.11	2.53	3.03	3.62	4.32	5.14	6.12	7.26	8.61	10.2	12.06	14.23	16.78	19.75	23.21	27.25	31.95
20	1.22	1.49	1.81	2.19	2.65	3.21	3.87	4.66	5.6	6.73	8.06	9.65	11.52	13.74	16.37	19.46	23.11	27.39	32.43	38.34
21	1.23	1.52	1.86	2.28	2.79	3.4	4.14	5.03	6.11	7.4	8.95	10.8	13.02	15.67	18.82	22.57	27.03	32.32	38.59	46.01
22	1.24	1.55	1.92	2.37	2.93	3.6	4.43	5.44	6.66	8.14	9.93	12.1	14.71	17.86	21.64	26.19	31.63	38.14	45.92	55.21
23	1.26	1.58	1.97	2.46	3.07	3.82	4.74	5.87	7.26	8.95	11.03	13.55	16.63	20.36	24.89	30.38	37.01	45.01	54.65	66.25
24	1.27	1.61	2.03	2.56	3.23	4.05	5.07	6.34	7.91	9.85	12.24	15.18	18.79	23.21	28.63	35.24	43.3	53.11	65.03	79.5
25	1.28	1.64	2.09	2.67	3.39	4.29	5.43	6.85	8.62	10.83	13.59	17	21.23	26.46	32.92	40.87	50.66	62.67	77.39	95.4
26	1.3	1.67	2.16	2.77	3.56	4.55	5.81	7.4	9.4	11.92	15.08	19.04	23.99	30.17	37.86	47.41	59.27	73.95	92.09	114.48
27	1.31	1.71	2.22	2.88	3.73	4.82	6.21	7.99	10.25	13.11	16.74	21.32	27.11	34.39	43.54	55	69.35	87.26	109.59	137.37
28	1.32	1.74	2.29	3	3.92	5.11	6.65	8.63	11.17	14.42	18.58	23.88	30.63	39.2	50.07	63.8	81.13	102.97	130.41	164.84
29	1.33	1.78	2.36	3.12	4.12	5.42	7.11	9.32	12.17	15.86	20.62	26.75	34.62	44.69	57.58	74.01	94.93	121.5	155.19	197.81
30	1.35	1.81	2.43	3.24	4.32	5.74	7.61	10.06	13.27	17.45	22.89	29.96	39.12	50.95	66.21	85.85	111.06	143.37	184.68	237.38
31	1.36	1.85	2.5	3.37	4.54	6.09	8.15	10.87	14.46	19.19	25.41	33.56	44.2	58.08	76.14	99.59	129.95	169.18	219.76	284.85
32	1.37	1.88	2.58	3.51	4.76	6.45	8.72	11.74	15.76	21.11	28.21	37.58	49.95	66.21	87.57	115.52	152.04	199.63	261.52	341.82
33	1.39	1.92	2.65	3.65	5	6.84	9.33	12.68	17.18	23.23	31.31	42.09	56.44	75.48	100.7	134	177.88	235.56	311.21	410.19
34	1.4	1.96	2.73	3.79	5.25	7.25	9.98	13.69	18.73	25.55	34.75	47.14	63.78	86.05	115.8	155.44	208.12	277.96	370.34	492.22
35	1.42	2	2.81	3.95	5.52	7.69	10.68	14.79	20.41	28.1	38.57	52.8	72.07	98.1	133.18	180.31	243.5	328	440.7	590.67
36	1.43	2.04	2.9	4.1	5.79	8.15	11.42	15.97	22.25	30.91	42.82	59.14	81.44	111.83	153.15	209.16	284.9	387.04	524.43	708.8
37	1.45	2.08	2.99	4.27	6.08	8.64	12.22	17.25	24.25	34	47.53	66.23	92.02	127.49	176.12	242.63	333.33	456.7	624.08	850.56
38	1.46	2.12	3.07	4.44	6.39	9.15	13.08	18.63	26.44	37.4	52.76	74.18	103.99	145.34	202.54	281.45	390	538.91	742.65	1020.7
39	1.47	2.16	3.17	4.62	6.7	9.7	13.99	20.12	28.82	41.14	58.56	83.08	117.51	165.69	232.92	326.48	456.3	635.91	883.75	1224.8
40	1.49	2.21	3.26	4.8	7.04	10.29	14.97	21.72	31.41	45.26	65	93.05	132.78	188.88	267.86	378.72	533.87	750.38	1051.7	1469.8

TABLE A.1 Interest Factors for Engineering Economy Compound Amount Factor Single Payment: To Find F Given P (F/P)

i% n Yrs	1%	2%	3%	4%	5%	6%	7%	8%	9%	10%	11%	12%	13%	14%	15%	16%	17%	18%	19%	20%
1	0.9901	0.9804	0.9709	0.9615	0.9524	0.9434	0.9346	0.9259	0.9174	0.9091	0.9009	0.8929	0.8850	0.8772	0.8696	0.8621	0.8547	0.8475	0.8403	0.8333
2	0.9803	0.9612	0.9426	0.9246	0.9070	0.8900	0.8734	0.8573	0.8417	0.8264	0.8116	0.7972	0.7831	0.7695	0.7561	0.7432	0.7305	0.7182	0.7062	0.6944
3	0.9706	0.9423	0.9151	0.8890	0.8638	0.8396	0.8163	0.7938	0.7722	0.7513	0.7312	0.7118	0.6931	0.6750	0.6575	0.6407	0.6244	0.6086	0.5934	0.5787
4	0.9610	0.9238	0.8885	0.8548	0.8227	0.7921	0.7629	0.7350	0.7084	0.6830	0.6587	0.6355	0.6133	0.5921	0.5718	0.5523	0.5337	0.5158	0.4987	0.4823
5	0.9515	0.9057	0.8626	0.8219	0.7835	0.7473	0.7130	0.6806	0.6499	0.6209	0.5935	0.5674	0.5428	0.5194	0.4972	0.4761	0.4561	0.4371	0.4190	0.4019
6	0.9420	0.8880	0.8375	0.7903	0.7462	0.7050	0.6663	0.6302	0.5963	0.5645	0.5346	0.5066	0.4803	0.4556	0.4323	0.4104	0.3898	0.3704	0.3521	0.3349
7	0.9327	0.8706	0.8131	0.7599	0.7107	0.6651	0.6227	0.5835	0.5470	0.5132	0.4817	0.4523	0.4251	0.3996	0.3759	0.3538	0.3332	0.3139	0.2959	0.2791
8	0.9235	0.8535	0.7894	0.7307	0.6768	0.6274	0.5820	0.5403	0.5019	0.4665	0.4339	0.4039	0.3762	0.3506	0.3269	0.3050	0.2848	0.2660	0.2487	0.2326
9	0.9143	0.8368	0.7664	0.7026	0.6446	0.5919	0.5439	0.5002	0.4604	0.4241	0.3909	0.3606	0.3329	0.3075	0.2843	0.2630	0.2434	0.2255	0.2090	0.1938
10	0.9053	0.8203	0.7441	0.6756	0.6139	0.5584	0.5083	0.4632	0.4224	0.3855	0.3522	0.3220	0.2946	0.2697	0.2472	0.2267	0.2080	0.1911	0.1756	0.1615
11	0.8963	0.8043	0.7224	0.6496	0.5847	0.5268	0.4751	0.4289	0.3875	0.3505	0.3173	0.2875	0.2607	0.2366	0.2149	0.1954	0.1778	0.1619	0.1476	0.1346
12	0.8874	0.7885	0.7014	0.6246	0.5568	0.4970	0.4440	0.3971	0.3555	0.3186	0.2858	0.2567	0.2307	0.2076	0.1869	0.1685	0.1520	0.1372	0.1240	0.1122
13	0.8787	0.7730	0.6810	0.6006	0.5303	0.4688	0.4150	0.3677	0.3262	0.2897	0.2575	0.2292	0.2042	0.1821	0.1625	0.1452	0.1299	0.1163	0.1042	0.0935
14	0.8700	0.7579	0.6611	0.5775	0.5051	0.4423	0.3878	0.3405	0.2992	0.2633	0.2320	0.2046	0.1807	0.1597	0.1413	0.1252	0.1110	0.0985	0.0876	0.0779
15	0.8613	0.7430	0.6419	0.5553	0.4810	0.4173	0.3624	0.3152	0.2745	0.2394	0.2090	0.1827	0.1599	0.1401	0.1229	0.1079	0.0949	0.0835	0.0736	0.0649
16	0.8528	0.7284	0.6232	0.5339	0.4581	0.3936	0.3387	0.2919	0.2519	0.2176	0.1883	0.1631	0.1415	0.1229	0.1069	0.0930	0.0811	0.0708	0.0618	0.0541
17	0.8444	0.7142	0.6050	0.5134	0.4363	0.3714	0.3166	0.2703	0.2311	0.1978	0.1696	0.1456	0.1252	0.1078	0.0929	0.0802	0.0693	0.0600	0.0520	0.0451
18	0.8360	0.7002	0.5874	0.4936	0.4155	0.3503	0.2959	0.2502	0.2120	0.1799	0.1528	0.1300	0.1108	0.0946	0.0808	0.0691	0.0592	0.0508	0.0437	0.0376
19	0.8277	0.6864	0.5703	0.4746	0.3957	0.3305	0.2765	0.2317	0.1945	0.1635	0.1377	0.1161	0.0981	0.0829	0.0703	0.0596	0.0506	0.0431	0.0367	0.0313
20	0.8195	0.6730	0.5537	0.4564	0.3769	0.3118	0.2584	0.2145	0.1784	0.1486	0.1240	0.1037	0.0868	0.0728	0.0611	0.0514	0.0433	0.0365	0.0308	0.0261
21	0.8114	0.6598	0.5375	0.4388	0.3589	0.2942	0.2415	0.1987	0.1637	0.1351	0.1117	0.0926	0.0768	0.0638	0.0531	0.0443	0.0370	0.0309	0.0259	0.0217
22	0.8034	0.6468	0.5219	0.4220	0.3418	0.2775	0.2257	0.1839	0.1502	0.1228	0.1007	0.0826	0.0680	0.0560	0.0462	0.0382	0.0316	0.0262	0.0218	0.0181
23	0.7954	0.6342	0.5067	0.4057	0.3256	0.2618	0.2109	0.1703	0.1378	0.1117	0.0907	0.0738	0.0601	0.0491	0.0402	0.0329	0.0270	0.0222	0.0183	0.0151
24	0.7876	0.6217	0.4919	0.3901	0.3101	0.2470	0.1971	0.1577	0.1264	0.1015	0.0817	0.0659	0.0532	0.0431	0.0349	0.0284	0.0231	0.0188	0.0154	0.0126
25	0.7798	0.6095	0.4776	0.3751	0.2953	0.2330	0.1842	0.1460	0.1160	0.0923	0.0736	0.0588	0.0471	0.0378	0.0304	0.0245	0.0197	0.0160	0.0129	0.0105
26	0.7720	0.5976	0.4637	0.3607	0.2812	0.2198	0.1722	0.1352	0.1064	0.0839	0.0663	0.0525	0.0417	0.0331	0.0264	0.0211	0.0169	0.0135	0.0109	0.0087
27	0.7644	0.5859	0.4502	0.3468	0.2678	0.2074	0.1609	0.1252	0.0976	0.0763	0.0597	0.0469	0.0369	0.0291	0.0230	0.0182	0.0144	0.0115	0.0091	0.0073
28	0.7568	0.5744	0.4371	0.3335	0.2551	0.1956	0.1504	0.1159	0.0895	0.0693	0.0538	0.0419	0.0326	0.0255	0.0200	0.0157	0.0123	0.0097	0.0077	0.0061
29	0.7493	0.5631	0.4243	0.3207	0.2429	0.1846	0.1406	0.1073	0.0822	0.0630	0.0485	0.0374	0.0289	0.0224	0.0174	0.0135	0.0105	0.0082	0.0064	0.0051
30	0.7419	0.5521	0.4120	0.3083	0.2314	0.1741	0.1314	0.0994	0.0754	0.0573	0.0437	0.0334	0.0256	0.0196	0.0151	0.0116	0.0090	0.0070	0.0054	0.0042
31	0.7346	0.5412	0.4000	0.2965	0.2204	0.1643	0.1228	0.0920	0.0691	0.0521	0.0394	0.0298	0.0226	0.0172	0.0131	0.0100	0.0077	0.0059	0.0046	0.0035
32	0.7273	0.5306	0.3883	0.2851	0.2099	0.1550	0.1147	0.0852	0.0634	0.0474	0.0355	0.0266	0.0200	0.0151	0.0114	0.0087	0.0066	0.0050	0.0038	0.0029
33	0.7201	0.5202	0.3770	0.2741	0.1999	0.1462	0.1072	0.0789	0.0582	0.0431	0.0319	0.0238	0.0177	0.0132	0.0099	0.0075	0.0056	0.0042	0.0032	0.0024
34	0.7130	0.5100	0.3660	0.2636	0.1904	0.1379	0.1002	0.0730	0.0534	0.0391	0.0288	0.0212	0.0157	0.0116	0.0086	0.0064	0.0048	0.0036	0.0027	0.0020
35	0.7059	0.5000	0.3554	0.2534	0.1813	0.1301	0.0937	0.0676	0.0490	0.0356	0.0259	0.0189	0.0139	0.0102	0.0075	0.0055	0.0041	0.0030	0.0023	0.0017
36	0.6989	0.4902	0.3450	0.2437	0.1727	0.1227	0.0875	0.0626	0.0449	0.0323	0.0234	0.0169	0.0123	0.0089	0.0065	0.0048	0.0035	0.0026	0.0019	0.0014
37	0.6920	0.4806	0.3350	0.2343	0.1644	0.1158	0.0818	0.0580	0.0412	0.0294	0.0210	0.0151	0.0109	0.0078	0.0057	0.0041	0.0030	0.0022	0.0016	0.0012
38	0.6852	0.4712	0.3252	0.2253	0.1566	0.1092	0.0765	0.0537	0.0378	0.0267	0.0190	0.0135	0.0096	0.0069	0.0049	0.0036	0.0026	0.0019	0.0013	0.0010
39	0.6784	0.4619	0.3158	0.2166	0.1491	0.1031	0.0715	0.0497	0.0347	0.0243	0.0171	0.0120	0.0085	0.0060	0.0043	0.0031	0.0022	0.0016	0.0011	0.0008
40	0.6717	0.4529	0.3066	0.2083	0.1420	0.0972	0.0668	0.0460	0.0318	0.0221	0.0154	0.0107	0.0075	0.0053	0.0037	0.0026	0.0019	0.0013	0.0010	0.0007

TABLE A.2 Interest Factors for Engineering Economy Present Worth Factor Single Payment: To Find P Given F (P/F)

i% n Yrs	1%	2%	3%	4%	5%	6%	7%	8%	9%	10%	11%	12%	13%	14%	15%	16%	17%	18%	19%	20%
1	0.990	0.980	0.970	0.960	0.950	0.940	0.930	0.930	0.920	0.910	0.900	0.890	0.880	0.880	0.870	0.860	0.850	0.850	0.840	0.830
2	1.970	1.940	1.910	1.890	1.860	1.830	1.810	1.780	1.760	1.740	1.710	1.690	1.670	1.650	1.630	1.610	1.590	1.570	1.550	1.530
3	2.940	2.880	2.830	2.780	2.720	2.670	2.620	2.580	2.530	2.490	2.440	2.400	2.360	2.320	2.280	2.250	2.210	2.170	2.140	2.110
4	3.900	3.810	3.720	3.630	3.550	3.470	3.390	3.310	3.240	3.170	3.100	3.040	2.970	2.910	2.850	2.800	2.740	2.690	2.640	2.590
5	4.850	4.710	4.580	4.450	4.330	4.210	4.100	3.990	3.890	3.790	3.700	3.600	3.520	3.430	3.350	3.270	3.200	3.130	3.060	2.990
6	5.800	5.600	5.420	5.240	5.080	4.920	4.770	4.620	4.490	4.360	4.230	4.110	4.000	3.890	3.780	3.680	3.590	3.500	3.410	3.330
7	6.730	6.470	6.230	6.000	5.790	5.580	5.390	5.210	5.030	4.870	4.710	4.560	4.420	4.290	4.160	4.040	3.920	3.810	3.710	3.600
8	7.650	7.330	7.020	6.730	6.460	6.210	5.970	5.750	5.530	5.330	5.150	4.970	4.800	4.640	4.490	4.340	4.210	4.080	3.950	3.840
9	8.570	8.160	7.790	7.440	7.110	6.800	6.520	6.250	6.000	5.760	5.540	5.330	5.130	4.950	4.770	4.610	4.450	4.300	4.160	4.030
10	9.470	8.980	8.530	8.110	7.720	7.360	7.020	6.710	6.420	6.140	5.890	5.650	5.430	5.220	5.020	4.830	4.660	4.490	4.340	4.190
11	10.370	9.790	9.250	8.760	8.310	7.890	7.500	7.140	6.810	6.500	6.210	5.940	5.690	5.450	5.230	5.030	4.840	4.660	4.490	4.330
12	11.260	10.580	9.950	9.390	8.860	8.380	7.940	7.540	7.160	6.810	6.490	6.190	5.920	5.660	5.420	5.200	4.990	4.790	4.610	4.440
13	12.130	11.350	10.630	9.990	9.390	8.850	8.360	7.900	7.490	7.100	6.750	6.420	6.120	5.840	5.580	5.340	5.120	4.910	4.710	4.530
14	13.000	12.110	11.300	10.560	9.900	9.290	8.750	8.240	7.790	7.370	6.980	6.630	6.300	6.000	5.720	5.470	5.230	5.010	4.800	4.610
15	13.870	12.850	11.940	11.120	10.380	9.710	9.110	8.560	8.060	7.610	7.190	6.810	6.460	6.140	5.850	5.580	5.320	5.090	4.880	4.680
16	14.720	13.580	12.560	11.650	10.840	10.110	9.450	8.850	8.310	7.820	7.380	6.970	6.600	6.270	5.950	5.670	5.410	5.160	4.940	4.730
17	15.560	14.290	13.170	12.170	11.270	10.480	9.760	9.120	8.540	8.020	7.550	7.120	6.730	6.370	6.050	5.750	5.470	5.220	4.990	4.770
18	16.400	14.990	13.750	12.660	11.690	10.830	10.060	9.370	8.760	8.200	7.700	7.250	6.840	6.470	6.130	5.820	5.530	5.270	5.030	4.810
19	17.230	15.680	14.320	13.130	12.090	11.160	10.340	9.600	8.950	8.360	7.840	7.370	6.940	6.550	6.200	5.880	5.580	5.320	5.070	4.840
20	18.050	16.350	14.880	13.590	12.460	11.470	10.590	9.820	9.130	8.510	7.960	7.470	7.020	6.620	6.260	5.930	5.630	5.350	5.100	4.870
21	18.860	17.010	15.420	14.030	12.820	11.760	10.840	10.020	9.290	8.650	8.080	7.560	7.100	6.690	6.310	5.970	5.660	5.380	5.130	4.890
22	19.660	17.660	15.940	14.450	13.160	12.040	11.060	10.200	9.440	8.770	8.180	7.640	7.170	6.740	6.360	6.010	5.700	5.410	5.150	4.910
23	20.460	18.290	16.440	14.860	13.490	12.300	11.270	10.370	9.580	8.880	8.270	7.720	7.230	6.790	6.400	6.040	5.720	5.430	5.170	4.920
24	21.240	18.910	16.940	15.250	13.800	12.550	11.470	10.530	9.710	8.980	8.350	7.780	7.280	6.840	6.430	6.070	5.750	5.450	5.180	4.940
25	22.020	19.520	17.410	15.620	14.090	12.780	11.650	10.670	9.820	9.080	8.420	7.840	7.330	6.870	6.460	6.100	5.770	5.470	5.200	4.950
26	22.800	20.120	17.880	15.980	14.380	13.000	11.830	10.810	9.930	9.160	8.490	7.900	7.370	6.910	6.490	6.120	5.780	5.480	5.210	4.960
27	23.560	20.710	18.330	16.330	14.640	13.210	11.990	10.940	10.030	9.240	8.550	7.940	7.410	6.940	6.510	6.140	5.800	5.490	5.220	4.960
28	24.320	21.280	18.760	16.660	14.900	13.410	12.140	11.050	10.120	9.310	8.600	7.980	7.440	6.960	6.530	6.150	5.810	5.500	5.220	4.970
29	25.070	21.840	19.190	16.980	15.140	13.590	12.280	11.160	10.200	9.370	8.650	8.020	7.470	6.980	6.550	6.170	5.820	5.510	5.230	4.970
30	25.810	22.400	19.600	17.290	15.370	13.760	12.410	11.260	10.270	9.430	8.690	8.060	7.500	7.000	6.570	6.180	5.830	5.520	5.230	4.980
31	26.540	22.940	20.000	17.590	15.590	13.930	12.530	11.350	10.340	9.480	8.730	8.080	7.520	7.020	6.580	6.190	5.840	5.520	5.240	4.980
32	27.270	23.470	20.390	17.870	15.800	14.080	12.650	11.430	10.410	9.530	8.770	8.110	7.540	7.030	6.590	6.200	5.840	5.530	5.240	4.990
33	27.990	23.990	20.770	18.150	16.000	14.230	12.750	11.510	10.460	9.570	8.800	8.140	7.560	7.050	6.600	6.200	5.850	5.530	5.250	4.990
34	28.700	24.500	21.130	18.410	16.190	14.370	12.850	11.590	10.520	9.610	8.830	8.160	7.570	7.060	6.610	6.210	5.850	5.540	5.250	4.990
35	29.410	25.000	21.490	18.660	16.370	14.500	12.950	11.650	10.570	9.640	8.860	8.180	7.590	7.070	6.620	6.220	5.860	5.540	5.250	4.990
36	30.110	25.490	21.830	18.910	16.550	14.620	13.040	11.720	10.610	9.680	8.880	8.190	7.600	7.080	6.620	6.220	5.860	5.540	5.250	4.990
37	30.800	25.970	22.170	19.140	16.710	14.740	13.120	11.780	10.650	9.710	8.900	8.210	7.610	7.090	6.630	6.220	5.860	5.540	5.250	4.990
38	31.480	26.440	22.490	19.370	16.870	14.850	13.190	11.830	10.690	9.730	8.920	8.220	7.620	7.090	6.630	6.230	5.870	5.550	5.260	5.000
39	32.160	26.900	22.810	19.580	17.020	14.950	13.260	11.880	10.730	9.760	8.940	8.230	7.630	7.100	6.640	6.230	5.870	5.550	5.260	5.000
40	32.830	27.360	23.110	19.790	17.160	15.050	13.330	11.920	10.760	9.780	8.950	8.240	7.630	7.110	6.640	6.230	5.870	5.550	5.260	5.000

Table A.3 Interest Factors for Engineering Economy Present Worth Factor Equal Payments: To Find P Given A (P/A)

i% n Yrs	1%	2%	3%	4%	5%	6%	7%	8%	9%	10%	11%	12%	13%	14%	15%	16%	17%	18%	19%	20%
1	1	1	1	1	1	1	1	1	1	1	1	1	1	1	1	1	1	1	1	1
2	2.01	2.02	2.03	2.04	2.05	2.06	2.07	2.08	2.09	2.1	2.11	2.12	2.13	2.14	2.15	2.16	2.17	2.18	2.19	2.2
3	3.03	3.06	3.09	3.12	3.15	3.18	3.21	3.25	3.28	3.31	3.34	3.37	3.41	3.44	3.47	3.51	3.54	3.57	3.61	3.64
4	4.06	4.12	4.18	4.25	4.31	4.37	4.44	4.51	4.57	4.64	4.71	4.78	4.85	4.92	4.99	5.07	5.14	5.22	5.29	5.37
5	5.1	5.2	5.31	5.42	5.53	5.64	5.75	5.87	5.98	6.11	6.23	6.35	6.48	6.61	6.74	6.88	7.01	7.15	7.3	7.44
6	6.15	6.31	6.47	6.63	6.8	6.98	7.15	7.34	7.52	7.72	7.91	8.12	8.32	8.54	8.75	8.98	9.21	9.44	9.68	9.93
7	7.21	7.43	7.66	7.9	8.14	8.39	8.65	8.92	9.2	9.49	9.78	10.09	10.4	10.73	11.07	11.41	11.77	12.14	12.52	12.92
8	8.29	8.58	8.89	9.21	9.55	9.9	10.26	10.64	11.03	11.44	11.86	12.3	12.76	13.23	13.73	14.24	14.77	15.33	15.9	16.5
9	9.37	9.75	10.16	10.58	11.03	11.49	11.98	12.49	13.02	13.58	14.16	14.78	15.42	16.09	16.79	17.52	18.28	19.09	19.92	20.8
10	10.46	10.95	11.46	12.01	12.58	13.18	13.82	14.49	15.19	15.94	16.72	17.55	18.42	19.34	20.3	21.32	22.39	23.52	24.71	25.96
11	11.57	12.17	12.81	13.49	14.21	14.97	15.78	16.65	17.56	18.53	19.56	20.65	21.81	23.04	24.35	25.73	27.2	28.76	30.4	32.15
12	12.68	13.41	14.19	15.03	15.92	16.87	17.89	18.98	20.14	21.38	22.71	24.13	25.65	27.27	29	30.85	32.82	34.93	37.18	39.58
13	13.81	14.68	15.62	16.63	17.71	18.88	20.14	21.5	22.95	24.52	26.21	28.03	29.98	32.09	34.35	36.79	39.4	42.22	45.24	48.5
14	14.95	15.97	17.09	18.29	19.6	21.02	22.55	24.21	26.02	27.97	30.09	32.39	34.88	37.58	40.5	43.67	47.1	50.82	54.84	59.2
15	16.1	17.29	18.6	20.02	21.58	23.28	25.13	27.15	29.36	31.77	34.41	37.28	40.42	43.84	47.58	51.66	56.11	60.97	66.26	72.04
16	17.26	18.64	20.16	21.82	23.66	25.67	27.89	30.32	33	35.95	39.19	42.75	46.67	50.98	55.72	60.93	66.65	72.94	79.85	87.44
17	18.43	20.01	21.76	23.7	25.84	28.21	30.84	33.75	36.97	40.54	44.5	48.88	53.74	59.12	65.08	71.67	78.98	87.07	96.02	105.93
18	19.61	21.41	23.41	25.65	28.13	30.91	34	37.45	41.3	45.6	50.4	55.75	61.73	68.39	75.84	84.14	93.41	103.74	115.27	128.12
19	20.81	22.84	25.12	27.67	30.54	33.76	37.38	41.45	46.02	51.16	56.94	63.44	70.75	78.97	88.21	98.6	110.28	123.41	138.17	154.74
20	22.02	24.3	26.87	29.78	33.07	36.79	41	45.76	51.16	57.27	64.2	72.05	80.95	91.02	102.44	115.38	130.03	146.63	165.42	186.69
21	23.24	25.78	28.68	31.97	35.72	39.99	44.87	50.42	56.76	64	72.27	81.7	92.47	104.77	118.81	134.84	153.14	174.02	197.85	225.03
22	24.47	27.3	30.54	34.25	38.51	43.39	49.01	55.46	62.87	71.4	81.21	92.5	105.49	120.44	137.63	157.41	180.17	206.34	236.44	271.03
23	25.72	28.84	32.45	36.62	41.43	47	53.44	60.89	69.53	79.54	91.15	104.6	120.2	138.3	159.28	183.6	211.8	244.49	282.36	326.24
24	26.97	30.42	34.43	39.08	44.5	50.82	58.18	66.76	76.79	88.5	102.17	118.16	136.83	158.66	184.17	213.98	248.81	289.49	337.01	392.48
25	28.24	32.03	36.46	41.65	47.73	54.86	63.25	73.11	84.7	98.35	114.41	133.33	155.62	181.87	212.79	249.21	292.1	342.6	402.04	471.98
26	29.53	33.67	38.55	44.31	51.11	59.16	68.68	79.95	93.32	109.18	128	150.33	176.85	208.33	245.71	290.09	342.76	405.27	479.43	567.38
27	30.82	35.34	40.71	47.08	54.67	63.71	74.48	87.35	102.72	121.1	143.08	169.37	200.84	238.5	283.57	337.5	402.03	479.22	571.52	681.85
28	32.13	37.05	42.93	49.97	58.4	68.53	80.7	95.34	112.97	134.21	159.82	190.7	227.95	272.89	327.1	392.5	471.38	566.48	681.11	819.22
29	33.45	38.79	45.22	52.97	62.32	73.64	87.35	103.97	124.14	148.63	178.4	214.58	258.58	312.09	377.17	456.3	552.51	669.45	811.52	984.07
30	34.78	40.57	47.58	56.08	66.44	79.06	94.46	113.28	136.31	164.49	199.02	241.33	293.2	356.79	434.75	530.31	647.44	790.95	966.71	1181.88
31	36.13	42.38	50	59.33	70.76	84.8	102.07	123.35	149.58	181.94	221.91	271.29	332.32	407.74	500.96	616.16	758.5	934.32	1151.39	1419.26
32	37.49	44.23	52.5	62.7	75.3	90.89	110.22	134.21	164.04	201.14	247.32	304.85	376.52	465.82	577.1	715.75	888.45	1103.5	1371.15	1704.11
33	38.87	46.11	55.08	66.21	80.06	97.34	118.93	145.95	179.8	222.25	275.53	342.43	426.46	532.04	664.67	831.27	1040.49	1303.13	1632.67	2045.93
34	40.26	48.03	57.73	69.86	85.07	104.18	128.26	158.63	196.98	245.48	306.84	384.52	482.9	607.52	765.37	965.27	1218.37	1538.69	1943.88	2456.12
35	41.66	49.99	60.46	73.65	90.32	111.43	138.24	172.32	215.71	271.02	341.59	431.66	546.68	693.57	881.17	1120.71	1426.49	1816.65	2314.21	2948.34
36	43.08	51.99	63.28	77.6	95.84	119.12	148.91	187.1	236.12	299.13	380.16	484.46	618.75	791.67	1014.35	1301.03	1669.99	2144.65	2754.91	3539.01
37	44.51	54.03	66.17	81.7	101.63	127.27	160.34	203.07	258.38	330.04	422.98	543.6	700.19	903.51	1167.5	1510.19	1954.89	2531.69	3279.35	4247.81
38	45.95	56.11	69.16	85.97	107.71	135.9	172.56	220.32	282.63	364.04	470.51	609.83	792.21	1031	1343.62	1752.82	2288.23	2988.39	3903.42	5098.37
39	47.41	58.24	72.23	90.41	114.1	145.06	185.64	238.94	309.07	401.45	523.27	684.01	896.2	1176.34	1546.17	2034.27	2678.22	3527.3	4646.07	6119.05
40	48.89	60.4	75.4	95.03	120.8	154.76	199.64	259.06	337.88	442.59	581.83	767.09	1013.7	1342.03	1779.09	2360.76	3134.52	4163.21	5529.83	7343.86

TABLE A.4 Interest Factors for Engineering Economy Compound Amount Factor Equal Payments: To Find F Given A (F/A)

i% n Yrs	1%	2%	3%	4%	5%	6%	7%	8%	9%	10%	11%	12%	13%	14%	15%	16%	17%	18%	19%	20%
1	1.0000	1.0000	1.0000	1.0000	1.0000	1.0000	1.0000	1.0000	1.0000	1.0000	1.0000	1.0000	1.0000	1.0000	1.0000	1.0000	1.0000	1.0000	1.0000	1.0000
2	0.4975	0.4950	0.4926	0.4902	0.4878	0.4854	0.4831	0.4808	0.4785	0.4762	0.4739	0.4717	0.4695	0.4673	0.4651	0.4630	0.4608	0.4587	0.4566	0.4545
3	0.3300	0.3268	0.3235	0.3203	0.3172	0.3141	0.3111	0.3080	0.3051	0.3021	0.2992	0.2963	0.2935	0.2907	0.2880	0.2853	0.2826	0.2799	0.2773	0.2747
4	0.2463	0.2426	0.2390	0.2355	0.2320	0.2286	0.2252	0.2219	0.2187	0.2155	0.2123	0.2092	0.2062	0.2032	0.2003	0.1974	0.1945	0.1917	0.1890	0.1863
5	0.1960	0.1922	0.1884	0.1846	0.1810	0.1774	0.1739	0.1705	0.1671	0.1638	0.1606	0.1574	0.1543	0.1513	0.1483	0.1454	0.1426	0.1398	0.1371	0.1344
6	0.1625	0.1585	0.1546	0.1508	0.1470	0.1434	0.1398	0.1363	0.1329	0.1296	0.1264	0.1232	0.1202	0.1172	0.1142	0.1114	0.1086	0.1059	0.1033	0.1007
7	0.1386	0.1345	0.1305	0.1266	0.1228	0.1191	0.1156	0.1121	0.1087	0.1054	0.1022	0.0991	0.0961	0.0932	0.0904	0.0876	0.0849	0.0824	0.0799	0.0774
8	0.1207	0.1165	0.1125	0.1085	0.1047	0.1010	0.0975	0.0940	0.0907	0.0874	0.0843	0.0813	0.0784	0.0756	0.0729	0.0702	0.0677	0.0652	0.0629	0.0606
9	0.1067	0.1025	0.0984	0.0945	0.0907	0.0870	0.0835	0.0801	0.0768	0.0736	0.0706	0.0677	0.0649	0.0622	0.0596	0.0571	0.0547	0.0524	0.0502	0.0481
10	0.0956	0.0913	0.0872	0.0833	0.0795	0.0759	0.0724	0.0690	0.0658	0.0627	0.0598	0.0570	0.0543	0.0517	0.0493	0.0469	0.0447	0.0425	0.0405	0.0385
11	0.0865	0.0822	0.0781	0.0741	0.0704	0.0668	0.0634	0.0601	0.0569	0.0540	0.0511	0.0484	0.0458	0.0434	0.0411	0.0389	0.0368	0.0348	0.0329	0.0311
12	0.0788	0.0746	0.0705	0.0666	0.0628	0.0593	0.0559	0.0527	0.0497	0.0468	0.0440	0.0414	0.0390	0.0367	0.0345	0.0324	0.0305	0.0286	0.0269	0.0253
13	0.0724	0.0681	0.0640	0.0601	0.0565	0.0530	0.0497	0.0465	0.0436	0.0408	0.0382	0.0357	0.0334	0.0312	0.0291	0.0272	0.0254	0.0237	0.0221	0.0206
14	0.0669	0.0626	0.0585	0.0547	0.0510	0.0476	0.0443	0.0413	0.0384	0.0357	0.0332	0.0309	0.0287	0.0266	0.0247	0.0229	0.0212	0.0197	0.0182	0.0169
15	0.0621	0.0578	0.0538	0.0499	0.0463	0.0430	0.0398	0.0368	0.0341	0.0315	0.0291	0.0268	0.0247	0.0228	0.0210	0.0194	0.0178	0.0164	0.0151	0.0139
16	0.0579	0.0537	0.0496	0.0458	0.0423	0.0390	0.0359	0.0330	0.0303	0.0278	0.0255	0.0234	0.0214	0.0196	0.0179	0.0164	0.0150	0.0137	0.0125	0.0114
17	0.0543	0.0500	0.0460	0.0422	0.0387	0.0354	0.0324	0.0296	0.0270	0.0247	0.0225	0.0205	0.0186	0.0169	0.0154	0.0140	0.0127	0.0115	0.0104	0.0094
18	0.0510	0.0467	0.0427	0.0390	0.0355	0.0324	0.0294	0.0267	0.0242	0.0219	0.0198	0.0179	0.0162	0.0146	0.0132	0.0119	0.0107	0.0096	0.0087	0.0078
19	0.0481	0.0438	0.0398	0.0361	0.0327	0.0296	0.0268	0.0241	0.0217	0.0195	0.0176	0.0158	0.0141	0.0127	0.0113	0.0101	0.0091	0.0081	0.0072	0.0065
20	0.0454	0.0412	0.0372	0.0336	0.0302	0.0272	0.0244	0.0219	0.0195	0.0175	0.0156	0.0139	0.0124	0.0110	0.0098	0.0087	0.0077	0.0068	0.0060	0.0054
21	0.0430	0.0388	0.0349	0.0313	0.0280	0.0250	0.0223	0.0198	0.0176	0.0156	0.0138	0.0122	0.0108	0.0095	0.0084	0.0074	0.0065	0.0057	0.0051	0.0044
22	0.0409	0.0366	0.0327	0.0292	0.0260	0.0230	0.0204	0.0180	0.0159	0.0140	0.0123	0.0108	0.0095	0.0083	0.0073	0.0064	0.0056	0.0048	0.0042	0.0037
23	0.0389	0.0347	0.0308	0.0273	0.0241	0.0213	0.0187	0.0164	0.0144	0.0126	0.0110	0.0096	0.0083	0.0072	0.0063	0.0054	0.0047	0.0041	0.0035	0.0031
24	0.0371	0.0329	0.0290	0.0256	0.0225	0.0197	0.0172	0.0150	0.0130	0.0113	0.0098	0.0085	0.0073	0.0063	0.0054	0.0047	0.0040	0.0035	0.0030	0.0025
25	0.0354	0.0312	0.0274	0.0240	0.0210	0.0182	0.0158	0.0137	0.0118	0.0102	0.0087	0.0075	0.0064	0.0055	0.0047	0.0040	0.0034	0.0029	0.0025	0.0021
26	0.0339	0.0297	0.0259	0.0226	0.0196	0.0169	0.0146	0.0125	0.0107	0.0092	0.0078	0.0067	0.0057	0.0048	0.0041	0.0034	0.0029	0.0025	0.0021	0.0018
27	0.0324	0.0283	0.0246	0.0212	0.0183	0.0157	0.0134	0.0114	0.0097	0.0083	0.0070	0.0059	0.0050	0.0042	0.0035	0.0030	0.0025	0.0021	0.0017	0.0015
28	0.0311	0.0270	0.0233	0.0200	0.0171	0.0146	0.0124	0.0105	0.0089	0.0075	0.0063	0.0052	0.0044	0.0037	0.0031	0.0025	0.0021	0.0018	0.0015	0.0012
29	0.0299	0.0258	0.0221	0.0189	0.0160	0.0136	0.0114	0.0096	0.0081	0.0067	0.0056	0.0047	0.0039	0.0032	0.0027	0.0022	0.0018	0.0015	0.0012	0.0010
30	0.0287	0.0246	0.0210	0.0178	0.0151	0.0126	0.0106	0.0088	0.0073	0.0061	0.0050	0.0041	0.0034	0.0028	0.0023	0.0019	0.0015	0.0013	0.0010	0.0008
31	0.0277	0.0236	0.0200	0.0169	0.0141	0.0118	0.0098	0.0081	0.0067	0.0055	0.0045	0.0037	0.0030	0.0025	0.0020	0.0016	0.0013	0.0011	0.0009	0.0007
32	0.0267	0.0226	0.0190	0.0159	0.0133	0.0110	0.0091	0.0075	0.0061	0.0050	0.0040	0.0033	0.0027	0.0021	0.0017	0.0014	0.0011	0.0009	0.0007	0.0006
33	0.0257	0.0217	0.0182	0.0151	0.0125	0.0103	0.0084	0.0069	0.0056	0.0045	0.0036	0.0029	0.0023	0.0019	0.0015	0.0012	0.0010	0.0008	0.0006	0.0005
34	0.0248	0.0208	0.0173	0.0143	0.0118	0.0096	0.0078	0.0063	0.0051	0.0041	0.0033	0.0026	0.0021	0.0016	0.0013	0.0010	0.0008	0.0006	0.0005	0.0004
35	0.0240	0.0200	0.0165	0.0136	0.0111	0.0090	0.0072	0.0058	0.0046	0.0037	0.0029	0.0023	0.0018	0.0014	0.0011	0.0009	0.0007	0.0006	0.0004	0.0003
36	0.0232	0.0192	0.0158	0.0129	0.0104	0.0084	0.0067	0.0053	0.0042	0.0033	0.0026	0.0021	0.0016	0.0013	0.0010	0.0008	0.0006	0.0005	0.0004	0.0003
37	0.0225	0.0185	0.0151	0.0122	0.0098	0.0079	0.0062	0.0049	0.0039	0.0030	0.0024	0.0018	0.0014	0.0011	0.0009	0.0007	0.0005	0.0004	0.0003	0.0002
38	0.0218	0.0178	0.0145	0.0116	0.0093	0.0074	0.0058	0.0045	0.0035	0.0027	0.0021	0.0016	0.0013	0.0010	0.0007	0.0006	0.0004	0.0003	0.0003	0.0002
39	0.0211	0.0172	0.0138	0.0111	0.0088	0.0069	0.0054	0.0042	0.0032	0.0025	0.0019	0.0015	0.0011	0.0009	0.0006	0.0005	0.0004	0.0003	0.0002	0.0002
40	0.0205	0.0166	0.0133	0.0105	0.0083	0.0065	0.0050	0.0039	0.0030	0.0023	0.0017	0.0013	0.0010	0.0007	0.0006	0.0004	0.0003	0.0002	0.0002	0.0001

TABLE A.5 Interest Factors for Engineering Economy Sinking Fund Factor Equal Payments: To Find A Given F (A/F)

i% n Yrs	1%	2%	3%	4%	5%	6%	7%	8%	9%	10%	11%	12%	13%	14%	15%	16%	17%	18%	19%	20%
1	1.0100	1.0200	1.0300	1.0400	1.0500	1.0600	1.0700	1.0800	1.0900	1.1000	1.1100	1.1200	1.1300	1.1400	1.1500	1.1600	1.1700	1.1800	1.1900	1.2000
2	0.5075	0.5150	0.5226	0.5302	0.5378	0.5454	0.5531	0.5608	0.5685	0.5762	0.5839	0.5917	0.5995	0.6073	0.6151	0.6230	0.6308	0.6387	0.6466	0.6545
3	0.3400	0.3468	0.3535	0.3603	0.3672	0.3741	0.3811	0.3880	0.3951	0.4021	0.4092	0.4163	0.4235	0.4307	0.4380	0.4453	0.4526	0.4599	0.4673	0.4747
4	0.2563	0.2626	0.2690	0.2755	0.2820	0.2886	0.2952	0.3019	0.3087	0.3155	0.3223	0.3292	0.3362	0.3432	0.3503	0.3574	0.3645	0.3717	0.3790	0.3863
5	0.2060	0.2122	0.2184	0.2246	0.2310	0.2374	0.2439	0.2505	0.2571	0.2638	0.2706	0.2774	0.2843	0.2913	0.2983	0.3054	0.3126	0.3198	0.3271	0.3344
6	0.1725	0.1785	0.1846	0.1908	0.1970	0.2034	0.2098	0.2163	0.2229	0.2296	0.2364	0.2432	0.2502	0.2572	0.2642	0.2714	0.2786	0.2859	0.2933	0.3007
7	0.1486	0.1545	0.1605	0.1666	0.1728	0.1791	0.1856	0.1921	0.1987	0.2054	0.2122	0.2191	0.2261	0.2332	0.2404	0.2476	0.2549	0.2624	0.2699	0.2774
8	0.1307	0.1365	0.1425	0.1485	0.1547	0.1610	0.1675	0.1740	0.1807	0.1874	0.1943	0.2013	0.2084	0.2156	0.2229	0.2302	0.2377	0.2452	0.2529	0.2606
9	0.1167	0.1225	0.1284	0.1345	0.1407	0.1470	0.1535	0.1601	0.1668	0.1736	0.1806	0.1877	0.1949	0.2022	0.2096	0.2171	0.2247	0.2324	0.2402	0.2481
10	0.1056	0.1113	0.1172	0.1233	0.1295	0.1359	0.1424	0.1490	0.1558	0.1627	0.1698	0.1770	0.1843	0.1917	0.1993	0.2069	0.2147	0.2225	0.2305	0.2385
11	0.0965	0.1022	0.1081	0.1141	0.1204	0.1268	0.1334	0.1401	0.1469	0.1540	0.1611	0.1684	0.1758	0.1834	0.1911	0.1989	0.2068	0.2148	0.2229	0.2311
12	0.0888	0.0946	0.1005	0.1066	0.1128	0.1193	0.1259	0.1327	0.1397	0.1468	0.1540	0.1614	0.1690	0.1767	0.1845	0.1924	0.2005	0.2086	0.2169	0.2253
13	0.0824	0.0881	0.0940	0.1001	0.1065	0.1130	0.1197	0.1265	0.1336	0.1408	0.1482	0.1557	0.1634	0.1712	0.1791	0.1872	0.1954	0.2037	0.2121	0.2206
14	0.0769	0.0826	0.0885	0.0947	0.1010	0.1076	0.1143	0.1213	0.1284	0.1357	0.1432	0.1509	0.1587	0.1666	0.1747	0.1829	0.1912	0.1997	0.2082	0.2169
15	0.0721	0.0778	0.0838	0.0899	0.0963	0.1030	0.1098	0.1168	0.1241	0.1315	0.1391	0.1468	0.1547	0.1628	0.1710	0.1794	0.1878	0.1964	0.2051	0.2139
16	0.0679	0.0737	0.0796	0.0858	0.0923	0.0990	0.1059	0.1130	0.1203	0.1278	0.1355	0.1434	0.1514	0.1596	0.1679	0.1764	0.1850	0.1937	0.2025	0.2114
17	0.0643	0.0700	0.0760	0.0822	0.0887	0.0954	0.1024	0.1096	0.1170	0.1247	0.1325	0.1405	0.1486	0.1569	0.1654	0.1740	0.1827	0.1915	0.2004	0.2094
18	0.0610	0.0667	0.0727	0.0790	0.0855	0.0924	0.0994	0.1067	0.1142	0.1219	0.1298	0.1379	0.1462	0.1546	0.1632	0.1719	0.1807	0.1896	0.1987	0.2078
19	0.0581	0.0638	0.0698	0.0761	0.0827	0.0896	0.0968	0.1041	0.1117	0.1195	0.1276	0.1358	0.1441	0.1527	0.1613	0.1701	0.1791	0.1881	0.1972	0.2065
20	0.0554	0.0612	0.0672	0.0736	0.0802	0.0872	0.0944	0.1019	0.1095	0.1175	0.1256	0.1339	0.1424	0.1510	0.1598	0.1687	0.1777	0.1868	0.1960	0.2054
21	0.0530	0.0588	0.0649	0.0713	0.0780	0.0850	0.0923	0.0998	0.1076	0.1156	0.1238	0.1322	0.1408	0.1495	0.1584	0.1674	0.1765	0.1857	0.1951	0.2044
22	0.0509	0.0566	0.0627	0.0692	0.0760	0.0830	0.0904	0.0980	0.1059	0.1140	0.1223	0.1308	0.1395	0.1483	0.1573	0.1664	0.1756	0.1848	0.1942	0.2037
23	0.0489	0.0547	0.0608	0.0673	0.0741	0.0813	0.0887	0.0964	0.1044	0.1126	0.1210	0.1296	0.1383	0.1472	0.1563	0.1654	0.1747	0.1841	0.1935	0.2031
24	0.0471	0.0529	0.0590	0.0656	0.0725	0.0797	0.0872	0.0950	0.1030	0.1113	0.1198	0.1285	0.1373	0.1463	0.1554	0.1647	0.1740	0.1835	0.1930	0.2025
25	0.0454	0.0512	0.0574	0.0640	0.0710	0.0782	0.0858	0.0937	0.1018	0.1102	0.1187	0.1275	0.1364	0.1455	0.1547	0.1640	0.1734	0.1829	0.1925	0.2021
26	0.0439	0.0497	0.0559	0.0626	0.0696	0.0769	0.0846	0.0925	0.1007	0.1092	0.1178	0.1267	0.1357	0.1448	0.1541	0.1634	0.1729	0.1825	0.1921	0.2018
27	0.0424	0.0483	0.0546	0.0612	0.0683	0.0757	0.0834	0.0914	0.0997	0.1083	0.1170	0.1259	0.1350	0.1442	0.1535	0.1630	0.1725	0.1821	0.1917	0.2015
28	0.0411	0.0470	0.0533	0.0600	0.0671	0.0746	0.0824	0.0905	0.0989	0.1075	0.1163	0.1252	0.1344	0.1437	0.1531	0.1625	0.1721	0.1818	0.1915	0.2012
29	0.0399	0.0458	0.0521	0.0589	0.0660	0.0736	0.0814	0.0896	0.0981	0.1067	0.1156	0.1247	0.1339	0.1432	0.1527	0.1622	0.1718	0.1815	0.1912	0.2010
30	0.0387	0.0446	0.0510	0.0578	0.0651	0.0726	0.0806	0.0888	0.0973	0.1061	0.1150	0.1241	0.1334	0.1428	0.1523	0.1619	0.1715	0.1813	0.1910	0.2008
31	0.0377	0.0436	0.0500	0.0569	0.0641	0.0718	0.0798	0.0881	0.0967	0.1055	0.1145	0.1237	0.1330	0.1425	0.1520	0.1616	0.1713	0.1811	0.1909	0.2007
32	0.0367	0.0426	0.0490	0.0559	0.0633	0.0710	0.0791	0.0875	0.0961	0.1050	0.1140	0.1233	0.1327	0.1421	0.1517	0.1614	0.1711	0.1809	0.1907	0.2006
33	0.0357	0.0417	0.0482	0.0551	0.0625	0.0703	0.0784	0.0869	0.0956	0.1045	0.1136	0.1229	0.1323	0.1419	0.1515	0.1612	0.1710	0.1808	0.1906	0.2005
34	0.0348	0.0408	0.0473	0.0543	0.0618	0.0696	0.0778	0.0863	0.0951	0.1041	0.1133	0.1226	0.1321	0.1416	0.1513	0.1610	0.1708	0.1806	0.1905	0.2004
35	0.0340	0.0400	0.0465	0.0536	0.0611	0.0690	0.0772	0.0858	0.0946	0.1037	0.1129	0.1223	0.1318	0.1414	0.1511	0.1609	0.1707	0.1806	0.1904	0.2003
36	0.0332	0.0392	0.0458	0.0529	0.0604	0.0684	0.0767	0.0853	0.0942	0.1033	0.1126	0.1221	0.1316	0.1413	0.1510	0.1608	0.1706	0.1805	0.1904	0.2003
37	0.0325	0.0385	0.0451	0.0522	0.0598	0.0679	0.0762	0.0849	0.0939	0.1030	0.1124	0.1218	0.1314	0.1411	0.1509	0.1607	0.1705	0.1804	0.1903	0.2002
38	0.0318	0.0378	0.0445	0.0516	0.0593	0.0674	0.0758	0.0845	0.0935	0.1027	0.1121	0.1216	0.1313	0.1410	0.1507	0.1606	0.1704	0.1803	0.1903	0.2002
39	0.0311	0.0372	0.0438	0.0511	0.0588	0.0669	0.0754	0.0842	0.0932	0.1025	0.1119	0.1215	0.1311	0.1409	0.1506	0.1605	0.1704	0.1803	0.1902	0.2002
40	0.0305	0.0366	0.0433	0.0505	0.0583	0.0665	0.0750	0.0839	0.0930	0.1023	0.1117	0.1213	0.1310	0.1407	0.1506	0.1604	0.1703	0.1802	0.1902	0.2001

TABLE A.6 Interest Factors for Engineering Economy Capital Recovery Factor Equal Payments: To Find A Given P (A/P)

Index

D

E

N

O

P

Q

R

CPSIA information can be obtained at www.ICGtesting.com
Printed in the USA
LVOW03*0407260515

439476LV00005B/8/P

9 780071 815369